THE FAMILY FARMERS' ADVOCATE

South Dakota Farmers Union
1914-2000

The Family Farmers' Advocate

South Dakota Farmers Union
1914-2000

BY LYNWOOD E. OYOS

The Center for Western Studies
Augustana College
2000

Published by
The Center for Western Studies
Box 727, Augustana College
Sioux Falls, SD 57197
605-336-4007
605-336-4999 (fax)
Publications catalog available upon request.

ISBN 0-931170-71-0 (cloth)
ISBN 0-931170-72-9 (paper)

The Center for Western Studies is an archives, library, museum, publishing house, and educational agency concerned principally with collecting, preserving, and interpreting prehistoric, historic, and contemporary materials that document native and immigrant cultures of the northern prairie/plains. The Center promotes understanding of the region through exhibits, publications, art shows, conferences, and academic programs. It is committed, ultimately, to defining the contribution of the region to American civilization.

Library of Congress Cataloging-in-Publication Data

Oyos, Lynwood.
The family farmers' advocate: South Dakota Farmers Union, 1914-2000 / Lynwood E. Oyos.
p. cm.
Includes bibliographical references and index.
ISBN 0-931170-710-(alk. paper) — ISBN 0-931170-729-(softcover: alk. paper)
1. South Dakota Farmers Union — History. 2. Agricultural laborers — Labor unions — South Dakota--History. I. Title.

HD6515.A292.S68 1999
338.1'06'073 21--dc21

Funding support for this book has been provided by the Farmers Educational Foundation, with assistance from the Harvest States Foundation.

Photo Credits:
All photos and illustrations are property of the South Dakota Farmers Union except for those on the following pages: pp. 4, 28, 96, 97, 116, 123, 211, 259 from Leo N. Rickertsen, *To Gather Together, CENEX, The First Fifty Years* (Minneapolis: Cooperative Printing Association, 1980; pp. 5, 44, 47, 85 from Charles and Joyce Conrad, *50 Years, The North Dakota Farmers Union* (n.p. 1976); pp. 6, 9 from *Economic Democracy for the Northern Plains: Cooperatives and North Dakota* (Jamestown: North Dakota Farmers Union, 1980); pp. 25, 52, 64, 75, 166 from South Dakota State Historical Society, State Archives Collection, Pierre, S.D.; pp. 33, 117 from *Cooperative League of USA*, *Yearbook, 1936*; p. 38 from Paul E. Huff and Raymond E. Novak, *A Rich Heritage of Community, A History of the NFU Insurance Companies* (Denver: NFU Property and Casualty Companies, 1955); pp. 42, 43 from the Center for Western Studies, Sioux Falls, S.D.; p. 49 from *The Depression Years*, Photographed by Arthur Rothstein (New York: Dover Publications, 1978); p. 154 from *East River Guardian*, 24 (May 1985); p. 196 from Marian Kleinsasser Towne, Indianapolis, Indiana; pp. 197, 198 from *One World, All States Manual* (Denver: NFU, 1948) courtesy of Marian Kleinsasser Towne; p. 156 from South Dakota Department of Tourism. Cover photography by Joel Strasser. Author photo by Augustana College.

Printed in the United States of America

PINE HILL PRESS, INC.
Freeman, S. Dak. 57029

CONTENTS

INTRODUCTION

In the latter part of the 19th century, economically distressed farmers sought succor through enrollment in a series of farm organizations. Within a few years after their inception, each of these associations failed or became politically impotent. Family farmers, however, found a new champion at the beginning of the 20th century when a new farmer-based organization emerged on the Prairie Plains. This new and lasting advocate for rural America was the Farmers' Educational and Cooperative Union of America, the Farmers Union. It is now the oldest active farm organization in the United States, excluding the Grange which became primarily a social organization. Three years hence, the Farmers Union will observe its 100th birthday, still carrying forward the goals and ideals set forth by Isaac Newton Gresham.

Newt Gresham, a farmer turned newspaperman, declared that family farmers needed a voice and an organization to fight for their rights and survival. Having experienced the realities of rural life, he argued that farmers were constantly being exploited by non-farmers. The credit, mortgage, processing and marketing systems were controlled by people who did not till the soil.Seeking to correct this imbalance in the American economy, Gresham organized the National Farmers Union (NFU) on August 28, 1902, at Point, Texas.

Because the problems faced by rural people on the northern Prairie Plains were similar to those confronting farmers in Texas, NFU organizers found willing converts in the Dakotas. At the turn of the century, North and South Dakota were considered colonial outposts to be exploited by corporate railroad and milling interests in Minneapolis, St. Paul and points east. In 1914, farmers in east-central South Dakota began to join Farmers Union locals in an effort to eradicate the unjust economic system. Three years later, in March 1917, the South Dakota Farmers Union (SDFU) received its charter. This book records the efforts of the Union over the following eight decades to improve and to maintain family farms in South Dakota.

In implementing the three basic programs set forth by Gresham and his associates — education, legislation and cooperation, the Farmers Union has carried out its mission during a century of radical change. Significant changes in technology made farm life easier and more efficient. These changes gave family farmers greater mobility and enabled them to become fully aware of what transpired in the greater society. Technology, however, had its downside. It contributed to a major decline in rural population and the growing urbanization of American society. The once powerful congressional farm bloc of the 1920s found its effectiveness diminished by the exodus of farmers to the cities. Technology enhanced global trade and farmers became dependent on healthy overseas economies. A decline in exports meant falling farm prices in the United States.

The quest for a "fair shake" from legislators was made more difficult by several presidential administrations that were hostile to any kind of safety net for farmers. Despite antitrust or anti-monopoly legislation early in the century, corporate greed always posed a threat to the economic survival of the family farmer. Diminished government regulation since the 1970s has permitted the almost unrestricted consolidation of corporations, in particular, mega-corporations that want to dominate all aspects of the farm economy,, production, processing, price fixing and marketing.

Readers of this work will discover how the South Dakota Farmers Union has confronted the foregoing changes in American society and how it has responded to the challenges of corporate dominance. The task has not been easy, but Union members have never wavered in their efforts to advocate and secure a better economic and social future for family farmers.

This is an account of farmers in action, as legislators, as lobbyists, as educators, as environmentalists and as co-op builders. Recognizing that an education program was needed if the Union were to survive, members of the SDFU did the pioneer work in Junior education, setting forth the organization's aims, principles and achievements before its youth. A strong educational program brought a steady flow of new strength into the SDFU. Inspired by the Rochdale cooperative in England, co-ops, like the crocus in late April, sprang up all across South Dakota in the 1930s and 1940s. Cooperation has always been a key to survival for rural people, and Union leaders recognized that the cooperative movement was necessary for the survival of farmers and small-town merchants.

The history of the South Dakota Farmers Union is also a story of democracy in action. Gresham emphasized that every member would have the right to participate in the Union's programs. Readers will discover there was not always union within the Farmers Union in South Dakota. Strong personalities had different views on how best to meet the farmers' needs. During the 20th century, the Union produced an abundance of strong, quality leaders. Several of them moved up to executive positions in the different components of the National Farmers Union. Others assumed leadership positions in state and national politics.

I have great admiration for family farmers. Having spent my formative years in a village dependent on the trade of area farmers, having worked for several summers on a farm, and having had numerous contacts with farmers throughout my academic career, I can only conclude that the Farmers Union has been the true advocate of the family farmer in South Dakota. The Union has worked diligently on behalf of those citizens who work the land. Necessary adaptations to changes in society have been made without compromising the original goals set forth by the NFU in 1902. After reading the following pages, I hope that you, too, will agree that given its past history and performance, the SDFU will continue to be the family farmers' advocate in the 21st century.

Chapter 1

THE PRECURSORS

"I've got a saddle and a mule and I'll start organizing tomorrow."
— Newton Isaac Gresham, September 2, 1902.

In the midst of the winter season, in January 1914, J. K. Weinmaster stepped off the train at the Mitchell, South Dakota, depot. He immediately introduced himself to a Mr. Anderson, standing near the railroad station. Weinmaster said that he was from Nebraska, and he indicated his interest in getting out into the country to talk to farmers about a new organization, the Farmers Union (FU). Shortly thereafter, an area farmer, Knute Strand, drove by on his way home after selling a load of hay in Mitchell. Anderson called to Strand, introduced him to Weinmaster, and after a brief conversation, the Nebraska resident was in Strand's hay wagon for the eight-mile ride to the Strand farm southwest of Mitchell.

After the chores were done and supper eaten, Weinmaster explained how the Farmers' Educational and Co-operative Union of America (FECUA) had been organized in Texas in 1902. Strand became engrossed in his guest's description of the the youth and vitality of this new farm organization, and its rapid spread northward through the Plains' states to Nebraska. Knute Strand, impressed by the organizer's zeal and common-sense talk, paid his dues and became the first paid-up Farmers Union member in South Dakota. The next morning, he loaned his team and buggy to Weinmaster, who spent several days organizing the Farmers Union in South Dakota. Included among the seventeen charter members that Weinmaster recruited were Knute and Carl Strand, Henry Oster, Herman Stahl, and Carl Lowrie. On February 6, 1914, they received a charter signed by Charles Simon Barrett, national president and A. C. Davis, national secretary, formally establishing Prosper Local No. 1. The first meetings were held in the Prosper Center School House, five miles south of Mitchell.

Prior to Weinmaster's arrival at Knute Strand's home in 1914, South Dakota farmers had already experienced over thirty years of economic struggle. Farm organizations had come and gone, and they had contributed little to improving farmers' income or standard of living. Because of exploitation by the railroads and flour millers in Minneapolis and St. Paul and perennially low farm prices, homesteaders and small farmers found it difficult to make an adequate living.

The first attempt to organize farmers into a major farm organization after the Civil War occurred when Oliver Hudson Kelley of Itasca, Minnesota, formed the National Grange of the Patrons of Husbandry, commonly known as the Grange, in 1867. Economic hard times in the Dakota Territory caused Henry L. Ferry to establish the first local grange in the winter of 1872 without assistance from the national organization.[1] A year later, on November 5,

1873, with national recognition, Territorial farmers formed the Dakota State Grange. In that year, there were twenty-five Grange locals and 2,000 members in the Dakota Territory.[2]

The efforts of Grange members to sponsor cooperatives ended in failure. The organization itself, after reaching its peak in 1875, began to decline with the return of good crops and competition from other farm organizations.

During the 1880s, economic unrest among farmers spread throughout the Prairie Plains. Deflation accompanied the resumption of specie payment in 1879. The ensuing decline in commodity prices, enhanced by increased production, meant decreased profits for Dakota farmers. It became more difficult to pay-off mortgages incurred during the "Dakota Boom" of 1878-1887.

Adding to the farmers' distress, crop yield was radically reduced by the lack of rainfall and periodic incursions of locusts (grasshoppers) between 1887 and 1894. Financial and natural adversity made the farmers more aware of their exploitation by credit agencies, the imposition of high interest rates, an unfair price structure determined by milling magnates and commodity firms in the Twin Cities and Chicago, arbitrary methods of grading and weighing at local elevators, abusive freight rates, and excessively high prices for farm implements and other rural consumer necessities. Wheat farmers, in particular, were the victims of high freight rates for short hauls. A wheat grower could expect to receive only about half the value of his product because of the transportation charges from Dakota Territory to Minneapolis.

The farmers' response to this continued and flagrant exploitation was the Farmers Alliance. Milton George, publisher of a Chicago newspaper for farmers, the Western Rural, organized the Alliance in April 1880. Within a year, Alliance chapters were chartered in Dakota Territory. The Northern Farmers Alliance issued a charter to a group of Yankton farmers in February 1881. The movement spread northward through the Territory and was well received in the wheat-growing area of what is now northern South Dakota. The price of wheat had dropped as low as eighty cents per bushel in 1884.[3]

The continued decline in grain and cattle prices led to the establishment of the Dakota Territorial Alliance at Huron on February 4, 1885. The farmers of Dakota Territory briefly enjoyed purchasing consumer products at a lower price when the Alliance incorporated as a joint-stock company with a capitalization of $200,000 in July 1887.[4] Alliance chapters in East River counties and a cooperative store at Watertown supplied members with binder twine, farm implements, furniture and coal at agreeable prices. The Dakota Alliance also provided fire, hail and life insurance for its members. Similar insurance programs were soon adopted by other state Alliances.[5] Unfortunately, these pioneer endeavors in cooperative work failed with the economic depression of 1893.

One of the leaders of the Farmers Alliance, Ignatius Donnelly, a former Minnesota Congressman, believed that the Grange had failed because it had not participated in politics. He argued that "you can do nothing, accomplish nothing, without the ballot box."[6] Convinced of the soundness of Donnelly's arguments, the Dakota Alliance entered into politics in the late 1880s. Led by Henry L. Loucks, who had homesteaded in Deuel County, the Alliance sought to achieve its objectives through the Republican Party when South Dakotans achieved statehood in 1889.

Finding Republicans uncooperative, Loucks formed an Independent Party in 1890 and entered the gubernatorial race. Loucks and members of the Alliance believed that their goals could only be achieved through legislative action. They sought the Australian (secret) ballot, equitable tax assessments for farmers, government control of railroads, the free and unlimited coinage of silver to increase inflation and higher prices for farm products.

Discord within the Alliance began to divide that organization in the late 1880s, and in 1889, the North Dakota and South Dakota units left the Northern Alliance and affiliated with the Southern Alliance, known as the National Alliance and Industrial Union. Coinciding with the dissension within the Alliance was the rise of a new farmer-dominated party, the People's or Populist Party. Organized in Kansas in 1890, it became a national party in 1891, integrating many small factions.

The ferment in Kansas can be described in terms of revivalism with the likes of the "Sockless Socrates of Kansas," Jerry Simpson, and the famous utterance of Mary E. Lease: "What you farmers need to do is raise less corn and more Hell!" At St. Louis, on February 22, 1892, embattled farmers had formally launched the People's Party. Assembled were 860 delegates, including all the leaders of the Prairie Plains and the South. Senator James H. Kyle of South Dakota was in this group. The delegates issued a call for a nominating convention to be held at Omaha on July 2, 1892.

Meeting in Omaha on July 4, the new People's Party broadcast its program to the world. The Omaha Platform, the first formal Populist document, read like a brief text in agricultural economics. Its three primary planks included: financial reform, government ownership of railroads and telegraph lines and the elimination of corporate and foreign ownership of land. The financial plank was the first and most comprehensive. It covered the longstanding grievances of the agrarian interests. Financially, the Populists demanded (1) the free and unlimited coinage of silver and gold at the current legal ratio of 16 to 1; (2) a graduated income tax; and (3) a federal postal savings bank for the safe deposit of the peoples' earnings and the facilitation of exchange.

If the federal government permitted the free and unlimited coinage of silver at the ratio of 16 to 1, there would be an adequate supply of cheap money available to enable rural debtors to ease their financial plight. Parity between gold and silver had been set at 16 to 1 prior to the Civil War. Because of the post-war mining boom in the West, more silver was available. The surplus silver, argued Populist leaders, should be used by debtors, who had incurred their debt through borrowing gold coins, to pay their creditors. Due to the abundance of silver, the ratio between the two metals probably approximated 30 to 1 by 1892. Thus, the true value of silver dollars was little more than fifty cents.

An additional set of resolutions, not written as such into the main platform, sought to enlist the support of a wider following. These planks included: a constitutional amendment for a single term for the President and the direct election of senators, and a recommendation to the states that they adopt the initiative, the referendum and the Australian ballot.

Andrew E. Lee, a Vermillion businessman, was nominated as the Populist gubernatorial candidate in 1896. With support from Democrats, Lee began the first of his two terms as governor. The Fusionists, Populists and Democrats, also dominated the South Dakota Legislature between 1896 - 1898 and won both seats in the U. S. House of Representatives. They were joined by one of South Dakota's leading politicians in the late 19th century, the Sioux Falls entrepreneur, Senator Richard F. Pettigrew.

Governor Lee achieved only one significant legislative success; the State constitution was changed to provide for the initiative and referendum. He proved ineffectual after 1898, as the Republicans, aided by abundant rainfall and a revived economy, controlled state offices. Populism's day in the South Dakota "political sun" was doomed with the

Original Equity Elevator built in St. Paul in 1915.

Republican's overwhelming triumph in the 1900 elections. Third party devotees returned to either the Democratic or Republican ranks.

As the ranks of the Populists dwindled, the American Society of Equity was founded by James A. Everitt in December 1902 in Indianapolis, Indiana. The Society's goal was to fix the prices of farm commodities, and it first spread southward into Kentucky, becoming popular among that state's impoverished tobacco growers. Organizing a tobacco growers association of some 40,000 members, the Society of Equity used "night riders" to bring uncooperative farmers into line and raised the price of tobacco from five to eleven cents a pound.

By 1906, the Society had a chapter among the wheat growers of South Dakota. Wheat was selling at seventy cents a bushel, and the Society of Equity sought to organize farmers to withhold wheat from the market until the price rose to $1.00 per bushel. When the withholding effort failed, Equity leaders turned to cooperative marketing as a means of controlling their product.[7]

After three years of raising money, the Society incorporated the Equity Cooperative Exchange in March 1911. With offices in Minneapolis and a warehouse in Moorhead, Minnesota, the Equity Exchange bought grain at its own country stations or handled it on consignment either for individual shippers or for country elevators. Because the Minneapolis Grain Exchange refused to accept the cooperative marketing agency as a member, Equity Exchange had to find a friendly commission firm to act as a selling agency. George Lofthus, general sales manager of Equity Exchange, unable to get a seat on the

Minneapolis Grain Exchange, moved the Equity office to St. Paul in 1914. Six years later, in 1920, the Equity Cooperative Exchange finally began construction of a $300,000 terminal at St. Paul, a terminal that would be used by its successor organization, the Farmers Union Grain Terminal Association (FUGTA).

None of the foregoing farm movements, the Grange, Farmers Alliance, Populists and the American Society of Equity, succeeded in their quest to ease the farmers' economic distress. Two of the organizations had not survived. Their efforts and ideas were not in vain. Emerging from their ranks would be the founder of one of the most active and influential farm organizations of the 20th century.

Isaac Newton Gresham, the founder of the Farmers Union, was born at Florence, Alabama, on February 20, 1858. Orphaned at ten, he grew up on tenant farms, sharing in the labor. Because of his work obligations, Gresham did not gain the education that he wanted. He was a grown man before he learned to read and write, attending school with neighbor children. He became obsessed with farm problems and would spend his adult life seeking answers. He asked why the farmers who had converted the countryside from wilderness into fertile land were hurting. Believing that a strong farm organization might be the solution, he had first joined the Grange, then the Agricultural Wheel and finally, the Farmers Alliance in the 1880s. In each case, as has been discussed, the influence and membership of the respective organizations declined.

Seeing no hope in the Alliance, Newt Gresham acquired a small newspaper, the *Graphic Truth*, through which he wanted to provide information to farmers. Subscriptions were often paid in farm products. He formulated ideas for a new farm organization, but he wondered how only one country editor could hope to get his ideas across to the farmers.

Gresham eventually moved to Point, Rains County, Texas, and about the turn of the century, he gained possession of another small newspaper, the *Point Times*. Using the town's only typewriter, he pecked out the outline for his proposed farm organization. Gresham bemoaned his limited education, as it made it difficult to relate what he wanted to say. He constantly reiterated that family farmers needed a voice and an organization to fight for their rights and survival. Farmers, Gresham argued, were continually being exploited by non-farmers. The marketing, production and distribution systems were owned and controlled by people who were not farmers. He saw a credit and mortgage system controlled by people who did not till the land. At the same time, this country editor also believed that there must be a permanent union of farmers with city laborers.

Newt Gresham. Founder and first president of the NFU, 1902-1906.

Some people in the community considered him an eccentric. Others refused to accept what they considered as his radical ideas. Soon Newt Gresham's meager funds were exhausted and he was refused credit for groceries. His convictions drove him on, and finally, nine other men in Rains County joined with him to secure a charter for a farm organization from the State of Texas on August 28, 1902. The charter included the following statements:

Original local of the FECU of America. Organized by 10 members near Point, Texas.

1. The name of the Corporation shall be the Farmers' Educational and Co-operative Union of America.
2. The purpose for which it is formed is to organize and charter subordinate Unions at various places in Texas and the United States, to assist them in marketing and obtaining better prices for their products, for fraternal purposes, and to cooperate with them in the protection of their interests; to initiate members, to collect a fee therefor.[8]

The officers, ten in number, were listed. Newt Gresham, Point, Texas, was General Organizer. The Farmers Educational and Co-operative Union of America (FECUA) had been founded by ten poverty stricken farmers. It would be built by other farmers who after a hard day's work in the field, unselfishly devoted themselves to the task of organizing and educating their neighbors.

Shortly after the receipt of the charter, the first meeting was scheduled for the Smyrna country schoolhouse, a few miles from Point. Concerned over high interest rates, manipulated markets and low farm prices, the farmers, who worked the sandy soil of Texas, packed the schoolhouse on the historic evening of September 2, 1902. They were there to join, but not all were eligible. The purposes of the organization were explained to the group by Gresham and some of his nine cohorts. The preamble to the organization's constitution read:

Point, Texas, Railway Station.

> In the course of modern industrial development, we find it necessary that the farmer not only apply the principles of scientific agriculture, but that he systematize the business by cooperation and apply the principles of scientific commerce. To enable farmers to meet these conditions, we have organized the Farmers' Educational and Co-operative Union of America.[9]

Furthermore, in its declaration of purposes, the constitution of the Farmers Union stipulated that the organization was designed:

1. To assist our members in buying and selling.
2. To teach farmers the classification of crops, domestic economy and the process of marketing.
3. To systematize the processes of production and distribution.
4. To eliminate gambling in farm products.
5. To secure and maintain profitable and uniform prices for farm products.[10]

Gresham argued that while the American people conducted their government along the lines of a political democracy, their economic life was ruled by a small economic elite. He believed that the only hope for agriculture rested with the introduction of democracy into the economic system by way of the organization and establishment of democratically controlled cooperative businesses. Only then would farmers be emancipated from the economic oligarchy that had dominated agriculture throughout the latter half of the 19th century.[11]

When asked how he would accomplish this seemingly impossible task, Gresham quickly replied: "I've got a saddle and a mule and I'll start organizing tomorrow."[12] Among those present at the Smyrna schoolhouse was John T. Garner, whose son, John Nance Garner, became vice president of the United States in 1933. Another of Gresham's nine colleagues was O. F. Dornblaser, "Uncle Dorn." Much of the credit for drafting the constitution and by-laws of the National Farmers Union (NFU) has been attributed to him.

"Uncle Dorn" had lived through the agrarian unrest of the 1880s and 1890s. He had been a member of the Farmers Alliance, the most promising farm organization in American history to that date. Dornblaser was a delegate to the national convention of the Farmers Alliance at which a representative from the Rochdale cooperatives of England was present. The Englishman pleaded with Alliance members to follow the lead of English cooperatives in building slowly but steadily a cooperative business system that would be owned and controlled by them. Unfortunately, delegates to the convention chose what they considered as a shorter and quicker way to emancipation — the road of partisan politics.

In the years that followed, "Uncle Dorn" saw the Farmers Alliance torn with political dissension, finally wrecked by self-seeking politicians who used the Alliance as a means to gain public office. Having witnessed this debacle, he placed in the original by-laws that partisan politics should be expressly forbidden in the Farmers Union, and that no man, if otherwise qualified, should be denied membership on the basis of either religion or politics.[13]

More farmers joined the organization than Gresham anticipated. The post office at Point, Texas, hired extra help to handle the large number of applications for membership. The initiation fee was $1.00, and dues were five cents per month. In addition to the low dues, the secret nature of the organization appealed to low income farmers. The movement provided a means for protest against the credit and mortgage system. Union organizers attempted to keep the membership strictly rural. It would be limited to dirt farmers, rural laborers and professionals working in the countryside such as doctors, ministers, teachers and sympathetic country editors.[14]

The Farmers Union quickly spread across the Cotton Belt of the South and became a national organization in 1905. Its membership, at that time, was estimated at 175,000 to 180,000.

Growth during the early years was hampered by dissension attributed to personality conflicts, poor business management and the reluctance of Texas Union officials to give the membership in other states greater autonomy. There were also objections to the inclusion of non-farmers in the organization and the partisan political activity of ex-Populists.[15]

The convention that created the National Farmers Union in December 1905 resulted from intense lobbying from Union groups in various states. Members from outside of Texas favored a federal system whereby governing powers would be lodged in a national council composed of the heads of various groups. The national constitution adopted at the convention closely resembled that of the founding body in Texas.[16]

In the early years, the Union established cooperatives and withheld crops from market until the minimum prices it set were met. The organization also named a new president. Following Newt Gresham's death in 1906, Charles Simon Barrett became the Union's leader, a position he would hold for the next twenty-two years. "Father Barrett," as old-time Union members called him, resided in Union City, Georgia, and for many years, it was the site of the national office. Although he was an able tactician, Barrett could not maintain the strength of the Farmers Union in the Deep South. Membership peaked in the Southern states in 1909, and by the end of World War I, the Union was in decline from Texas eastward to Alabama. Many Union-sponsored cooperative enterprises failed, some state organizations fell victim to improper business management, and many members lost money by unwisely withholding cotton from the market because they planned on higher prices that never materialized.

Charles S. Barrett,
President of the NFU, 1906-1928.

As the Farmers Union waned in the South; it spread north up the Mississippi and Missouri River valleys into the states where the Union would have its greatest strength. In 1907, charters were issued for Illinois, Kansas and Missouri. The surge continued northward across the Prairie Plains during World War I and culminated with the organization of the North Dakota Farmers Union in the late 1920s.

Following the granting of a charter to Nebraska in 1914, Union organizers from Nebraska such as the previously named J. K. Weinmaster, John M Burdick, and yes, even "Uncle Dorn" (O. F. Dornblaser), now residing in Iowa, began to organize locals in South Dakota. Having considered the roots of the Union, one must now turn to the role played by South Dakotans in the Farmers Union. The FECUA became a major force in rural America during the 20th century and the most effective voice of the old agrarianism.

Chapter 2

GETTING UNDERWAY 1914 - 1920

Pick the best man for the job ...
not necessarily the richest, strongest, ablest, most successful
or even the most popular ... but ... one whose leadership
the others are willing to accept.
— W. C. Lansdon in a manual "for the use of organizers." c. 1921.

As Farmers Union organizers drove the dirt and gravel roads of southern Davison County in 1914, the farmers of South Dakota were enjoying the best economic conditions they had experienced in many years. From 1900 to 1920, the pressure on American agriculture was temporarily relieved. Mechanization on a large scale had not yet come to the farms. The output of farm products had not increased in proportion to the output of other consumers goods or the increase in urban population. The increased urbanization, accompanied by rising wages and employment, meant a larger share of farm produce was domestically consumed. Agriculture found itself in the happy position of being limited to domestic market needs, primarily cereals and meat products, without the consequent disorganizing effects of world prices.

Some students of rural life consider the years between 1900 and 1914 as a "golden age" of American agriculture. It was a time when both market prices and land values rose. The years between 1910 and 1914 became known as the "parity years" when supply and demand stood in effective equilibrium.

Agricultural stability ended with the onset of World War I. During the war years, a large number of American farmers actually became affluent. The Allied demand for foodstuffs and cotton, the requirements of U. S. armed forces beginning in 1917, and the immediate needs of post-war Europe between 1918 and 1920 caused the prices of crops and agricultural lands to soar. Beginning in 1918, grain production reached into the more arid sections of the Great Plains. Wheat production rose to an average of approximately 870 million bushels while corn production remained relatively stable.

Using the 1910-1914 base as 100, farm prices in 1918 stood at 208 and in 1920 at an unprecedented 211, more than double 1915 returns. Lulled by a sense of security, many farmers pushed out their financial horizons. In their new found prosperity, land, equipment and buildings were mortgaged to buy more and more farmland and machinery. Land prices rose to 160 percent of the prewar average.[1] Farmers also clamored for the extension of social ser-

Horse and Buggy Recruiting.

vices by state and county governments, to the accompaniment, of course, of an ever-mounting tax rate.

Although farmers were heartened by a revived agrarian economy, there remained dark undercurrents of abuse and exploitation. Conditions were such that it made it imperative for Dakota farmers to join and unify in an organization to right the economic injustices that had prevailed since the latter decades of the 19th century. It was ironic that South Dakotans, with the greater portion of the population being composed of farmers and ranchers, were still being manipulated by out-of-state oligarchical grain and railroad interests. For example, the Minneapolis Chamber of Commerce, an association of big grain traders, terminal elevators, commission houses and the like, determined the wheat farmers' economic lot. Members of the Chamber set the prices of wheat and determined the grading and buying conditions. Twin Cities bankers and railroad magnates were allies of the grain traders.

Many farmers continued to pay-off mortgages, and they had to contend with banking firms that charged as much interest as the traffic would bear. The farmer was, almost completely, at the mercy of the local banker upon whom he was dependent for operating capital. Freight rates, like interest rates, remained comparatively high. In 1915, in South Dakota, it cost $24.00 to ship a carload of hogs one hundred miles. This amount compared unfavorably with Minnesota and Iowa where the costs for shipping the same car the same distance was $20.48 and $16.50 respectively.[2]

As Farmers Union organizers took to the road throughout southeastern South Dakota, they enjoyed the hospitality of farm people, who provided their visitors with coffee and good farm fare. The organizers had plenty of evidence to show their gracious hosts that

East House of Farmers Union Coop Association, Mitchell.

farmers living on the Prairie Plains were the victims of out-of-state interests, depriving them of their just earnings. In addition to providing information, Union organizers emphasized the advantages of cooperative bargaining in the market place.

Language sometimes proved a problem for organizers. There were many immigrant farmers and homesteaders who spoke little English and mistrusted what they could not understand. Being able to speak German, Norwegian or other European languages, was an important attribute for an organizer. The Farmers Union recruiters, who could not speak the language, tried to get a neighbor of the immigrant to accompany him to the latter's farm. A familiar bilingual person proved to be a valuable asset to the organizers as farmers trusted their neighbors. The effective organizer was the man who located the neighborhood's leader — the person the area farmers looked to for guidance.

During the nine months that followed the chartering of Prospor Local No. 1 on February 6, 1914, John M. Burdick, now designated as state organizer, and Owen Dornblaser, "Uncle Dorn" from Iowa, organized fourteen other locals in Davison County.[3] Sixty-four delegates from these locals met at Mitchell for an organizational meeting on November 24, 1914. The first regular county meeting, called to order by state organizer Burdick at 10:00 A.M., was an all-day affair.[4] Meeting in the Odd Fellows Hall, the delegates elected David Lovinger, president; E. E. Halfhill, vice president; A. J. Frantz, secretary-treasurer; J. E. French and R. C. Stiles as members of the Board of Directors; and Charles Mooney, purchasing agent.[5] The Davison County unit of the Farmers Union received its charter on December 5, 1914. It was the first county organization in the state.[6]

From the first day of organization, the first cooperative project of the members was to buy farm and household supplies through their locals. Later on, purchases were made through county organizations, cooperatives or purchasing associations. Charles Mooney, the Davison County purchasing agent, bought coal and flour by train carload, and members of the locals enjoyed substantial savings on their purchases. Other products purchased in wholesale quantities and sold in smaller lots by the local purchasing agent were apples, brooms, honey, oil, potatoes, twine and vinegar.[7]

In addition to cooperative purchasing of consumer products, a co-op elevator and a co-op produce house were established at Ethan. Soon thereafter, a cooperative supply was built in Mitchell as well as a mercantile co-op at Mount Vernon.[8]

What was happening in Davison County was in accordance with the primary goals of the National Farmers Union during the war years: the unification of organized agriculture and the development of cooperatives. A step toward fulfilling the first objective occurred in 1915 with the creation of the National Board of Farm Organizations as a lobbying group in Washington, D.C. It was composed of representatives from several farm organizations. The Union had been a leading advocate for this lobbying body, and C. S. Barrett was named its chairman.[9]

Almost all of the first locals organized at country schoolhouses and took the name of the particular school. One such local, Dodd Local No. 16, was founded on January 1, 1915, the first local established after the chartering of Davison County the prior December. Only men attended the business meetings at the schoolhouse. They would take a kerosene lantern and walk to the evening meetings if they lived close enough. Later, social meetings were held in member homes once a month for the whole family. At a time when commercial entertainment was minimal, these residential get-togethers were a social highlight and they were anticipated by all member families. At times, forty to fifty adults and children would crowd into a home.[10]

The early locals in the state were chartered by the National Farmers Union and operated under the NFU constitution and by-laws. Article III, Section 1 of the 1909 by-laws, stipulated that no local union could be organized with less than five male members. No charter could be issued until the fee, which in no state should be less than $15.00, was paid.[11] Member dues for Dodd Local No. 16 and other Davison County locals were first set at $2.50 per year.[12] Ritualism played an important part in local and county meetings. A small handbook prepared by the NFU in 1910 gave complete and detailed information covering many ritualistic ceremonies. From the opening of a session to its closing, a fixed procedure was followed by both officers and members.

A member's dues had to be paid before he was allowed to attend meetings. In the spirit of early lodge meetings, farmers applying for membership in a local were subject to a vote by current members. Voting was done with black and white marbles. An applicant was rejected if three black marbles were found in the drop-box.[13] A doorkeeper asked each arriving member for the current password to the meeting. The whole procedure was taken quite seriously and much was made of knowing the password — be it planting, corn, wheat or barley. A conductor presided over initiation ceremonies; there were signs of recognition and early Union members often addressed one another as "brother." There were directions for the closing ceremony and a prescribed burial service was available.

Many first-generation local members regretted the passing of the pageantry that accompanied ritualism but the more pragmatic Union members believed it meant more time for

One of the oldest Farmers Union locals in South Dakota is Buckeye Local 23. The hall, built in 1917, was still in use during the 1950s. It was originally located 7 miles northwest of Plankinton.

discussion, study, plans and work. Not too many years would pass before younger members knew little, if anything, about ceremonial-style meetings.

The secretary played a very important role in the local. He attended every meeting and usually was the last person to leave. Prior to the meeting, he collected dues, distributed membership cards and explained the activities of the Brokerage House. The secretary took minutes and transcribed them. He handled the locals' correspondence and deposited its money in a designated bank.[14]

Local Union gatherings soon became more than business and ritual gatherings. Hope Chapel Local No. 3, organized on February 20, 1914, began with meetings every week, later the members decided to meet every two weeks, and in 1920 they voted to meet only once a month, making the meeting both a business and social event.[15] The presidents of the locals were reminded by the state office to start meetings on time. Meetings scheduled for 8:00 P.M. sometimes did not begin until 9 or 9:30 P.M. because members first completed their field work and chores for the day. As the meeting dragged on into the night, the women waited for the men to finish the business meeting in order to serve lunch. It was often very late before the families returned home and the children put to bed. Parents and children found it difficult to arise early enough the following morning for the children to be at school on time.[16]

There were oyster suppers, a great many oyster suppers. One-act plays were staged, and pie and box socials were used to raise funds for the local. Oscar Fosheim and Gustav Moseson used to stage debates at meetings of their Miner County local.[17] The precedent for local and county Union picnics, many scheduled for July 4, occurred during the first year of

On February 24-25, 1915, the South Dakota Farmers Union was organized in a meeting at Mitchell. S.W. Ricords, Fulton, was elected president.

the Davison County Farmers Union. A picnic was held in the Strand grove four miles south of Mitchell with decorations, a band, a whole roasted beef, a wrestling match, many games, a concession stand and a dance in the evening. The Farmers Union was not only an organization dedicated to improving the economic lot of the family farmer; it was a community.

One of the well-known locals in east-central South Dakota, Local No. 447, was organized in the mid-twenties. Members of the local constructed a meeting site northeast of Forestburg in Sanborn County. The women members were asked to provide a name for the local and the building, thus, Harmony Hall was dedicated on January 6, 1926. The building became a landmark for communities activities and political gatherings. Area residents would give directions based on the location of the Hall.

Harmony Hall Local held meetings twice a month in the years immediately following its organization. Dramatic presentations by local members, including a three-act play, The Eyes of Love, were among the forms of entertainment that followed business sessions. Harmony Hall, during the 1920s and 1930s, was also a distribution point for foodstuffs, purchased in bulk from the Farmers Union Brokerage at Sioux Falls.[18]

Union leaders soon realized that a centralized, state organization was needed to bring together the diffused number of Union locals. To establish such an organization, a state had to have 5,000 dues paying members, the minimum requirement for a state charter under the prevailing by-laws of the National Farmers Union. From Davison County, the Union had spread rapidly to surrounding counties in east central South Dakota. In the midst of an "old-time blizzard," an organizational meeting of the South Dakota Farmers Union was held in

Farmers Union State Convention at City Hall, Mitchell, March 16-17, 1917. The state organization received its charter.

Mitchell on February 24-25, 1915. Members gathered on the second floor of the Odd Fellows building on West 4th Street. S. W. Ricords of Fulton was elected president of the nascent farm organization.[19] By 1917, 134 locals had been established. On March 16-17, 1917, 268 delegates from fourteen counties gathered in Mitchell to receive a state charter from the National Farmers Union. Meeting in the city hall, the convention was called to order by the national secretary, A. C. Davis, Gravette, Arkansas.[20] Also present were John M. Burdick, the veteran organizer, and National Farmers Union president, C. S. Barrett; the latter addressed the delegates.[21]

Hanson County farmer S. W. Ricords was elected the first state president. Chosen as vice president was Clair D. Johnson, Kimball, and R. B. Warne, Loomis, was named secretary-treasurer. The first state directors (Executive Board) included:

E. H. Everson	Gregory County
Frank Bradley	Hutchinson County
O. J. Murphy	Lyman County
J.Glenn Richards	Aurora County
Henry Turner	Davison County.[22]

Approximately four months after the convention, on July 9, 1917, the Executive Board received bids for the publication of the *Union Farmer*. Having received bids from the *Fulton Advocate* and the *Mitchell Gazette*, the Board arranged with C. W. Downey, editor of the *Gazette* to print about 3,000 issues per month and have them sent to all members in good standing.[23] J. Glenn Richards, Plankinton, became the first editor and manager of the *Union Farmer*. A change in both the publisher and editor of the state's Union media outlet occurred after the 1918 convention. The Executive Board accepted the bid of L. W. Kreidler, owner of the *Fulton Advocate*, to print the *Union Farmer* for thirty-five cents per copy per year. John W. Batcheller became editor-in-chief and Kreidler served as editor and manager.[24]

Hoping to gain greater recognition and new members, the Executive Board arranged for the Union's presence at the South Dakota State Fair. A tent was erected on the State Fairgrounds from which Union officials dispensed information and met with farmers interested in joining the organization.[25]

S.W. (Sid) Ricords, President, SDFU, 1915-1918.

Less than a month after the organizing convention adjourned, the United States entered World War I on April 6, 1917. The war had an immediate impact on the newly established South Dakota Farmers Union (SDFU). R. B. Warne was soon called to military service, and before the next convention in April 1918, many of the young members of the State Union were in uniform.

None of the state officers appeared at the opening of the 1918 convention. S. W. Ricords had resigned earlier to enter politics. In 1919-1920, he served in the State House of Representatives. Vice President Clair D. Johnson succeeded to the presidency for only a short time as he had entered the garage business in Mitchell, disqualifying him for Union membership. Johnson was persuaded to come to the convention and call it to order. The undaunted delegates quickly elected a new set of officers that included John W. Batcheller, Gayville, president; E. H. Everson, vice president; and Albert Stevens, secretary-treasurer. Batcheller had been a long-time activist, and Everson served on the Executive Board. Stevens had been secretary of Local No. 2, eight miles southwest of Mitchell for several years.[26]

Stevens established the state office in a room at the Merchants Hotel on Main Street, Mitchell. His salary was to be $100 per month.[27] The office remained in Mitchell until December 1921 when E. H. Sauber, Gregory County, was elected secretary-treasurer. Sauber moved the state office to Yankton, a location more accessible for Batcheller and him.[28] One of Stevens' biggest problems when he held the office involved maintaining a correct mailing list as many local secretaries were careless about providing the full and correct addresses of the their members. Stevens said: "It provided a lot of paid-up members with the opportunity to 'rake us over the coals.'"[29]

The South Dakota Farmers Union was incorporated on December 15, 1919, for one million dollars. Shares in the organization cost $50.00 each, and they bore a 10 percent interest rate. The Executive Board also approved a motion to prepare by-laws for the incorporation of the State Union.[30]

Batcheller, the State Union's third president, was one of the pioneers of early Dakota Territory, having been born in Yankton County in 1868. Educated in Territorial schools, he attended Cornell College, Mount Vernon, Iowa, and received a degree in theology from Northwestern University. After serving in the ministry for several years, he concentrated on farming, bringing him to the forefront in championing the farmers' lot. Two of his sons, George and Paul, managed the Sioux City and Sioux Falls Brokerages respectively. George Batcheller was manager of the Sioux City wholesale house from 1925 to 1936, and Paul Batcheller oversaw the Sioux Falls outlet from 1919 to 1937.[31]

Batcheller presided over the South Dakota Farmers Union from 1918 to January 1, 1929. These were difficult and formative years for the organization. During his ten years at the helm, he piloted the organization through some very troubled waters. Factional and political movements threatened to tear the organization apart. As a member of and organizer for the South Dakota unit of the American Society of Equity, he had witnessed the detrimental aspects of factionalism at the Society's convention in Indianapolis, Indiana, in 1907.[32] Although he wanted to avoid similar disunity within the SDFU, his administration was fraught with controversy. The Union faced competition from farm organizations on both the left and right. On the left was the Nonpartisan League, attempting to attract members with socialistic legislation. On the right was the newly founded American Farm Bureau, committed to a conservative farm policy. Towards the end of his tenure, Batcheller felt threatened by the very formidable Northwest Organizing

John W. Batcheller, President, SDFU, 1918-1928.

Committee. On top of this struggle to save their newly born organization, Union members were caught up in a new economic crisis after 1920. Through it all, Batcheller adhered to his primary goal: "The Union — it must be preserved."[33]

Chapter 3

BUILDING STRENGTH THROUGH ADVERSITY 1920 - 1928

I have had to fight the enemy, both without and within our ranks, the records will show that I have refused to compromise, have never had to slink away licked, have been on the square with the Union....
— J. W. Batcheller, *Union Farmer*, August 29, 1928.

If a cattleman sold a steer
and they would let him weigh all the mortgages
that was on the steer with him,
he would weigh 50 pounds heavier.
— Will Rogers

During World War I, prices rose as Europeans needed grain and the federal government sustained artificially high prices for wheat. On the Northern Plains, agriculturalists applied new technology, heavy gasoline powered machinery, to increase crop production. Farmers mortgaged their farms to invest in more land and machinery, and in 1919, they planted more acres that ever before. Land prices rose to 160 percent of the pre-war average.

PRICES DOWN — PRODUCTION UP

Alas, the boom times were all too brief, and the bubble of the farmer's content was pricked with a suddenness and completeness that left him shaken to his depths. In 1920, the bottom fell out of the grain market; it began with wheat selling for $2.96 per bushel. Within two years, the price dropped to ninety-two cents a bushel on the Minneapolis market. What had happened? Prices plummeted when the government withdrew its supports on wheat. At the same time, the foreign market collapsed as overseas competitors returned to production. Because of the forementioned mechanization and the need to compensate for lower prices, farmers continued to overproduce cereal crops. With the transition from animal to mechanical power, the number of acres that could be harvested by a single worker in South Dakota increased from 33.2 in 1920 to 100 by 1929. Land used for crop production had increased from 16,441,000 acres in 1924 to 19,003,000 acres in 1929. In their efforts to place more acres in production, many farmers cultivated submarginal land, thereby endangering the delicate ecological balance. Within a few years, they would be shown the folly of their ways as a prolonged drought, high wind, and an unrelenting sun created the horrendous dust

O.H. Bochman's, Threshing Rig, 1920s, Tyndall, South Dakota

storms of the early 1930s. Farmers overproduced because as the price of farm commodities fell, the goods they purchased cost more. For example, in 1913 a farmer could buy a suit for twenty-one bushels of wheat. In 1923, the same suit cost him the equivalent of thirty-one bushels of wheat.

The farmer's role, as a participant in the consumer economy, continued to shrink as the decade progressed. Worst of all, land values declined while taxes on farm land did not decrease and mortgage payments remained the same.

All farmers did not overproduce every year during the 1920s. Central South Dakota experienced poor crops from 1924 to 1926. Mrs. David Lovinger of Ethan described 1926 as a bone-dry year. Roadside ditches, small streams and several of the lakes were all dry. The lack of rainfall led to empty grains bins and cribs, half-filled hay mows and empty fruit jars. Farmsteads in the James River Valley could have used the surplus corn dumped on the market a few years earlier.[1]

Throughout the decade, conservative politicians in South Dakota refused to provide succor for the state's largest industry. When the State Legislature failed to provide any tax relief for beleaguered farmers, Batcheller called a special convention of the SDFU at Pierre on February 17-18, 1927. He hoped to create a greater public awareness of the farmer's plight and to encourage consideration of tax relief legislation. Almost 500 people attended the session, including 227 voting delegates.

O.H. Bochman on steam-driven tractor, 1920s, Tyndall, South Dakota.

The *Pierre Daily Dakotan* probably spoke for the conservative-dominated Legislature in publishing a scornful editorial response to the Union's special session. In condescending and haughty terms, the paper's editor wrote:

> The Farmers Union. These brawny tillers of the soil who have sprung into organization at the cost of four dollars per, have crusaded to the capital to instruct the legislature after the date for introduction of bills and measures has gone. They apparently have nothing to offer at this time aside from their desire to show that there is such an organization.... Blindly following, willing to be led, seeking relief - from what?[2]

In this instance, the editor was obviously blind to the economic conditions existing in South Dakota's leading industry.

Another era of farm foreclosures and bank failures had begun. As economic conditions in the rural areas deteriorated, bank failures accelerated. Nine banks in South Dakota failed in 1922, thirty-six in 1923, and by January 1925, 175 state banks had closed their doors. The federal government gave temporary relief to rural banks through short-term loans from the Agricultural Credit Corporation, but these loans did not stem the tide of bank failures.

Cooperation, legislation, and education formed what the Union called the "triangle" of its functions. Leadership at both the national and state levels believed that legislation was the best approach to solving the critical farm situation during the 1920s.[3] In South Dakota, a legislative approach was preferred over any cooperative solution proposed by President Batcheller and his successor, E. H. Everson, through the end of the latter's tenure in 1934. Working with other farm organizations, the Union tended to give its support to a bipartisan farm bloc in the Congress that included Senator Peter Norbeck, and in the late twenties, Senator William H. McMaster.

Although the farm bloc made farm legislation one of the crucial issues of the 1920s, governmental programs during the decade were concerned only with details. There was little or no effort made to come to grips with the fundamental problems of high costs, heavy fixed charges and the contraction of the market for agricultural goods. Advocates of the cooperative movement appreciated the passage of the Capper-Volstead Cooperative Act in February 1922. This law exempted, once and for all, agricultural associations or cooperatives from attack under the Sherman Law. Cooperatives were now granted the right to process, prepare, handle and market their wares in interstate commerce.

Maintaining debt-ridden farmers as dues-paying members of the South Dakota Farmers Union and lobbying for federal legislation to improve the farmers' economic plight were serious matters; another problem was the appearance of other farm organizations that vied for the Union's membership.

THE NONPARTISAN LEAGUE

One of these organizations was the Nonpartisan League (NPL), of which President Batcheller was a member. Interested in any cause that might promote the small farmer, he had joined both the Farmers Union and the Nonpartisan League after his affiliation with the Equity. Late in his life, in an interview with Robert Thompson, Batcheller said that he had been reluctant to join the League because of his existing membership in the Union. He was persuaded to become a League member when one of its organizers was physically attacked in western South Dakota by League opponents. Batcheller argued that he joined the NPL out of sympathy, nothing more.[4]

The Nonpartisan League began to gain prominence in 1915 in North Dakota. Arthur C. Townley, its founder, was determined that the League would be more than a farm organization. It would become a political party. He and his organizers recruited members by telling farmers that their only party obligations were to pay their dues and to vote in local and state elections. He said the League leadership would solve their problems. Believing that political elections were won in the precincts, Townley wanted to establish strong local organizations. Above all, the NPL should avoid becoming a third party. Through the direct primary, the Nonpartisan League sought to gain control of one of the two major political parties, the Democrats or the Republicans. Once the League members achieved political dominance, they proposed to implement a program of public ownership and control as a solution to the farmers' economic ills. Townley called for state ownership of terminal elevators, flour mills, packing houses and cold storage plants. The state would supervise inspection of grain and grain dockage, establish rural credit banks operated at cost, and exempt farm improvements from taxation.

By 1916-17, League organizers began operating in neighboring states. In South Dakota, Batcheller was one of many farmers holding membership in both the SDFU and the NPL. Why not? Both were dedicated to improving the lot of the small farmer. Batcheller, nevertheless, was worried about the ever-increasing presence of the League. Another very concerned person was Governor Peter Norbeck (1917-1921).

Governor Peter Norbeck, 1917-1921.

Norbeck took preemptive action to quash the League's effectiveness in South Dakota. He countered Townley's program of state enterprises and intent to control one of the two major political parties. Norbeck pledged to support constitutional amendments that would permit the state to enter into business enterprises if voters concurred. He asked the Legislature to consider state hail insurance on crops, state-owned and controlled coal mines, terminal grain elevators, flour mills, stockyards and a cement manufacturing plant.

The centerpiece of his program was a Rural Credits system that would enable the State of South Dakota to loan money to farmers at low interest rates. Following approval of Rural Credits early in 1917, Norbeck persuaded the Legislature to fund a marketing commissioner, believing that the business of marketing farm products was in the public interest and subject to control by the state. As a progressive, the Governor firmly believed in the use of governmental powers to alleviate social and economic distress. Norbeck's Rural Credits program never had the full support of the Farmers Union. The Union approved of the principle but disliked the way in which it was administered in association with private banks. The SDFU said Rural Credits should be administered entirely by the state.[5]

Viewing the Nonpartisan League as ideological and leader controlled, Norbeck took additional, but questionable, steps to make it politically impotent in South Dakota. After the United States entered World War I, the Governor questioned the patriotism of League members and tried to discredit them. At one time, he issued an order prohibiting Arthur Townley from speaking in the state, rescinding the order three days later. A month after American entry into the war, Norbeck appointed a Council of Defense to facilitate South Dakota's role in the war. Local councils took it upon themselves to break up political meetings of the NPL and to drive organizers out of communities such as Britton.

Batcheller's concern was not the patriotism of League members, but that their growing numbers in the Union could lead to factionalism. He had witnessed, first hand, what divisiveness had done to the American Society of Equity. He went to the FU state convention at Mitchell in October 1918 to alert the delegates to the dangers of factionalism. Batcheller indicated that divisiveness was evident upon his arrival in Mitchell. He believed that his speech to the convention convinced the anti-Nonpartisan League delegates to place his name in nomination for state president. Because the anti-NPL element had a slim majority, they had been able to elect him as the third president of the SDFU. At the time of the vote, neither faction was aware of his dual membership.[6]

According to the newly elected president, the NPL group, later in the day, discovered that he was a paid-up member of the League. When his election supporters were confronted with this revelation, they felt deceived by Batcheller and the League. By 1923, the NPL was no longer a significant organization, but the Union president's former membership in that organization tended to hinder his policy-making endeavors for several years. Unlike Governor

Norbeck, Batcheller practiced tolerance in relations with League members. He believed that his neutral position toward the NPL represented "the unbiased and impartial attitude that the Farmers Union should have toward the League."[7]

Other SDFU leaders were not as tolerant. Shortly after Batcheller took office, the state organizer, Ed Martly, in a sense, declared war on those farmers who held dual membership. Without authorization from the state office, he informed pro-League locals that they were expelled from the Farmers Union. When Batcheller became aware of Martly's activities, he relieved the organizer of his duties. Martly continued his association with the Union and found an ally when E. H. Everson succeeded Batcheller as president in 1928. Everson openly declared his opposition to the few NPL members who were still members of the Union.[8]

THE AMERICAN FARM BUREAU

At the time the Nonpartisan League was collapsing, a more powerful and enduring competitor was emerging in the eastern United States. Post-World War I industrial unrest, the prospect of a farmer-labor alignment, remote though it was, and a general fear of radicalism caused certain vested interests to consider a more conservative farm organization.[9] In 1919, the American Farm Bureau was organized.

The Farm Bureau became a strong supporter of the county agent system which, in turn, played an important role in its organizing activities.[10] Putting its major energies into marketing and legislative efforts, the Farm Bureau became strongly entrenched in the Old Midwest, the states of Michigan, Ohio, Indiana and Illinois. It nearly wiped out the Farmers Union in the South, severely crippled the Grange, and devastated the American Society of Equity except in Wisconsin. The Bureau, which became an anathema to President Batcheller and other SDFU leaders, moved quickly through Iowa and into South Dakota.

As of 1922, National FU President Charles Simon Barrett enunciated the official stance of the Union toward the new farm organization. Speaking to the delegates of the national convention of the Farmers Union, he said:

> For the last two or three years a powerfully financed movement has been attempting to do the farmer's thinking for him .... Generally speaking, the organization known as the American Farm Bureau Federation has, throughout its state units, been officered and led by the "complacent" farmer and the banker or politician farmer, and it has been in the activities of the farm bureau movement that one could best see the efforts of . . . financial, manufacturing and political magnates to direct and to mold the destiny of American agriculture . . . .[11]

The fiery and forceful Milo Reno, president of the Iowa FU during the 1930s, told a convention audience in 1934 that the Farm Bureau was "born a bastard of the Railroads and the Chamber of Commerce. No farmer had anything to do with organizing [it]."[12] Other Union leaders likened the Farm Bureau to a company union versus a genuine labor union.

Batcheller, the former minister, could also engage in evangelical rhetoric, and he and the paper he edited, the *South Dakota Union Farmer*, supported the indictment made by Barrett. The *Union Farmer* claimed that the Farm Bureau originated in the Binghamton, New York, Chamber of Commerce, and that the first county Farm Bureau was organized in Brooms County, New York. In charge was a county agent, an organizer for the Bureau. The

SDFU press further stated that the Chicago Board of Trade gave a cash grant of $1000 to each of the first 100 Farm Bureau units formed, beginning with the one at Binghamton.[13]

On May 8, 1914, Congress had passed the Smith-Lever Act. This legislation provided for a system of agricultural extension work based on cooperation between the Department of Agriculture and the land-grant colleges. Federal grants-in-aid were to be matched by state appropriations in carrying out the program. Local farm bureaus, created to sustain the nationwide system of county agents authorized by Smith-Lever, were incorporated into the American Farm Bureau Federation after 1919. Batcheller believed there was an unholy alliance between the county agents, higher education, and the Farm Bureau. He argued vigorously that the federal government was spending hundreds of millions of dollars not to educate the farmer but to make him profitable to someone else.[14]

When President C. W. Pugsley, South Dakota State College, invited the Farmers Union to hold its annual convention on the Brookings campus, Batcheller delivered a stinging reply. In his letter of response in June 1926, he based his refusal on the inconvenience of the location and his conviction that "the social, commercial and political atmosphere would be as foreign to us as Babylon was to the captive Jews,...." He informed Pugsley that his "institution [State College] is trying to make the world safe for county agents." The Union president believed that a massive conspiracy of the "educational, professional, political and commercial institutions of the state, together with the embryonic South Dakota Chamber of Commerce" were plotting the Union's finish. Batcheller began a discussion to circulate petitions to abolish extension work.[15]

The Union also associated the much detested compulsory tuberculosis eradication movement with county agents, State College and government bureaucrats. Beginning with the 1924 state convention, and at several ensuing conventions, members of the SDFU passed resolutions expressing their bitter opposition to the program. They believed that the program was unnecessarily extravagant and a useless expenditure of public funds.[16] It was pointed out that less than 2 percent of the cattle in the state were infected with tuberculosis, and the test was being forced on farmers who did not need or want it. The accuracy of the tests was questioned. Veterinarians were accused of becoming inspectors, because they were unable to earn a living in a competitive field.[17] Lora C. Little spoke for Union members when she wrote: "Enough! We have known all along that the thing is unnatural and opposed to common sense; but we have been awed by humbug and pretense masquerading under high sounding names, and too many of us have helped swell the cause."[18] The opposition to the testing program did not dissipate until the mid-1930s when a more moderate leadership took over direction of the State Union.

The Farm Bureau quadrupled its family memberships in South Dakota between 1920 and 1925, but after the mid-1920s, its membership declined rapidly, and by 1935, only 556 farm families were members of the Bureau.[19] During the same time period, the SDFU lost less than 700 family memberships; there were 5,303 members in 1919 and 4,660 in 1933.[20] Despite the farm crisis and the competition, the Union had held its own through strong leadership and a vigorous political action program supplemented by developing education and cooperative programs.

Perceptions of the Farm Bureau expressed in the 1920s and '30s by Batcheller, Barrett, Reno, John Simpson and others did not change in the following decades. The Union, at both the national and state levels, continued to assert that the Farm Bureau usually aligned itself with business groups against other farm organizations and Rural Electrification Administration cooperatives.

JOUSTING WITH THE NORTHWEST COMMITTEE

Batcheller always argued against factionalism in the Union, but he contributed to a rancorous struggle within the organization that began in the latter half of the twenties and continued through 1936. As in many Union disputes, certain personalities played a major role. In this case, Batcheller's nemesis was A. W. Ricker, a Farmers Union pioneer, champion of the cooperative movement and editor of the *Farmers Union Herald*.

The controversy found its roots in another Townley-inspired farm organization. In February 1923, during the final days of the Nonpartisan League, the North Dakota radical organized the National Producers Alliance with the help of old League organizers. Farmers were wary of Townley's promises and many were reluctant to join the Alliance. Townley lost interest in his new brainchild, and early in 1924, A. W. Ricker became the national manager of the Alliance. The Alliance enrolled approximately 40,000 members, primarily in the spring wheat area of the Northern Plains. In its short life, it remained an amorphous organization.

There existed among the wheat farmers that had survived the Nonpartisan League — the Equity Cooperative Exchange (ECE), which had been organized in 1911. It had continued to market grain through its parent organization, the American Society of Equity, until the latter fell victim to the Farm Bureau and other farm organizations. The Equity Exchange had grown rapidly after 1914, establishing a terminal grain marketing agency in St. Paul. During World War I, the ECE controlled some eighty local elevators in Minnesota and North Dakota.The depressed farm economy of the 1920s forced the Equity Exchange into receivership in 1923. Its economic problems can also be attributed to the lack of good business management and the opposition of the private grain trade.

Members of the Alliance and the Equity were present when National Farmers Union President C. S. Barrett called a meeting of all farm organizations operating on the Prairie Plains at Des Moines, Iowa, on May 25, 1925. They were in Iowa's capital city to discuss ways through which farmers could improve their economic status. Delegates from the National Producers Alliance included Ricker and Ed Overvold of South Dakota.

A.W. Ricker,
Editor of the *Farmers Union Herald*.

The upshot of the Des Moines meeting was the formation of a new organization, the Corn Belt Committee or Grain Belt Confederation, composed of twenty-four separate farm groups. The cost of production as a basis for price became the official program of the Confederation. To achieve cost of production, it was decided that a dynamic organization was needed to control marketing and to secure passage of tariff legislation. In the years that followed, farm speakers and organizers constantly reiterated the necessity for cost of production. After all, as one speaker said: "The farmer plants with faith, harvests with hope and markets with charity."[21]

In a jab at the Ricker-led Alliance, the delegates insisted that the new body include only farmer-member groups. Many in the assemblage expressed their disenchantment with cooperatives because of cooperative failures in the ECE that had been accompanied by bad publicity and the loss of money. Barrett, nevertheless, extended an invitation to the Equity Cooperative Exchange and the National Producers Alliance to merge with the Farmers Union. Don Eliason, another Alliance delegate from South Dakota, was appointed to the merger committee.

On January 14-15, 1926, Equity stockholders and Alliance members met in joint-convention at Fargo, North Dakota, and decided, with minimal opposition, to have their entire farmer membership join the Farmers Union and give it their active support.[22] Following the two-day meeting, the officers of the Alliance, receivers for the Equity Exchange, President Barrett and the National Secretary of the NFU, studied the realities of the new situation. Alliance members were given a year to transfer their membership in the various states, including South Dakota, to the Farmers Union. The NFU further agreed to have all dues turned over to a Northwest Fund to help pay the costs of organizing farmers in at least three states, including North Dakota.

The SDFU played host as the National Farmers Union held its annual convention at Mitchell on November 17, 1925. "Uncle Reuben," NFU President Charles S. Barrett, led the parade through downtown Mitchell. He was unanimously reelected president before the delegates debated the merger proposal.[23] Members of the SDFU were cool towards the proposed merger, but they did not oppose it. In South Dakota, the Alliance had built a membership that was larger than the Farmers Union, and Batcheller and the Board of Directors feared dominance by the Alliance.[24] In what was really a *pro forma* vote, convention delegates approved the merger resolution.

During the first months of 1927, FU leaders made preparations for making the greatly expanded organization even more effective. The merger brought skilled, aggressive leadership and ideas to the Union. *The Farmers Market Guide*, published by the Alliance, and the *Cooperators' Herald*, published by the Equity Exchange, were combined into a new publication, the *Farmers Union Herald*. In March 1927, NFU and cooperative officials planned their strategy for the coming months. They decided to group Wisconsin, Minnesota, North Dakota and Montana into a "Northwest Division" and to name a committee to guide organizational work in these states. M. W. Thatcher of Minnesota and A. W. Ricker were named chairman and secretary respectively. They, along with C.C. Talbott of North Dakota, made up the executive committee.

Ricker, an articulate spokesman for the Northwest Organizing Committee, assessed the merger with the following statement: "When the merger of the Alliance, Equity and Farmers Union came in 1926 ... it was the Alliance which furnished the organizers, the Farmers Union the good name and history, but it was the old Equity philosophy and spirit which provided the stickers in our organization."[25]

Although the merger generated much enthusiasm, lines of battle were being formed within the Union. Three definite points of view emerged. First, there were those members who wanted farmers to develop strength by using the collective bargaining methods developed by organized labor. Others, including Ricker and Thatcher, favored cooperative marketing and buying. A third group pushed for political action on the national level for farm legislation.

The full Northwest Committee avidly backed certain business activities, including the newly organized Farmers Union Terminal Association. The Terminal Association was the former Equity Exchange operation at St. Paul. It had been purchased by the Farmers Cooperative Grain Commission, a temporary organization, to bring about the transition to the FU Terminal Association in 1926.[26] In 1930, when the work of the Committee was brought to a close, some 60,000 new FU members had been added in four states. The organizers had been financed in large part by the profits of the Union.[27]

Batcheller became a bitter opponent of Ricker's aggressive policies. He accused Ricker of being an autocrat and of allying with managers and business firms. Saying that the SDFU was strictly a farm group, the Gayville farmer rejected what he called "Autocratic Commercialism." The leaders of the Northwest Committee were not true farmers. The controversy was made public as the letters exchanged between the two men were published on the front page of the *South Dakota Union Farmer*. The two men "pulled no punches." In fairness to Batcheller, it must be noted that he aired both sides in the Union paper. He made it a policy to print the primary correspondence he received and to offer comments on the various letters. What occurred was an interesting dialogue in print. In May 1927, the SDFU president wrote: "... I do feel that our programs are diametrically opposed, and that it is now up to us to engage in a fight to the finish in order to determine whether autocratic or popular control, is to determine the policy of the Farmers Union."[28] In an effort to undermine Ricker's reputed success, the SDFU Board of Directors, on August 5, 1927, appointed E. H. Everson to document the financial standing of the FU Terminal Association in St. Paul.[29]

Ricker used his positions as editor of the *Farmers Union Herald* to accuse President Batcheller of obstructing organizational activity in South Dakota counties where there were hundreds of paid-up members ready and anxious to join the Union.[30] In a blistering editorial that followed Batcheller's announced retirement in 1928, Ricker wrote: "South Dakota has been a bone of contention in the Union for the past several years, because its state president, John Batcheller, has been out of step and out of harmony with the balance of the Union."[31] The St. Paul editor said the trouble began at the Farmers Union convention at Hot Springs, Arkansas, in 1927, when Batcheller tangled with Iowa's Milo Reno. The South Dakota Farmers Union president had failed to be elected as a national board member and had been an obstruction in the Union at the state and national levels ever since.[32]

At the annual state convention at Miller on October 4, 1927, Batcheller was reelected president, but he had opposition. The incumbent received 162 votes, while E. H. Everson had the approval of sixty-three delegates. Everson was then elected vice president.[33]

The invective became so bad that the SDFU Executive Board, on March 14, 1928, formally expressed its concern over the personal attacks being issued from the state office. The Board said that the attitude of the state office was wrecking "our national program of marketing and financing in South Dakota." In addition, the Produce House was suffering because of the split in South Dakota's Union membership.[34]

Another controversial issue during the latter half of the 1920s was the ill-fated McNary-Haugen bill. It was the most important farm legislation proposed in the 1920s. The bill made use of traditional devices, including a high tariff, to protect American producers from foreign competition. Reminiscent of the Grange's export debenture plan, McNary-Haugen provided for a two-price system, one for the domestic market, another for the world market.[35] Championed by the farm bloc in the United States Senate, including Norbeck and McMaster of South Dakota, and by America's farm organizations, farmers mobilized enough strength to force the measure through both houses in February 1927, and again in May 1928. In neither instance did the farm bloc have margins large enough to override President Calvin Coolidge's veto. The bill, in its final form, provided for a federal farm board.[36] The board was to have authority to make loans to cooperative associations for the purpose of assisting them in controlling seasonable surpluses in excess of domestic needs. The board was to collect an equalization fee from the grower of each staple sufficient to cover the losses incurred in the marketing of the surplus, due to the difference between the domestic price and the world price. A revolving fund of $400,000,000, to finance these operations, was to be established.

Although the National Farmers Union spent much energy in support of the McNary-Haugen bill, Batcheller opposed it. In 1927, he said the SDFU was against the bill because they [the farmers] were at the base of the pyramid as the producers of the new wealth that must finance the whole show. The bill would help only capitalist farmers.[37] He went on to argue that if the farmer can't or won't organize, control his products, and protect himself through collective buying and selling, he had a dark future.[38]

When McNary-Haugen cleared both houses in May 1928, the SDFU president sent Coolidge a telegram asking him for a second veto of the bill. Batcheller had also insisted that a statement in the *Congressional Record* that the SDFU was for the measure be corrected.[39] He believed that the proposed farm board would create new government machinery, self-promoters, and enlarge the Department of Agriculture. President Batcheller also asserted that the board could not maintain the domestic price of wheat at forty cents above the world-price level unless it was prepared to buy every bushel of the crop that was offered for sale.[40] The passage of the Smoot-Hawley tariff and the resulting paralysis in international trade rendered the McNary-Haugen plan obsolete.

The beleaguered Batcheller announced his retirement at the 13th annual state convention of the Union, held in the American Legion Hall at Winner, October 9-11, 1928. As the convention began, he believed that he had sufficient votes for reelection, but because of ill-health and the divisions within the Union, it was best to step down. He felt there were lines of battle at the convention between the "enterprise control" [Ricker] and "membership control" [Batcheller] groups. His retirement, he was convinced, created great jubilation among some people. In addressing the delegates he plead for unity. "Folks," he said, "this Union is bigger than any man and bigger than this convention. That there is a division in our camp is well known and the thing we must look to is the holding together of this organization and building it for greater service."[41]

In his 1948 interview with Robert Thompson, Batcheller claimed that prior to the convention the Northwest Committee, in their efforts to remove him as president, had planned to have three candidates nominated. The nominees included Batcheller, E. H. Everson and the Northwest Committee's candidate, Joe Atkins of Wessington Springs. The withdrawal of the SDFU president from the election and his endorsement of Everson, he argued, threw the opposition into disarray.[42]

In the hotly contested election, described by one delegate as an awful fight, Everson defeated Atkins 149-120.[43] Everson had earlier gained the attention of the delegates with various recitations and a series of jig dances.[44] E. H. Sauber was again elected state secretary-treasurer. Another highlight of the convention was the observance of "Father" Barrett's birthday. According to Douglas Murphy of Woonsocket, who attended the 1928 convention, the delegates paid tribute to President Barrett by throwing nickels and dimes on the stage.[45]

What occurred between 1926 and 1928 was more than a dispute between Batcheller and Ricker. Within the ranks of the National Farmers Union, there was a growing difference over whether cooperatives or legislation was the best approach to the farm problem. Batcheller and his ilk saw the value of cooperative effort. After all, cooperatives had been formed in South Dakota in the 1920s, but under present conditions the power of cooperation was so limited that it would not bring the needed economic relief. At that particular moment in history, he believed that a major emphasis on cooperatives would only weaken the Union. Such emphasis would detract from more valuable programs, primarily those that were legislative in nature. In the event of a cooperative's fiscal failure, the SDFU leader believed the Union would be criticized rather than the management of the cooperative endeavor.[46]

THE FIRST COOPERATIVES

The SDFU, nevertheless, joined in several cooperative endeavors during the 1920s. By the time Batcheller resigned in 1928, the State Union was involved in three major cooperative enterprises, an insurance company, livestock commission and a brokerage firm that did wholesale as well as retail business.

In 1923, the SDFU had absorbed the Rosebud Mutual Insurance Company. The Rosebud firm had filed articles of incorporation on May 20, 1914, but it had failed to grow and expand as envisioned by its stockholders. By the early 1920s, the stockholders sought reorganization at the very time the State Union was considering entering the insurance business.[47]

During the annual convention in 1922, a committee was chosen to establish a mutual insurance company. In a meeting with representatives of Rosebud Mutual in the summer of 1923, Rosebud officials offered to sell the company with no cash being involved. As its part of the bargain, the SDFU agreed to assume all obligations on condition that the Union receive all the good will and assets owned by the company. On March 1, 1924, the Rosebud Mutual's policy holders approved an amendment to the articles of incorporation, changing the name of the company to the Farmers Union Mutual Insurance Company of South Dakota and changing the number of directors from five to thirteen.[48]

The renamed company continued to have its headquarters at Burke, South Dakota. P. G. Eckhoff, Platte, was president of the firm. As of 1924, the company insured farmers' homes, barns, grain and livestock against fire, lightning, and tornadoes.

One of the most successful cooperative ventures of the SDFU through its first half-century of existence was the sale of cattle through livestock commissions. Many farmers joined the State Union because of the financial benefits they could reap through the cooperative enterprise. From 1903 to 1920, the American Society of Equity and the FU took the lead in furthering the livestock cooperative movement.[49] The first FU Livestock Commission on the Northern Plains was established at Omaha on April 1, 1917. The Nebraska Farmers Union loaned the firm $2,000 for its first capital.[50] The Omaha operation was the first successful farmer-owned terminal livestock marketing agency in the United States even though it was denied membership in the Omaha Livestock Association because it issued patronage dividends. It served as a model for cooperative firms later established at Sioux City and Sioux Falls.

Shortly after the creation of the FU Livestock Commission at Sioux City in 1918 by the Nebraska Union, the SDFU Executive Board appointed a committee to investigate affiliation with the Sioux City firm.[51] Batcheller was intrigued by the Commission's goal of working

Sioux City Branch, SDFU Brokerage Co. The Home office of the SDFU Cooperative Wholesale until 1936.

toward cost of production plus some profit. Because of Sioux City's location, South Dakota's FU members began sending their livestock to this northwest Iowa community. In its first year of operation, the FU Livestock Commission did 2 percent of the business at the yard. In 1924, it did 12.5 percent, one-eighth of the total business at the Sioux City yard. Batcheller believed that the Farmers Union should be doing several times that amount of business. With typical Batcheller hyperbole, he said that some members were doublecrossing and playing into the hands of the enemy. "A man, "he said, "is either a Union man or he is nothing, and it is about time we begin to draw the line or close up shop."[52] Plans to establish a FU Livestock Commission at Sioux Falls were set in motion by the Executive Board on November 11, 1924. A committee of two, F. A. Uttecht and Martin Ellingson, was appointed to investigate the feasibility of such a venture.[53] By 1925, South Dakota Union members furnished approximately 45 percent of the livestock being processed through the firm in Sioux City.[54] The State Union's leadership believed it was imperative that the power and efficiency of the Union's business institutions must be broadened. Attention must be given to both producer cooperatives (selling farm produce) and consumer cooperatives (the purchasing and sale of supplies to members).[55] The conclusion was reached that it was time either to incorporate with the firm in Sioux City or, at least, to establish a branch livestock house at Sioux Falls.[56]

Over three years passed before the SDFU Executive Board, on February 1, 1929, decided to take $1000 from the State Union treasury to finance the opening of a livestock marketing agency at Sioux Falls.[57] Within two weeks, C. H. O'Hern, at $350 per month, was employed

—3—

FARMERS UNION STATE EXCHANGE, OMAHA, NEB.

SPECIALS

SUGAR

A-5771—Beet sugar, f. o. b. Omaha, net cash..............$7.40
A-5770—Cane sugar, f. o. b. Omaha, net cash............$7.50
A-5776—Powdered sugar, 25-lb. bag, f. o. b. Omaha, net cash, per pound .... .11
A-5775—Light brown sugar, f. o. b. Omaha, net cash........ 8.65
A-5773—Medium brown sugar, f. o. b. Omaha, net cash..... 8.45
A-5774—Dark brown sugar, f. o. b. Omaha, net cash........ 8.40

A5723 Oyster Shells, 5 bag lots or more $1.20

Prices Quoted Subject to Change.

BAKING POWDER

	Per Can	Per Doz.
A-5012—Calumet, 10c size	.09	$0.97
A-5013—Calumet, 20c size	.18	1.90
A-5011—Calumet, 1-lb. cans	.30	3.25
A-5014—Calumet, 5-lb. cans, cases of six	1.20	6.30
A-5015—Farmers Union, per pound	.30	.25
Farmers Union, 5-lb. cans, ea....	1.10	...
A-5017—K. C. 15c size....	.14	1.40
A-5018—K. C. 25c size....	.23	2.35
A-5019—Royal, 10c size...	.09	.97
A-5020—Royal, 6-oz. can.	.25	2.70
A-5021—Royal, 12-oz. can	.47	5.20

BEANS

Packed in 100-lb. bags.

	Per lb.	Bag
A-5026—Lima Beans	.10	9.00
A-5029—Michigan Hand Picked Navy Beans......	.07	5.50

BLUING

	Each	Case
A-5033—Liberty Ball, 36 pkgs. to case............	.04	.85
Mrs Stewarts, 24 bottles	.15	3.00
A-5034—Packed regularly 2 doz. to the case		
A-5036—Packed 6½ doz. to case, billed as 6 dozen.		
A-5037—Packed 14 doz. to case, billed as 12 doz.		
A-5035—Paddle bluing, 12 to the case..............	.05	.50

BROOMS

	Each	Doz.
A-5044—Exchange Brand	.65	7.50
A-5047—Farmers Union Special	.45	5.00

WASHING POWDER

	Per Pkg.	Per Case
A6236—Lux, 100 pkgs to case...........	.12	10.75
A6237—Lux, 50 pkgs. to case......................	.12	5.50
A6235—Rinso, 100 pkgs. to case....................	.07	5.75
A6225—Ammonia, powered, 36 pkgs. to the case	15.	4.25
A6226—Gold Dust, large, 24 pkgs...	.30	6.90
A6227—Gold Dust, small	.05	4.05
A6228—Old Dutch Cleanser.........	.11	4.75
A6229—Rub-No-More, small	.05	4.50
A6231—Sani-Flush	.22	2.25
A6232—Sunbright Cleanser, 50 to cs.	.05	2.15
A6233—Seafoam, 16 4-lb. pkgs to case	.29	4.00
A6234—Seafoam, 100 8-oz. pkgs to the case	.05	4.15
A6230—Rub-No-More, large	.23	4.50

SPECIAL DEAL

We are offering for this month only the following shipment of Kellogg's Products, freight prepaid. This is an unusual deal and will be worthy of your attention.

	Per case
1 case Kellogg's Corn Flakes, large, 24 pkgs..............	$3.50
1 case Kellogg's Corn Flakes, small, 36 pkgs..............	3.50
1 case Kellogg's Corn Krisps, large, 24 pkgs................	3.20
1 case Kellogg's Krumbled Bran, 12 pkgs..................	2.25
1 case Kellogg's Shredded Krumbles, 24 pkgs...............	2.75

We will ship the above in any quantity or way you want it but there must be at least 5 cases in order to make a prepaid shipment.

Let us have your orders now.

PLEASE MENTION OUR NUMBER WHEN ORDERING

Price List, Farmers Union Exchange, Omaha, Nebraska - 1920s.

as the manager and salesman of the new cooperative enterprise.[58] At its opening, the firm was equipped to handle only hogs. Within a month-and-a-half, O'Hern announced that the Commission was also prepared to process cattle and sheep. The establishment of the Commission in Sioux Falls created a competitive market with other buyers. In addition to receiving dividends on commissions, by directing their livestock through the member agency, they forced other buyers to compete for their stock.[59]

One other major cooperative venture supported by Batcheller and the SDFU membership that survived the crisis years of the early 1930s was a brokerage business. A Farmers Union Brokerage and Produce Commission had been organized at Sioux City in October 1924. Beginning business in January 1925, at 412 Pearl Street, the Brokerage provided farm machinery, binder twine, coal, salt, wire, posts, minerals, coffee, flour, feeds, oil, gasoline, tires and other necessities to FU members. It returned all profits back to the Union farmers who furnished the business on which the profits were made.

Although located in Sioux City, the cooperative had been formed and licensed in South Dakota in January 1925. Over a year later, on April 5, 1926, it received a certificate of incorporation from South Dakota.[60] The Brokerage had an uncertain beginning. Plagued by poor management, there was an almost immediate investigation of its operations. On January 6-7, 1926, the SDFU Executive Board acted to dispose of stock on hand in the Produce House and converted the firm into a strictly brokerage business.[61] The manager of the Brokerage resigned on April 1, 1926, and was succeeded by one of his subordinates.[62] Within eighteen months, the Brokerage was solvent, having not only monies available to repay all funds borrowed from the SDFU but a small operating surplus. The Brokerage had survived the onset of the Great Depression, because it had quickly developed a sound economic base.[63]

One other landmark event occurred in the twilight of Batcheller's administration. The first serious effort to unite organized labor with the SDFU happened in 1927. Theodore Reise, secretary-treasurer of the South Dakota Federation of Labor, invited President Batcheller to address the Federation's eighth annual convention on August 7. Reise suggested that both organizations had much in common, and they would have a better understanding of each other if it was possible to exchange ideas.[64] Batcheller readily accepted the invitation, and when the State Union held its convention in October, the delegates elected three people to be a part of the six-person permanent committee to develop a basis of cooperation that would be profitable for both groups.[65]

The SDFU's third and first long-term president was a man of strong convictions, perhaps too many. He personified the individualism of the first inhabitants of the Dakota Territory. Despite his inability to compromise, Batcheller successfully led the Union in its infancy through grim times for American agriculture. At the time of his retirement, paid-up membership in 1928 totaled 7,017. There were 333 locals and thirty-five county organizations, with Bennett, Todd and Minnehaha counties being organized since the last convention.[66] The organization was in place, and with no regional or national help, the State Union had made significant strides in both the cooperative and educational areas. Organization and unity were essential for survival as Prairie Plains farmers confronted their worst ecological and economic crisis to date.

Chapter 4

DIRECT-ACTION TACTICS 1928-1936

Let's call a Farmers' Holiday
A Holiday let's hold
We'll eat our wheat and ham and eggs
And let them eat their gold.
— Iowa *Union Farmer*, February 10, 1932

What had been bad for farmers and the the Union became worse as the 1920s drew to a close. Commodity prices dropped radically following the collapse of the stock market in October 1929. Drought for several years coupled with the "high winds" and "high sun" of the Great Plains caused fields once covered with lush vegetation to become barren and brown. The deepening economic crisis now accompanied by ecological disaster led to a tidal wave of farm foreclosures. The Farmers Union, at both the national and state levels, was rife with factionalism, and there was the possibility that a divided Union might give birth to two separate farm organizations.

THE ROOTS OF FACTIONALISM

The dispute between the pro-legislative and pro-cooperative factions within the Union deepened in 1929. The faction that championed an active legislative program believed that the answer to the farmers' problems was guaranteed cost of production prices for farmers. This panacea had emerged out of the deliberations of the Corn Belt Committee at Des Moines in May 1925. It called for the itemizing of the average expenses of producers of each commodity, including a respectable wage for the producer. Using these averages as guidelines, a price would be determined for all products consumed domestically that would guarantee to the average operator his costs, labor, and a reasonable profit.[1]

Federal law would require processors to purchase at this price the percentage of the year's crop consumed on the home market. It was the farmer's responsibility to deal with any surplus. He could sell it at world market prices, or he could store it, hoping for higher prices.

The pro-cooperative element pegged their hopes on the Agricultural Marketing Act, approved by Congress on June 15, 1929. This legislation drew the line between the political activists and the cooperators. The Marketing Act established a Federal Farm Board of eight members plus the Secretary of Agriculture. Authorized to support new and existing farm cooperative marketing associations, the Board promised real benefits to grain cooperatives. The Farm Board encouraged the large grain cooperative associations, including the Farmers Union Terminal Association (FUTA), to consolidate into the Farmers National Grain Corporation. C. E. Huff, president of the National Farmers Union, 1928-1930, and a close friend of M. W. Thatcher, became its president. The FUTA was the largest cooperative in the new corporation.

Congress also appropriated $500 million for a self-sustaining revolving fund. The so-called "stabilizing corporations" owned by the cooperative would draw upon the fund to purchase the surplus grain and provide price supports. For the first time in the history of American agriculture, albeit indirectly, the federal government was committed to stabilizing farm prices.

Unfortunately, the price support program did not succeed because farmers were reluctant to reduce acreage. Those who opposed the Marketing Act and large bureaucracies, including SDFU President E. H. Everson, believed it was ironic that while the Federal Farm Board exhorted and cajoled in an effort to obtain acreage reduction, the Department of Agriculture and the states, through their educational agencies and experiment stations, continued teaching the farmers how to grow bigger and better crops.

The Hawley-Smoot tariff, approved by Congress on June 17, 1931, contributed to the Farm Board's dilemma. By raising the tariff on agricultural raw materials from 38 percent to 49 percent, Hawley-Smoot made the sale of commodities overseas almost impossible. Foreign countries needed to sell products in the United States to get dollars to buy our grain. The Farm Board itself had to buy grain, and when grain prices plummeted, the Board incurred heavy financial losses. In 1931, the Farm Board terminated its purchasing program. After spending more than $180 million, it went out of existence in 1933.

The Northwest Organizing Committee, led by Ricker and Thatcher, was an enthusiastic supporter of the Federal Farm Board. C. E. Huff, a Kansan, who succeeded Barrett as NFU president in 1928, had allied himself with the pragmatic Northwest Committee. As a moderate, Huff was willing to accept the half-a-loaf represented by the Farm Board. Later on, he would help shape the New Deal's farm programs.

The pro-cooperative faction lost the FU presidency in 1930 when John A. Simpson, president of the Oklahoma Farmers Union, succeeded Huff. Huff, a quiet idealist, gained a reputation as a business leader as the head of the Farmers National Grain Corporation following his failed bid for reelection.

C.E. Huff, President, NFU, 1928-1930.

Later, he managed the National Farmers Union's insurance program.[2] Simpson's victory meant that the NFU was committed to a program of political action. He would demand monetary reform and cost of production legislation. Simpson was a seasoned veteran of farm struggles and farm organization politics. An attorney by profession, he gained a reputation as a spell-binding orator. During World War I, he built the Oklahoma Union from 231 to 23,000 members.[3]

There had been no change in leadership policies in South Dakota. When E. H. Everson, St. Charles, succeeded Batcheller as president of the SDFU on January 1, 1929, he continued to oppose the position of the Northwest Committee with the same fervor as his predecessor.

When "Ed" Everson first homesteaded in South Dakota in 1908, he lived in a dugout. He later remarked, "I lived in a hole in the ground the first winter, because I was too poor to build a shanty." Six years later, in 1914, Everson heard about the Union and traveled to Nebraska to talk with its leaders. Upon his return home, he helped organize the Union in his area. As a young farmer, he struggled to make ends meet, but he was probably not poorer than his neighbors.

Everson, Simpson and Milo Reno, president of the Iowa Farmers Union, all had great verbal facility. Simpson, as the president of the NFU, was in heavy demand as a speaker, "not only because of what he said but also the manner in which he said it."[4] He could easily hold an audience for two hours, and there were instances where he spoke for three hours. Everson could run the full scale of emotions in discussing Union issues, from calmness and moderation to indignation, and close by vigorously sounding the tocsin for the American farmer. There is no doubt that personalities played a major role in the early 1930s, and many farmers joined a leader more than an organization.[5]

On the divisive issue of whether the Union should stress cooperativism or lobbying, Simpson and Everson came down on the side of lobbying. Their strongest support within the NFU came from Iowa, Michigan, and until 1934, from South Dakota. First and foremost in the legislative program of the so-called radical faction was federal approval of cost of production. They believed that the Federal Farm Board detracted from that all-consuming objective. Questions were raised as to the effectiveness of the Agricultural Marketing Act. After all, the prices of farm commodities continued to tumble. The Farm Board, they argued, was a colossal failure. It could not help all farmers. Cooperative techniques could best be applied to non-perishables such as grain and wool. How much assistance could the program give to corn and livestock producers?, they asked.

C.E. Huff interviewed by Eileen Paradies at All-States Camp (1956).

Much of their ire was directed at the Board's attempt to reduce cotton and wheat acreage. The Board argued that overproduction because of overplanting was the real barrier between the farmer and a fair price.[6] Unwilling to admit that basic changes were taking place in agriculture, and American society in general, the radicals argued that any attempt to restrict the farmers was a blow to traditional agrarian freedom and individualism. Farmers should not have to endure the tyranny of government bureaucrats. The government should guarantee farmers the cost of production for that part of their crops consumed domestically.

Union members generally agreed with the policies advanced by Simpson, Everson and Reno. Each man continued to be reelected by his respective constituencies into the 1930s. Simpson was unanimously reelected national president at the annual convention at Des Moines, Iowa, in November 1931. E. H. Everson spoke for the delegates when he said, "Farmers have been relieved of everything but their mortgages."[7]

The moderate or pro-cooperative element refused to go away. Despite Simpson's annual election victories, state Unions were not unanimous in subscribing to cost of production. The Northwest Committee had its own cadre of skilled leaders and its own newspaper, the *Farmers Union Herald*. Unfortunately, there was a lot of name-calling on both sides, and even some of Simpson's supporters believed several of his statements were out-of-line.

His actions in supporting Everson for vice president of the NFU in 1933, outraged many Nebraskan Union members. One of their own, H. G. Keeney, had served the previous three years as vice president under Simpson. At the same annual convention at Omaha, Simpson refused to admit A. W. Ricker to an executive session, claiming he did not have proper credentials. Following the session, they met in the hotel lobby. News accounts stated that Ricker called Simpson a liar, and Simpson responded with a fist to Ricker's jaw.[8]

The magnitude of the economic crisis should have fostered the Union's growth. The ideological struggle, however, divided the organization's activities, and the factions could not agree on how to go about serving the interests of farm families. Meanwhile, the tillers of the soil in South Dakota were entering the grimmest period in the state's agricultural history.

THE WORST OF TIMES

Will Rogers, the cowboy humorist from Oklahoma, aptly described the rural crisis of the 1920s when he said: "Stock on the Exchange in Wall Street was never higher, stock on a ranch was never lower." The lack of government regulation and the the ease with which stocks could be bought on margin caused speculation to run wild on the floor of the New York Stock Exchange. All of that came to an end with the collapse of the stock market on October 29, 1929, when 16 million shares of stock were sold. A universal Great Depression was truly underway.

E.H. Everson,
President of SDFU, 1928-1934.
President of NFU, 1934-1937.

Following the Wall Street crash, commodity prices fell faster and further than any other period in the history of American agriculture. Hog prices stood at \$11.36 per head in 1931; at \$6.14 in 1932; and at \$4.21 in 1933. The market price of a bushel of wheat sank from \$1.03 in 1929, to sixty-seven cents in 1930, and to a low of twenty-four to twenty-eight cents per bushel in 1932. It was the lowest price for both products since the mid-1890s.[9] A person with a wry sense of humor related the story of the champion wrestler who broke his back attempting to lift fifty cents worth of oats. Speaking over WNAX radio, President Everson, in February 1930, used statistics from the Department of Agriculture for the period between 1890 and 1929 to indicate the dwindling purchasing power of the farmer's dollar.

YEAR	PURCHASING POWER IN CENTS
1890	.83
1895	.85
1898	.88
1904	.93
1909	.100
1913	.100
1918	.112
1920	.96
1922	.89
1923	.58.4
1926	.62.2
1928	.65.6
November 1929	.68[10]

By the early 1930s, Dakota's farmers, once again, were reminded how dependent they were on the vicissitudes of nature. During World War I, farmers had overcropped the Plains as they produced millions of bushels of wheat and corn and fed not only America but many overseas nations. Slogans such as "Food will win the war" and "Plow to the fence for national defense" encouraged an increase in production. Wheat acreage increased from 1913 to 1919. Nationwide, wheat production increased 27 percent and the price per bushel nearly tripled. The overcropping caused the plowing up of thousands of acres of grassland. In prior decades, the limited amount of rain that fell on the sea of grass was trapped by the vegetation's tough roots, resulting in soil that was nourished and held in place. Then, in 1930, years of drought commenced. Summer sun baked the land; any rain that fell did not penetrate the soil. High winds, synonymous with the region, swept over the Plains. It picked up the parched, dry soil that had nothing to hold it down and blackened the sky with dust. *The Union Farmer*, on July 16, 1930, urged farmers not to rush their grain to market, because with millions of unemployed and the high tariff, the demand would be comparatively light.[11]

At the annual SDFU convention at Mitchell in October 1931, President Everson wondered what would be the fate of the Union membership because of the extended drought and grasshoppers that had ravaged crops in almost a dozen counties. He had been appointed to a governor's committee to work out plans for farm relief.[12]

The dusters reached their intensity in 1933 and 1934. Hardest hit in South Dakota were the counties located between the 98th and 100th meridians. R. D. Lusk, editor of the *Evening Huronite*, provided a vivid description of the impact of the "great black blizzard" of November 11, 1933, on the Karnstrum farm in Beadle County. Lusk wrote:

> By mid-morning a gale was blowing, cold and black. By noon it was blacker than night, because one can see through night and this was

Soil blown from fields into farmyard, near Gregory, South Dakota. (1934)

> an opaque black.... Inside the house the Karnstrums soaked sheets and towels and stuffed them around window ledges, but these didn't help much.
>
> When the wind died and the sunshine shone forth again, it was on a different world. There were no fields, only sand drifting into mounds and eddies that swirled in what was now an autumn breeze.... In the farmyard, fences, machinery, and trees were gone, buried. The roofs of sheds stuck out through drifts deeper than a man is tall.[13]

As the drought intensified throughout two-thirds of South Dakota, there was a general air of despair and hopelessness. During the dust storms, rural Dakotans placed water-soaked sheets and towels over windows at night, making it easier to breathe while attempting to sleep. Fields once covered with lush vegetation were now blown barren. Ditches and fence lines were filled with drifted dust. Tumbleweeds and thistles blew across prairie and roads until they came to rest against the side of a building or fence line. Cattle became gaunt as there was no fodder for feed. There were times when farmers found it difficult to sleep, listening to the lowing of hungry cattle. As of 1933, cattle were being transported from the state to truly greener pastures. The sale of cattle did not even pay the cost of shipping them to market because they sold for five cents per pound.

The forces of nature were relentless. The dry conditions provided an excellent environment for grasshoppers to lay their eggs and to multiply. Periodically,throughout the early 1930s, they struck at fields and orchards that contained some vegetation. Ruth Loriks, wife

One of South Dakota's "Black Blizzards" near Gregory, S.D. (1934)

of former SDFU president Emil Loriks, recalled that on driving from their farm home near Arlington to Aberdeen in 1933, that a sunny sky suddenly turned dark. A huge swarm of grasshoppers had obscured the sun.[14] They made quick work of cornfields, devoured the leaves and bark off of trees, and consumed whole gardens, chewing beans and other vegetables down to the ground.

In this hostile economic and natural environment where the daily lunch for many school children was lard, mustard or syrup sandwiches, the Farmers Union labored valiantly to secure some form of government assistance for the beleaguered farmers.

THE FARMERS' HOLIDAY ASSOCIATION

The Wall Street crash of 1929 had a dire effect on the farmer's credit. Following October 1929, insurance companies and banks were forced to call upon their assets, and this meant pressing farmers to pay their debts or forfeit their land. Land prices had declined so radically in the 1920s that the proceeds from a forced sale did not cover the full amount of the debt. The debtor was left with a much-hated deficiency judgment to cover the balance.

Encumbered farmers unable to meet their credit obligations sought legislation to prevent foreclosures. In reviewing the legislative record, they found almost no victories for the tillers of the soil since 1920. Fearing the loss of their land, farmers displayed great interest in the proposal of the Union, led by Reno and Simpson, to call a farm strike.[15] Most members of the Union, however, believed that other avenues of farm relief should be explored. At the 1931

The Frazier Bill.

Cost of Production.

NFU convention, the members rejected by a vote of over three to one (68-19) Milo Reno's motion that the Union join in calling a general farm strike for January 1, 1932. Reno had said the strike would end once cost of production legislation had been approved.

The Union, in February 1932, made an all-out effort to get its legislative agenda approved by Congress. One FU measure, sponsored by Senator Lynn Frazier (North Dakota) called for the refinancing of farm mortgages by the federal government at 1.5 percent yearly payment on principal and a 1.5 percent yearly interest. At the same time, Senator Elmer Thomas (Oklahoma) introduced legislation to abolish the Federal Farm Board and to secure for the farmer cost of production.[16] Simpson and other FU lobbyists made a veiled threat that unless favorable legislation was forthcoming, farmers would resort to more drastic measures.

The Seventy-Second Congress was unmoved by either appeals or threats, and the decision was made to declare a "holiday." Banks had done it in the past, why not farmers? Plans for a withholding movement were laid by Reno and John Bosch, a FU leader from Minnesota. Realizing that within the Union there were many who opposed the political program and desiring not to embarrass the business operations of the FU, it was decided to create an organization independent of the Union, the Farmers' Holiday Association. Homer Ayres, a rancher near Zeona, a Union member whose political views were far left of center, said that the "Holiday Association was kind of manufactured by the Farmers Union because the Farmers Union had some cooperatives, and they were quite timid about these cooperatives. They were afraid somebody might sue them you know, [they were] the only real assets that they might have."[17] Ayres wanted greater involvement of the SDFU in the withholding action, and he believed direct-action tactics to be the best approach to solving the farmers' problems. Oscar Brekke, the president of the Grant County Holiday Association, when asked if the Holiday movement was sponsored by the Farmers Union, was more direct. In an interview with William C. Pratt on September 26, 1987, he referred to the Holiday as "the army and the navy of the Farmers Union."[18]

The organization was formally instituted when approximately 2,000 farmers, including the Executive Board of the SDFU, gathered at Des Moines on May 3, 1932. It was reiterated that the withholding action was designed to secure cost of production. What began as a peaceful withholding action soon became the most extreme expression of the farmers' belief that they were disadvantaged. The Farmers Holiday became the most aggressive agrarian upheaval of the 20th century.

Several South Dakota farmers met at Yankton on July 20 to establish a state Holiday Association. Approximately three weeks later, representatives from thirty-eight counties met at Mitchell to formalize the state organization and to develop plans for organizing the remaining counties.[19] Barney McVeigh, Republican Speaker of the State Legislature, was elected president. Chosen for the Board of Directors were W. C. Lusk, Yankton, the president of the state Chamber of Commerce; D. D. Brisbine, Woonsocket, president of the South Dakota Banker's Association; E. H. Everson, president of the SDFU; H. B. Test, president of the state Farm Bureau, and J. J. Martin, master of the South Dakota Grange.[20]

Emil Loriks, Arlington, was named executive secretary of the South Dakota Holiday group. Prior to the meeting at Mitchell, he had been organizing counties for the Association. A group of farmers had asked Loriks to come to Egan to assist in organizing not only Moody County but adjoining counties. He believed they called him because he was a state senator and committed to easing the farmers' plight. The Sunday following the Egan meeting, Loriks called key men in eight counties and the organizing process formally began.

He and his fellow organizers borrowed techniques from North Dakota's now defunct Nonpartisan League. They toured the countryside asking farmers to sign pledges of support for the forthcoming withholding action. Organizers did not have to serve coffee or doughnuts or have door prizes to secure signatures, people were desperate. Many were glad to have an organization to fight for them.[21] After all, number two yellow corn, shelled, sold for two cents per bushel while ear corn had a negative market value.

Swamped with demands for speakers, the Holiday Association organized a state speaker's bureau. They had more than seventy volunteer speakers on the roster, including many business and professional men. Loriks indicated that he drove over 10,000 miles at his own expense to speak at meetings. During the harvest season, he hired people to run his threshing machines as he was on the road.[22]

Loriks' office in Arlington was like a command post. Using the mail and the telephone, he tried to prod along the various county chairmen. He had a large map in his office and as each county was organized, it was colored in. The greatest organizational success was located east of the Missouri River. The Executive Secretary urged E. N. Hammerquist, Farmingdale, South Dakota's representative on the Holiday's national board of directors, to renew efforts to organize the West River counties. Hammerquist's primary success was in Pennington County.[23] With the emphasis primarily on ranching, the West River area was not a fruitful recruiting ground. The Holiday movement received its greatest support in regions and states with a mixed farm economy, including Wisconsin, Iowa and eastern South Dakota.

In early autumn, the State Association further refined its organizational structure. The Association created an executive committee in every organized county and advisory committees at the town and township level. The committees had a threefold purpose: 1) to per-

Farmers Holiday Movement.

suade the general populace not to buy commodities below the cost of production; 2) to urge farmers to exchange commodities among themselves; and 3) to prevent such commodities from entering marketing channels where ever possible.[24]

The strike or withholding action had begun on August 8, 1932. At its beginning, however, it was a spontaneous and undisciplined movement. Farmers in northwest Iowa formed picket lines and blockaded the roads into Sioux City. Violence ensued as the pickets attempted to stop non-cooperating farmers attempting to pass through the blockades of logs, barrels, old tires, and machine belts containing spikes. By the end of August, there was violence on the roads leading into Omaha, Nebraska. The "farm holiday" was not proceeding as its planners had conceived it. In South Dakota, roads outside of Sioux Falls, Watertown and Yankton were temporarily blocked, but pickets had not resorted to the type of force displayed around Sioux City.

During the evening of August 26, 1932, approximately 10,000 South Dakota farmers gathered in the city park at Huron for a rally with Barney McVeigh presiding. After several telephone calls, Loriks had persuaded a rather timid Governor Warren Green to speak to the gathered throng. Green's remarks had little impact upon the crowd as he did not speak of reform and relief, but he confined his remarks to the importance of law enforcement and keeping the roads open. The arrival of John Simpson, following his speaking engagement at Redfield, electrified the crowd. The national FU president spoke for only ten minutes, but when he had finished and the meeting formally adjourned, the aroused mass of farmers continued to shout "We want Simpson! We want Simpson!," and he returned to the podium to speak again. Simpson had prefaced his remarks with a statement that Loriks felt was too strong. The NFU president had said that "When Constitution, laws and court decisions stand in the way of human progress, it's time they be scrapped." Loriks would have preferred that this revolutionary statement have been altered to conclude with the words that "it's time they be changed."[25]

Governor Green, who had not proved effective at the Huron meeting, did support the farmers' goal of cost of production, and he hinted that he might invoke legal sanctions for an orderly farm embargo. At his suggestion, the governors of Iowa, Minnesota, North Dakota, along with representatives from four other states, agreed to meet at Sioux City on September 9, 1932, and to hear the appeals of the Holiday leaders. Milo Reno and other Association leaders demanded that the governors order a moratorium on all foreclosure proceedings and that states enforce embargoes against the sale of farm products at less than the cost of production. The governors believed the last demand to be highly unfair as it would have compelled all farmers to keep produce at home.[26] In a more-or-less symbolic gesture, the unmoved governors prepared a series of moderate and ineffectual resolutions to be sent to President Herbert Hoover.

Because of the growing violence, Reno had declared a temporary truce in the farm strike prior to the governors' conference. With no relief forthcoming from their elected government officials, it was decided to resume the withholding. Speaking over WNAX Radio, Yankton, on September 20, Loriks announced that a "holiday" would be effective the following morning. Citing the record number of bankruptcies, mortgage foreclosures and tax sales, he said that "YOU FARMERS throwing your farm commodities on the market below cost of production are responsible for the situation."[27] The embargo would begin at the farm gate and would apply only to grain and livestock. The "holiday" would continue until the cost of production was received. If not procured in thirty days, member farmers would apply the embargo to all farm products, including poultry, dairy and perishable products.[28]

John Simpson,
President of NFU, 1930-1934.

A limited number of farmers participated in the renewed withholding action. Joseph Trudeau, Jefferson, president of the Union County Association, reported to Loriks on September 27 that livestock was moving to market in large quantities and that members of the Holiday were waiting

Cars lined up for an auction sale.

the word to picket. He urged clamping down on the movement of all livestock and grain into the markets at Sioux Falls, Sioux City, Huron and Mitchell.[29] President Everson emphasized that farmers would only succeed if they remained organized and acted in unison.[30]

Throughout the last three months of the year, farmers were preoccupied with the campaign for president. Although the Union did not take a formal stand, many of its members prepared to vote for the Democratic candidate, Franklin D. Roosevelt, hoping that a new administration would bring legislative relief to farmers.

There were infrequent spontaneous blockades, such as the one that occurred around Sioux Falls on October 5-6. The action was immediately disavowed by the Minnehaha County Holiday president. Withholding activities for 1932 culminated with a Farm Holiday demonstration at Sioux Falls on December 7. Following a parade with banners and slogans, and an educational program, participating members gathered at the Coliseum to hear talks by Reno and Everson. Acknowledging that it was too cold for picketing highways, it was minus 20 degrees, the speakers asked the audience to inaugurate a new withholding movement and to hammer away at Congress for favorable farm legislation.[31] Holiday members carried their message to the biennial session of the State Legislature in 1933. They displayed a banner with red letters that read: "We buy together, we sell together, we vote together." A senator from Lake County referred to them as "a bunch of Bolsheviks and anarchists."[32]

The South Dakota Farmers' Holiday Association had greater success in combatting farm foreclosures. The farm strike and the road blockades were spectacular but not as significant as the "penny" or "Sears-Roebuck" sales that forced creditors to consider carefully the circumstances before beginning foreclosure proceedings.[33] The anti-foreclosure movement, as with picketing, was a spontaneous, unplanned activity. The first "penny auction" probably occurred in Wright County, Iowa, several months before the Holiday Association was instituted at Des Moines in May 1933. The sales directly associated with the Holiday movement

Auction Sales, too many, too often.

began in Madison County, Nebraska. A widow's property at Elgin, Nebraska, was being sold at foreclosure. With several hundred farmers orchestrating the proceedings, cattle, horses, chickens, and furniture were all sold for five cents each, there being no competing bids. At the close of the sale, the property was returned to the distressed widow. The total proceeds of the sale were $5.86.[34] Many farm women participated in the foreclosure They increased the size of the crowd and added to the volume of the protest. Women were known to be more vocal than their male counterparts at these sales.[35]

The foregoing tactic spread to neighboring states, including South Dakota. In late September 1932, an extreme leftist farm organization, the United Farmers League(UFL) picketed a foreclosure sale at Frederick, South Dakota. Holiday members adopted the tactic, and if action was justified, farmers would proceed *en masse* to the site of the sale. Hand painted signs, such as the one that appeared in Aberdeen on February 27, 1933, "Farm Foreclosure -Robbing a Farmer at Court House this afternoon - Come at Once" were sufficient notice to bring a crowd to the auction scene.[36] The Holiday battle cry became "Neither buy nor sell and let the taxes go to hell."

Although they admired the enthusiasm of the participants in such a gathering, the spontaneity of such events concerned Union and Holiday leaders. SDFU President Everson, a member of the Holiday board, confined his actions largely to getting mortgage relief, not calling strikes and resorting to violence. He and other Holiday leaders instructed county Holiday Associations to take a debtor's case only after careful investigation convinced them that justice rested with the debtor, and there was likelihood that he or she would make the debt good.

Loriks believed that events at a foreclosure sale in early September 1933 "really put the brakes on future foreclosure sales."[37] Four to five hundred members of the Holiday showed up at a farm sale near Milbank, bidding for the livestock and machinery at the usual low prices; a horse was secured for eight cents.[38] During the sale one of the deputy sheriffs irresponsibly discharged a gun containing a tear gas canister at one of the protesters. Some violence ensued with the deputy being hit over the head with a blunt instrument. Other farmers quickly disarmed the remaining deputies.[39]

Oscar Brekke, a native of Albee and the Holiday leader in the area, a tall, erect man with considerable presence, quickly got up on a manure spreader and made a passionate speech against violence. "We've seen some violence here," Brekke said, "We don't like it. These deputies coming in here and starting a shooting match."[40] Following the incident, the presiding officials were forcibly compelled to continue with the sale.

The sale netted $6.30. A collection was taken up of over $16.00 which paid for the pennies spent by the bidders and for lunch for the visiting guests from Minnesota, Deuel, Roberts and Codington counties. Later efforts by Grant County authorities to set the sale aside and to make an example of the Holiday men who confronted the deputies were totally unrealistic given the tenor of the times.[41]

Shortly after the much publicized incident, Brekke wrote Loriks a long letter recommending that the Holiday establish a code relating to foreclosures that would have all of the essentials found in a moratorium bill introduced in Congress by Senator Lynn J. Frazier, North Dakota. Brekke recommended no evictions and giving the owner, being foreclosed on, the possession of the farm as a renter until proper financing was available. It would give the dispossessed virtually the same protection that they would have under a moratorium, a blow against unjust convictions.[42]

Looking back at the Milbank incident from the perspective of four decades, Loriks commented that after that time, the Association had no problem stopping sales. If Holiday leaders heard of a foreclosure sale they considered unjustified, those handling the sale were contacted and, very often, it was called off. The most depressing aspect of the foreclosure sales was the farmer's possessions being sold for little or nothing and the person's loss of equity for all of the years he had worked.[43] The anti-foreclosure movement peaked in 1933 and in the remaining years of the Holiday's life, the number of "penny auctions" nationwide dwindled to nine in 1936.[44]

Most farmers closely watched how the new Democratic administration would deal with the farm problem in the early months of 1933. During this period of watchful waiting, South Dakota experienced its first fatality associated with the Holiday movement. On February 4, 1933, R. D. Markell, sixty-eight, an Elk Point farmer, was fatally shot when he and his two sons attempted to run a blockade placed on Highway 77 near Sioux City. Markell left his truck containing milk to remove planks from the highway when pickets opened fire. His two sons returned fire and by the time the shooting concluded, six persons had been struck by bullets, the two sons, three pickets and the fatally wounded Markell. One of the pickets, an Iowa native, was sentenced to the South Dakota state penitentiary for three years.

Farmers Union leaders were convinced that the Democratic platform of 1932 committed the newly inaugurated Roosevelt administration to cost of production. Unfortunately, Roosevelt, ever the wily politician, had made ambiguous statements on agriculture during the campaign. Wanting to make sure that their message was heard in the nation's capital, 3,000 delegates came together at a special national convention at Des Moines, Iowa, on March 12, 1933, to present a united front to Congress. A majority of the delegates resolved that if the farmer did not receive legislative justice by May 3 a nationwide strike would be called. The South Dakota delegation did not support any immediate strike move believing there was an excellent chance for the enactment *"of our cost of production* legislation by Congress."[45]

Union members were alarmed because the new Secretary of Agriculture, Henry A. Wallace, an Iowa native, had presented a bold new farm relief bill to the President that did not include the cost of production. The Administration seemingly favored a domestic allotment plan that had been endorsed by the Farm Bureau and the Grange. Farm surplus would be controlled by making benefit payments to farmers who agreed to regulate their production in accordance with the government plan. The New Deal's objective was to restore the farmer's purchasing power to parity level, during that period between 1900 and 1914 when supply and demand for farm products stood in effective equilibrium.

The farm relief bill went to Congress on March 16, 1933. It had five major components. The Department of Agriculture would: 1) agree to acreage reduction contracts with individual farmers to lease land withdrawn from production; 2) float price sustaining loans on stored crops (namely wheat, cotton, tobacco and corn); 3) enter into marketing agreements with processors; 4) levy a processing tax on distributors; and 5) license processors, if necessary, in order to enforce compliance.

Everson, Reno and Simpson bitterly opposed the legislation. They argued that cost of production placed no limits on production and that it measured farm prices on immediate costs, not the parity level. Their proposals would provide guaranteed cost of production prices, inflation of currency, and low credit rates for farmers.[46] It was their contention that domestic allotments would not work. Government mandated reduction of acreage would only cause the farmer to crop more intensively, and his output would actually increase. In addition, they raised the spectre of an army of bureaucrats to carry out Secretary Wallace's proposals.

A Farmers Union inspired amendment to the Farm Bill, authorizing the Secretary of Agriculture to fix farm prices at the cost of production level, was rejected by the House of Representatives by a decisive majority on May 9, 1933, and later deleted by the Senate on May 12, 1933. The Agricultural Adjustment Act (AAA), the Wallace bill, passed Congress on May 12, 1933.

South Dakota's FU members were surprised and disappointed that cost of production had been stricken from the Farm Bill at the insistence of Secretary Wallace. Many believed that the removal of Wallace from the cabinet was justified.[47] At that moment, Loriks believed that the passage of cost of production legislation was the only alternative to a farm strike. An impulsive Milo Reno responded to the Congressional action by notifying state presidents and secretaries of the Holiday Association that a general farm strike would be called for May 13, 1933. Sounding an optimistic note, he said: "If all farmers would refrain from buying or selling for ten days, I am firmly convinced the question would be settled."[48] Shortly before the deadline, the mercurial Holiday leader cancelled the strike, deciding that farmers should wait and see how effective the new farm legislation would be. In addition, Reno was bothered by recent outbursts of violence in northwest Iowa.

The National Executive Committee of the Holiday, with John Batcheller, Everson and Loriks present, assembled in Des Moines to make one last effort to work through legislative channels in early fall.[49] They asked President Roosevelt to order currency inflation and, once again, guaranteed cost of production prices. There was a new item on their request list. Having observed the fanfare connected with the "Blue Eagle" or National Industrial

Recovery Act (NIRA), June 16, 1933, they also asked for a production code for farmers like merchants and manufacturers enjoyed under the NIRA.

"The time has come when patience ceases to be a virtue" was the theme of Loriks' newsletter to Holiday members on October 3, 1933. Because the Administration in Washington, D.C. had failed to respond in any concrete way to the Holiday's latests requests, he alerted the membership to prepare for a strike at a moment's notice. The disparity between the farm dollar and the industrial dollar was widening; the farmer had less purchasing power than ever. Regarding Wallace, Loriks said that prior to the Secretary's "ascending to his present high and exalted position," he was an ardent advocate of cost of production principles.[50]

The strike called for October 21, 1933, was used to get the attention of the region's governors. Reno asked Governor Clyde Herring of Iowa to call a meeting of his counterparts in neighboring states to discuss alternatives to the strike. Prior to their coming together, Governor Tom Berry sent a telegram to Loriks on October 27, expressing his hope that the Farmers Union, in convention, would work out a farm code under the NIRA. The Governor felt that a reasonable code would be accepted by President Roosevelt, and it would accomplish much more than a strike.[51]

On October 30, Herring gathered with four other governors, including Tom Berry. Berry, who was really sympathetic to the New Deal farm program, played a minimal role in the conference. He was noticeably silent during most of the meeting. Although Loriks claimed that Berry was a member of the Holiday, except for moratorium legislation, the Governor usually engaged in symbolic gestures to appease Association members.[52] Everson charged that the Berry administration was not fulfilling the Farmers' Union program. In fairness to the Governor, he had not opposed several ameliorable acts of legislation. The State legislature reduced the maximum rate of interest from 10 to 8 percent. A bill was passed for a two-year moratorium on mortgage foreclosures, and the legislators repealed the judgment deficiency law. They also extended the time for redemption of property under tax sales to four years, reducing the interest rate and the penalty on the same.[53]

The five governors agreed to carry the following Association demands to Washington, D.C.: 1) An N.I.R.A. code for agriculture to be administered by representatives of farm organizations; 2) Cost of production price levels; and 3) Revision of farm loan policies by reducing interest rates and extending coverage to 75 percent of the value of the farm.[54] The parting words to the governors were "If you fellows ... go to Washington, get a farm program — a decent workable farm program — we'll settle for that."[55]

The President, supporting his Secretary of Agriculture, did not accede to the demands. Governor Berry, in his reluctance to offend the Administration, told his South Dakota constituents: "I think he's [Roosevelt] doing everything he can for us .... He may not have the right program for some things but he's sincere."[56] Everson said that FDR had not fulfilled his promises. Nothing had been done to enact cost of production and to refinance

Governor Tom Berry, 1933-1937.

farm mortgages over a long period of time at low rates of interest. The SDFU president concurred with Simpson that international bankers still controlled Washington.[57]

Angered by Roosevelt's unwillingness to alter his basic farm policy, Reno ordered continuation of the strike. Between November 6-9, pickets were able to prevent the majority of livestock shipments to markets at Watertown and Sioux Falls, but in a few days, receipts were back to normal. Other markets throughout the state were unaffected.

MILITANCY WANES

The Holiday movement in South Dakota had peaked in its effectiveness. Ever hopeful leaders made two further attempts to secure their objectives. On November 10, 1933, they asked Governor Berry to declare a South Dakota farm embargo. Giving lip-service to the strike, he rejected the request and indicated there was growing sentiment against the strike among farmers. He did, nevertheless, proclaim November 20, 1933, as Justice for Agriculture Day. All communities in the state were asked to protest against the continued debasement of agriculture. All producers and buyers were asked to cease business for the day. Loriks stated that the one-day holiday was effective. Receipts, or lack of same, indicated that the Holiday had almost 100 percent cooperation at livestock terminals in Sioux City, Sioux Falls and Watertown.[58] He was of the opinion that a general embargo on farm products for a dozen main breadbasket states would be effective and that Prairie Plains governors should approve such action if Congress continued to ignore the Holiday program.[59]

At the NFU convention in Omaha on November 21-22, John A. Simpson and E. H. Everson were reelected president and vice president respectively. After the convention adjourned, a few members of the National Farmers' Holiday were called into session. They voted a temporary suspension of the farm strike at the request of the region's governors. The latter group planned another conference to formulate plans for advancing the Holiday's program in Congress.[60] By this time, even Reno despaired of the Administration acquiescing to any of the Holiday's demands. He expected the Wallace gang, operating through the Farm Bureau, "to make a supreme effort to prevent any action being taken that will interfere with their program."[61]

Although the Farmers' Holiday Association had operative units in Iowa and Minnesota as late as 1937, the South Dakota Holiday Association held its last state convention in June 1934. Data on the organization's numbers at its peak is limited. A report dated June 7, 1933, indicated that $800 had been collected. Thus, the state probably had about 1,600 Holiday members.[62] Dues for Association members were fifty cents, half of the amount went to the state's national dues while the remaining twenty-five cents was retained in the county of the dues-paying member.

John Simpson died unexpectedly of a heart attack on March 15, 1934, elevating E. H. Everson, SDFU president, to the presidency of the National Farmers Union. Everson, a disciple of Simpson, called the deceased president "the greatest champion of human rights in America today."[63] With the death of Simpson, Milo Reno, who no longer had a close liaison with the National Farmers Union, was further isolated from that organization.[64] Loriks, the new president of the SDFU, relinquished his duties as primary director of Holiday activities, because his time was occupied with Farmers Union activities. Writing to L. M. Peet, editor of the *Farm Holiday News* on April 17, 1934, he said others would have to "carry the torch" but that he would continue to be a Holiday booster.[65]

Emil Loriks stated on many occasions that the South Dakota Holiday Association was much more conservative that the national body.[66] He and Oscar Fosheim, a Howard resident, the second president of the South Dakota unit, maintained good relations with much of the business community and discouraged picketing in numerous instances. Loriks found this noteworthy in the Northeast Regional Holiday of South Dakota. Under Oscar Brekke's leadership, the Northeast Regional was so well organized that "they bottled up everything air-tight without ever picketing."[67]

Because of the presence of the extreme leftist United Farmers League in the northeast part of the state, Holiday organizers were invited into Britton and Sisseton by more conservative elements in those communities. Ultimately, these newly enrolled farmers transferred their membership to the Farmers Union, resulting in a significant increase in membership for the Union.

In five years time, the Farmers' Holiday Association disappeared back into the Farmers Union, from which it had sprung in 1932.[68] Although its legislative agenda, including the cost of production, never became national law, the organization's militancy aroused the attention of politicians in Washington, D. C. and served as a catalyst in the development of farm relief legislation during the first hundred days of the New Deal. The New Deal was the catalytic agent of time and change that contributed to the defeat of agrarian militancy in the 1930s. The agricultural program removed rural discontent.[69] In November 1933, an aide to Harry Hopkins made the following comment on the diminished fervor of the Holiday movement:

> Apparently one thing that is contributing largely to its failure is the arrival of wheat allotment checks. I have that from a chap named McCandless, farm reporter for the *Omaha World Herald*. I believe our CWA program will also do a lot to calm them down. Quite a few farmers will be getting jobs out of it. It will give them something to do and a little money.[70]

In addition, the direct-action tactics of the Holiday caused state legislatures to introduce moratorium and tax relief legislation, providing distressed farmers with some succor at a time when it was most needed.

The Holiday Association was the last stand of the old Populist principles of economic individualism and opposition to bureaucracy and academic elitism. The United States was undergoing a significant economic and social transformation during the 1920s and 1930s. The New Deal was an important part of this change, and many members of the Farmers Union, as shall be noted in Chapter Five, who first bitterly opposed New Deal farm legislation, changed their minds by 1936.

Chapter 5

LEGISLATION — NEGATIVE AND AFFIRMATIVE 1929 - 1937

The fight is on - Oh Union forces
Be strong and in your might hold fast
The Ore Tax before us,
The Union banner o'er us
We'll sing a victory song at last!
— *South Dakota Union Farmer*, 17 October 1934

The Farmers Union national leadership placed a heavy emphasis on legislation during the first half of the 1930s. In South Dakota, the FU tended to take a negative approach to most federal legislation until 1934 when a change in the state Union's leadership occurred. The death of John A. Simpson and the elevation of E. H. Everson to the national presidency permitted younger men, at the state level, with progressive convictions, to introduce new ideas and policies. This change in direction was reflected in the Union's changing view of New Deal farm legislation.

RADICALS VERSUS PROGRESSIVES AND THE NEW DEAL

Until his death in 1934, Simpson led the charge against Secretary of Agriculture Henry A. Wallace and the Democratic administration. There would be no compromise with the New Dealers. Simpson said that New Deal policies wasted valuable natural resources and restricted the farmer's freedom. He believed that by restricting production and setting farm prices, the Agricultural Adjustment Act (AAA) fixed by law the growing economic disparity between agriculture and industry. Addressing a national radio audience in 1933, a bitter Simpson stated:

> The brain trust when they prepared a bill for farmers did not give us what had been given every other group, cost of production. They said we will give you parity prices and a processors tax. We are not American citizens according to their views and not entitled to what other American citizens get.[1]

As one of Simpson's disciples, Ed Everson continued the assault upon the New Deal until his ouster from the national presidency in 1937. Speaking to the delegates at the National FU convention in 1934, he said that the AAA and the NIRA "were conceived not in the secret chambers of the Roosevelt Braintrust, but in the staid and conservative Chamber of Commerce."[2] He charged the Farm Bureau and the Grange with deserting Brother Simpson and supporting Wallace's scheme which Everson believed would reduce buying power more rapidly than the AAA would reduce production. Everson accused the government of developing a famine in order to bring about better farm prices. He objected to the wholesale slaughter of cattle and the plowing under of cotton fields when millions of people lacked the very things that were being destroyed.[3] He conceived of the AAA as a relief program only for the many thousands of government employees required to administer the act.[4] It is questionable whether Everson fully understood the Agricultural Adjustment Act as he viewed it not only as a scheme to keep farmers from securing cost of production but as a plot to compel them to accept the parity prices of 1900 to 1914 for their products.[5] He viewed the NIRA, popularized through a blue eagle symbol, as another means for exploiting farmers. In his regular column in the *Union Farmer*, he inserted the following poetic jab at the National Industrial Recovery Act:

N.R.A.

That bluebird says "We do our Part,"
But we would like to know
Just what it is they do to start
Our farm incomes to grow.
We know that things that we must buy
Are higher than before.
But when we sell, we heave a sigh
For our prices are still lower.
And then we wonder if that bird so wise
Is not a big black crow,
Just masquerading in disguise
To get from us more dough.[6]

Many South Dakota farmers agreed with Everson. They, too, saw the NIRA as a device to raise industrial prices at the expense of agriculture, and they viewed it as symbol of regimentation and a threat to individual values.

As late as 1934, the foregoing attacks upon the New Deal agrarian program expressed the views of most Union members in South Dakota. When E. E. Kennedy, secretary of the NFU, addressed delegates to the annual state convention in Rapid City in October, he received an enthusiastic response to his jibes at the AAA. There were only two programs, he said, the program of the Farmers Union (cost of production) and the program of the farm exploiters. Kennedy, another very articulate speaker who combined wit with cynicism, said that "the AAA was a program of exploitation originated in Wall Street. Henry [Wallace] Harriman met with the US Chamber of Commerce officials and prepared the AAA in 1931, when 'Sir Herbert' was president. It was turned over to [Rex] Tugwell and [Raymond] Moley in 1932." Farmers who participated in the AAA had signed a confession that they were responsible for the Great Depression.[7]

Emil Loriks, an ardent supporter of the New Deal later on, was one of the AAA's severest critics a year after the New Deal's "One Hundred Days." He accused the administrators of the AAA of proposing "to drive two million farmers from the land, while another agency

[NIRA] ... seeks to settle two million of the city unemployed on subsistence farms." The AAA, he said, subjects farmers to the virtual dictatorship of a bureaucracy over which they had no control. It seeks to gain prosperity through starvation. "We therefore declare," he went on, "that the AAA is a failure and a fraud, an insult to every self-respecting farmer and a crime against society and civilization."[8]

Loriks believed there was corruption in Wallace's hog buying program. He asked why bankers were buying thousands of hogs beforehand and disposing of them on the opening days of the government's purchasing program.[9] It was his opinion that Roosevelt had not driven the money changers from the temple, but he had provided them with a safe harbor during the economic storm. Loriks vented his anger at the press for its attacks on Robert S. LaFollette and Huey Long for their interest in the welfare of humanity. Such people, he said, "invariably have to wear the crown of thorns, often crucified by the very people they're trying to help."[10]

Within the ranks of the Farmers Union there was the moderate Northwest Organizing Committee, which actively supported the New Deal farm program. They adopted the attitude that the FU should attempt to get from government everything possible in the way of help and constructive benefits for farmers. Union leaders from the spring wheat areas were the primary backers of the AAA. The first commodity program prepared under the New Deal legislation was for wheat, including a crop insurance feature.

Both radical and moderate Union members agreed to support the Frazier-Lemke bill to refinance farm indebtedness and to enact a moratorium on farm mortgages. William Lemke, destined to be a congressman from North Dakota, became associated with John A. Simpson in 1928. Attracted by Lemke's views on monetary legislation, Simpson hired him to engage in speaking tours for the NFU between 1928-1932. When Roosevelt campaigned for the presidency in 1932, Lemke became the contact man between FDR's advisers, the Farmers Union, and the remnants of the Nonpartisan League.[11] Elected to the House of Representatives in 1932, Lemke was one of those farm-state legislators who felt betrayed by the new Democratic administration's emphasis on production control. Simpson, once again, arranged for the Farmers Union to finance Lemke as a spokesman for the FU, this time on the radio. The Congressman used this forum to advocate cost of production and bankruptcy legislation favorable to farmers.[12] He was a featured speaker at the annual conventions of the SDFU through 1935.[13]

Emil Loriks,
President of SDFU, 1934-1938.

Lemke's bankruptcy proposal encountered stiff opposition from President Roosevelt and his point-men in the House of Representatives. Through persistence and petition signatures, the legislation was brought to the floor of the House for a vote. The Farmers Union used the lobbying powers at its disposal to get the bill approved. The SDFU Board of Directors made annual pleas to South Dakota's Congressional delegation to support the legislation proposed by Lemke and Lynn J. Frazier. Everson and the *South Dakota Union Farmer* throughout the early months of 1934 reported on the growing support for the proposed legislation. Lemke's proposal was combined with a bill authored by Senator Lynn J. Frazier, North Dakota, that provided a moratorium on farm mortgages.[14] With Senator Huey Long leading the way, the Frazier-Lemke Farm Bankruptcy Act was approved by Congress on June 18, 1934.[15] President Roosevelt, reluctantly, signed the measure ten days later. After months of having their legislative proposals rebuffed on Capitol Hill, the Union old guard finally had a victory. Loriks described the enactment of the Farm Bankruptcy Act as a red-letter day for the Farmers Union.[16]

The legislation, designed to prevent foreclosures, provided additional relief to farmers by enabling them to secure credit extensions. Provision was made whereby farmers could repurchase their properties at a newly appraised value, with small annual payments at an interest rate of 1 percent, distributed over a six-year period. In the event creditors opposed such a settlement, farmers could retain possession of their property on fair and reasonable terms for five years, during which time bankruptcy proceedings were suspended.[17]

Senator Long had been a favorite of many SDFU members prior to his staunch support of the bankruptcy law. He had been elected governor of Louisiana in 1928, campaigning under the slogan, "Every man a king, but no one wears a crown." Two years later, he won election to the United States Senate. As New Deal policies unfolded in the months after Roosevelt's inauguration, Long found himself more and more at odds with the new administration. In January, 1934, he organized the Share-Our-Wealth Society, with chapters in every state. Long proposed that every American family be given a $6,000 "grub stake" or homestead allowance. His scheme would be financed by confiscating the largest American fortunes through taxes.

John W. Batcheller, a former president of the SDFU, lauded Long. In writing to the Louisiana Senator, Batcheller said: "Let the good work go on, until you are dubbed 'the man with one idea' for without radical action where the real trouble is, we are doomed. The trouble is the centralization of wealth and the rackets that have made said concentration possible...."[18] John A. Simpson declared that Long was the best friend the Union had in Congress. He was valued because he focused national attention on the farm crisis and worked vigorously for the passage of ameliorative legislation. As late as 1970, Emil Loriks believed Long to be a great liberal. He had some misgivings about his style, believing that Long used ruthless means to accomplish his goals, like William Langer, governor of North Dakota in the 1930s, but he was basically well intentioned.[19]

Senator Long was scheduled to speak at the SDFU's annual convention in October 1935. When he died of an assassin's bullet on September 10, 1935, Loriks, then president of the SDFU, asked "Is the assassin's bullet, the cross, or the hemlock ever [sic] going to be the reward for faithful service to humanity?" In eulogistic terms, he described Long as a great champion of the common people who was misrepresented, maligned, and persecuted by the powerful interests that ruled the nation.[20]

By 1935, the younger and more progressive-minded within the SDFU, including Loriks, were shifting their support to Wallace's farm policies. The progressives were impressed with the dramatic changes in New Deal policies. New emphasis was placed on welfare, on planning and on regulating business. Massive employment programs, such as the Works Progress Administration and the National Youth Administration, had significant economic impact in the rural areas of South Dakota.

The radicals, however, continued their vendetta with FDR. Ed Everson, still fixated on the cost of production as late as 1936, asked why the government did not perform its proper functions rather than "engaging in hopeless and impossible experiments that can only result, at best, in temporary relief, while at the same time they impose added and increased burdens upon the future?"[21] His formidable ally was E. E. Kennedy of Illinois, national secretary of the FU from 1931 to 1936 and editor of the *National Union Farmer* as of 1934. Following the death of Simpson in 1934, Kennedy played a leading role in formulating Union policies.[22] His decision to back William Lemke, candidate of the Union Party, in the presidential election of 1936, was fatal to the radicals. As described earlier, Lemke had been a favorite of Simpson and co-sponsor of the Frazier-Lemke Farm Bankruptcy Act. As a third party candidate, he found much of his support coming from fringe groups in American politics, including Father Charles Coughlin. Kennedy openly supported Father Coughlin's political program. Coughlin's Social Justice movement appealed to depression-stricken Union members, but his radio diatribes against Jews and the New Deal made him controversial. Kennedy's overt support of Lemke violated the Union's constitution and its precedent of, at least, nominal nonpartisanship.[23]

Loriks believed it was Kennedy's ambition to "tie the Union on the tail of Father Coughlin's political kite." He was correct as Kennedy admitted that he made 134 speeches before the election for the Union Party. Everson also made a speech supportive of the same cause.[24]

Lemke discovered that, although he had the verbal support of radical spokesmen, the rank and file of the Farmers Union backed FDR in the 1936 campaign. By associating with Coughlin, he denied himself further assistance from the Union. After the election, in which he received only 891,886 votes, he was no longer an effective leader of the agrarian bloc.[25]

The moderates or progressives in the Union had bided their time, while the Simpson-Everson-Kennedy faction was dominant. They had used the *Farmers Union Herald* as the medium for expressing their views. Realizing that the tide had now turned in their favor, they moved to oust Kennedy as national secretary at the 1936 NFU convention. Ousted by a narrow margin, Kennedy attempted to have his progressive successors recalled. The national board, now dominated by the progressives, ruled the recall petitions void, and the recall request was never voted on by the membership.[26]

The final major encounter between the Union's factions came at the national convention in 1937. The radicals still dominated the Iowa Farmers Union, and they advanced an amendment to elect the national officers and board by referendum. Delegates rejected the amendment by a 3-1 vote. The progressives, their base of power being the wheat growing states and Oklahoma, elected a compromise candidate, John Vesecky, former head of the Kansas Union, as president. Former president Ed Everson left Union politics permanently, and former secretary Kennedy tried, in vain, to establish a rival farm organization.

Differences within the Farmers Union did not end in 1937, but the serious possibility that the factionalism and infighting of the 1930s would lead to two farm organizations no longer existed. The convictions of the two groups and the balance between them helped explain why the NFU presidency changed hands five times from 1928 to 1940.[27]

The Union remained the voice of the small farmer, but it become a firm supporter of the AAA. New Deal policies initiated a new approach to preserving the family farm. Benefit payments, acreage allotments, and price supports became a necessary and accepted mode of operation. The New Deal helped remove rural discontent as did the return of rainfall to the Prairie Plains and the recovery of commodity prices because of World War II. The primary shortcoming of the AAA and other New Deal farm legislation was that it failed to slow the gradual trend toward consolidation of farms. The program was an "adjustment" policy only. It soon became apparent that government benefit payments served to cushion and to assist the larger farm operator more than the family farmer.[28]

THE UNION AND THE GOVERNOR

As the sense of crisis deepened in 1932, Americans had the opportunity to go to the polls. Perhaps the machinery of American politics could be used, as it had been at times of crisis in the past, to solve the people's problems. There was the opportunity for new leaders to emerge, who would offer new programs and new solutions.

In the gubernatorial election, South Dakota voters could choose between the Republican incumbent, the modest and soft-spoken dirt farmer from Hazel, Warren E. Green, and the Democratic candidate, Tom Berry, a rancher from near Belvidere. Initially, Democratic party regulars had offered to support Emil Loriks if he would run for the position, but the state senator from Kingsbury County, since 1926, rejected the offer. He believed the odds were too formidable to justify mortgaging his farm to fund a campaign.[29]

Very few people had perceived the magnitude of the Democratic victory in November 1932. Roosevelt defeated Hoover by almost a two-to-one margin in traditionally conservative South Dakota and carried in all but one of the Democratic statewide candidates. Only Senator Peter Norbeck, the Progressive Republican, escaped the onslaught. Berry defeated Governor Green in a landslide, 158,058 to 120,473. For the first time in the state's history, the Republicans became onlookers in the Legislature as the Democrats had a 29-15 majority in the Senate and a 70 to 34 dominance in the House. Eleven of the senators and about forty-five members of the House belonged to the South Dakota Farmers Union.[30]

The ideological stance of the new governor was not well received by many Union members. In his inaugural address, Governor Berry expressed his opposition to several Progressive programs, dating back to the World War I era. Berry proposed to eliminate several state-directed entities including a coal mine, hail insurance and the bonding department. He recommended an end to the state primary law. Worst of all, in the eyes of Farmers Union members, the new governor made no recommendations for a debtor relief program.

A radical drop in income had caused the farmer's debt burden to increase dramatically. Cash income from the state's crops decreased from $17 million in 1929 to $6 million in 1932. The income from livestock had a simultaneous decline from more than $150 million to less than $45 million. From 1921 to 1932, a total of 34,419 farm foreclosures were inaugurated, involving 19.6 percent of the farm acreage on the assessment rolls. Nearly one-third of the foreclosures occurred during 1931 and 1932.[31]

A schism between South Dakota's Democrats became apparent during the legislative session of 1933. Berry and the party regulars favored few changes in the state's budget. They sought only enough money to operate the state government. Ironically, the Governor was a staunch supporter of the New Deal on the national level. Speaking for the progressive faction of the Democratic party were Union leaders Oscar Fosheim and Emil Loriks. Favoring their position was Lieutenant Governor Hans Ustrud, a LaFollette progressive.[32]

Senator Loriks and Representative Fosheim used their position as co-chairmen of the Joint-Appropriations Committee to try to obstruct several of Berry's proposals. Because the state appeared ready to default on the interest charges of its rural credit bonds, the Governor sought to end the exemption on gasoline for farm purposes and to divert from the highway fund half of the four cents per gallon gasoline tax to meet obligations on the rural credit debt. The Farmers Union, through its elected representatives, fought to retain the exemption and to use the diversion from the highway fund to lower the property tax. The Governor had his way in both houses of the Legislature.

It was the first of several defeats for the SDFU. The biggest battle was over Berry's proposal for a gross income tax to replace the property tax as the primary source of state revenue. Farmers Union spokesmen argued that such a tax was unfair. A flat 10 percent rate on incomes disregarded both financial losses and the ability to pay. The Governor's response was that a graduated net income tax, a Union proposal, would not raise enough money. Progressives answered by saying that a gross income tax was regressive, similar to a sales tax. The Governor was shifting the tax burden from the rich to the poor. The acrimonious debate alienated Berry from the SDFU, but he had the votes to carry the measure through both houses.[33]

The state's chief executive also had his way in eliminating state hail insurance, the state coal mine, the state bonding department, and the Child Welfare Commission. The SDFU had no position toward the coal mine and cement plant, but it never supported the state hail insurance program. The Union's two primary objections to the Norbeck sponsored program were (1) the farmer's participation in the program was voluntary; the Union believed it should be compulsory on 100 percent of the tillable land in South Dakota; (2) the program had never established uniform premium rates throughout the state, instead it stipulated that rates were to be determined sectionally by the farmers themselves.[34] Taxpayers had lost approximately $265,000 in the insurance venture. Another financial loser had been the coal mine. It was estimated that South Dakota lost nearly $175,000 in the fourteen years of its operation. Only the state cement plant proved a success.[35]

Berry was able to persuade Democratic regulars to carry out such measures at a time when the national spirit was becoming more progressive for the following reasons: (1) a desire to maintain party harmony, (2) only about one-third of the legislators had any experience in the state capital, and (3) the Governor's inordinate use of the patronage system. Upon taking office in 1933, the Governor dismissed many perennial Republican officeholders. He assessed the Democratic replacements 1 percent of their salaries under the patronage system. Berry wanted to create a strong party organization, but the assessment was self-serving as he planned to collect a war chest for the 1934 gubernatorial campaign.[36] His use of patronage to control votes was effective only during his first term. Oscar Fosheim's observation with regard to these new officeholders was that "a lot of them are good boys until they get their jobs."[37]. Fosheim, Loriks and their Union colleagues in the Legislature were indifferent to patronage. They identified more with the Farmers Union and the small farmer than the Democratic party. In retaliation, the Governor said he would not assist any of Loriks' associates, because the Senator did not support his administration.[38]

Oscar Fosheim, President of SDFU, 1944-1947.

The Governor's conservative philosophy was also expressed in his reluctance to do something about mortgage foreclosures. He hoped creditors would not foreclose on mortgages, but he denied a Holiday request for a moratorium. Critical of "penny auctions," he asked people to respect foreclosures of mortgages that were in arrears for more than two years. Berry blamed foreclosures on bad management by individuals rather than the unfavorable economic and natural environment.

The progressives in the Legislature, led by the SDFU, succeeded in enacting a moratorium law that provided an extra year for the redemption of foreclosed farms. Property owners, who were unable to pay their real estate taxes, had the option to redeem their property within two to four years of the date of the sale. The progressives, knowing that the moratorium would expire before the next biennial session of the Legislature would have a chance to extend it, failed in their efforts to secure a two-year moratorium.[39]

The Farmers Union did not come away empty-handed from the legislative session. On the first day, Loriks had introduced four bills designed to secure tax and interest relief for farmers. He called for cutting the penalty interest on delinquent taxes from 12 to 6 percent. The Legislature ultimately set the penalty rate at 8 percent, also agreeing to extend the time for taking tax deeds from two to four years. Furthermore, the legislators agreed to the Union's request for reducing interest rates from 10 to 8 percent, reducing legal interest from 7 to 6 percent and abolishing the compounding of interest.[40] Fosheim played a leading role in guiding the reduction in the interest rate through the Legislature. Knowing the final vote would be close, he got an avid supporter of the bill, Senator Hans Erdahl, Lake County, out of his sick bed, had a taxi transport him to the Capitol where he made an excellent supportive speech, and returned him immediately to his room. The bill passed by one vote, and Governor Berry reluctantly signed the measure.[41]

Fosheim also secured a rule change that enabled the progressives to get more bills out of committee and on to the floor for general debate. The number of signatures required on petitions to force bills out of committee was reduced from thirty-five to twenty.[42]

As chairmen of the Joint-Appropriations Committee, Fosheim and Loriks were able to carry out the greatest reduction in appropriations in the history of the state. Their economy measure translated into a 25 percent reduction in the state budget.[43] Traditional conservatives reacted with bitterness to the tax and debt relief legislation. Edward Oddy, editor of the *Woonsocket News*, typified this reaction. He accused Loriks of going around the state advocating socialist doctrines.[44] Oddy went on to say: "Senator Loriks with his radical legislation would ruin the state of South Dakota. He would cut the penalty interest on taxes, extend the time for taking tax deeds, when everybody knows the way to get taxes paid is to increase the penalty." Loriks was particularly sensitive to Oddy's editorial because other weekly newspapers in the state reprinted the editor's attack.[45]

The Union's progressive vanguard could not forget that they had failed to secure enactment of an ore tax and a graduated net income tax during the 1933 legislative session. The membership of the SDFU and the Grange conducted a petition drive to prevent the gross income tax from going into effect on July 1, 1933. Twenty-three thousand signatures were collected requesting a referendum on the issue, but the State Supreme Court on June 30 denied the legality of the referendum petition by a three-to-two decision.[46]

"TAX GOLD, NOT RUSSIAN THISTLES"

A new spirit permeated the SDFU in 1934 when Emil Loriks was elevated to the presidency following Ed Everson's succession to the top executive position of the NFU after the death of John A. Simpson. Loriks believed that the Union had to be constructive; it should be for something. The first issue of the South Dakota Union Farmer under his administration praised the state legislators. The Farmers Union had earlier criticized legislators for their expense allowance. Now the Union's stance was that they should be paid more.[47] Many of the men who went to Pierre in the 1930s, had to hitchhike to the capital and had to use the same method to return home.[48]

Loriks became president of the SDFU on March 16, 1934. He began another chapter in his long career as a prairie populist and farm leader. He was born on July 18, 1895, at the family homestead in Kingsbury County. The son of Swedish immigrants, he first learned English when he enrolled in a country school at age five. After farming with his father, Loriks took pre-college and college courses at nearby Eastern Normal (presently Dakota State University) and later transferred to the University of Nebraska, Lincoln, where he completed his bachelor's degree. Following a year of teaching at Alma, Nebraska, he enrolled in the U.S. Army Signal Corps where he trained as a biplane pilot. Before he could be assigned to overseas duty, World War I ended.

Following his discharge from the military, he returned home and began farming. The continued decline of commodity prices and the value of farm land during the 1920s caused him to enter politics. Believing that agriculture was decidedly unrepresented in South Dakota, Loriks was elected a Democratic state senator from Kingsbury County in 1926. At the age of thirty-one, he was the youngest member of the Legislature. He was reelected to three additional terms, and following the New Deal sweep of 1932, which gave Democrats control of the Legislature, he became co-chairman of the Joint-Appropriations Committee.

At the time of his election to the Senate, he was a member of the Farm Bureau. While attempting to "smoke out" a bill from a committee, Loriks caucused with several Farmers Union people at the St. Charles Hotel in Pierre. The more he became impressed with the activism of the Union members, the more he became disillusioned with the Farm Bureau. Farm Bureau farmers were "bigger, wealthier, and of the 'big shot variety.'"[49]

In 1930, he joined the Farmers Union and rapidly became one of its foremost leaders, being elected vice president in 1933. By this time, as discussed earlier, he had played a key role as executive secretary of the South Dakota Farm Holiday. From the moment he entered the activist ranks in 1926, he used the debating skills acquired in high school and college to be acknowledged as one of the state's all time great agrarian populist orators.

Quickly discovering that his duties as the fifth president of the SDFU would demand his full attention, Loriks, on August 27, announced that he was no longer a senatorial candidate in the election campaign of 1934. He had also tendered his resignation as executive secretary and treasurer of the Farm Holiday on June 7, 1934.

Something had to be done soon to meet the needs of drought ravaged farmers. Writing to H. A. Cowden, president of the Union Oil Company, on June 20, 1934, the new head of the SDFU predicted:

> There will be no crop in South Dakota this year. Forage and fodder is about all we can expect even under the most favorable circumstances from

Emil Loriks.

Rep. Oscar Fosheim.

"Tax gold, not Russian thistles."

> now on. The people of our state are just simply crucified to the cross of poverty and destitution by reason of the succession of crop failures.[50]

Nature was playing the leading role in Secretary Wallace's efforts to reduce crop reduction. The impact of the drought was a constant theme in Loriks' periodic addresses over WNAX radio, Yankton. He said the state faced the most serious drought in its history. On the Northern Plains, the harvest of 1934 yielded the smallest corn crop in forty years, oats and rye production was the smallest in more than fifty years, and the wheat crop was only 50 percent of its average size.[51]

In the spring of 1934, President Loriks and his progressive colleague, Oscar Fosheim, announced that the Farmers Union would conduct a statewide petition campaign to pressure the 1935 Legislature to enact an ore tax. Their goal was to secure 30,000 names. President Roosevelt had raised the price of gold from $20.67 to $35.00 per ounce, and this was a bonanza for the Homestake gold mine, controlled by the W. R. Hearst family. Realizing that Homestake was prospering and that almost all of its profits were leaving the state, the Farmers Union wanted to tax ore at its source — a tax on the production of gold.

The Union would, once again, have to contend with Tom Berry as he was reelected governor in 1934. His win came without the blessing of the state's leading progressive, Senator Peter Norbeck. Shortly before the election, he had said that Berry had "the viewpoint of big business and yielded too much to the demands of Democratic spoilsmen...."[52] The Democrats maintained their majority in the Legislature, but Homestake, the designated tax target, had several senior legislators "in its pocket," and they controlled key committees.

One of the strong supporters of an ore tax was Henry Mundt, a Sioux Falls attorney. He was a cousin of Karl Mundt, later to become a prominent South Dakota politician. Henry Mundt helped in the petition drive which culminated in 28,000 signatures. Mundt, a member of the House of Representatives, and Senator Earl Mumford, of Howard, did the legal work in preparing the ore tax bill. The bill that Fosheim introduced was based on the Minnesota iron ore tax, legislation that had been sustained by the U. S. Supreme Court.[53]

The Farmers Union presented the giant initiatory petition to the Secretary of State the day before Fosheim introduced the bill to the 1935 Legislature. The Secretary of State said she would have to turn the petition over to the Attorney General to verify the signatures. Fosheim displayed a copy of the state code and informed her that there was nothing in the law requiring such verification. Accompanied by Loriks, he told her that if she referred the petition to the Attorney General "... we'll consider it an unfriendly act." She complied with their request.[54] Another hurdle had been overcome as it was well known that the Attorney General was a Homestake man and very much for Governor Berry, who had already expressed his opposition to an ore tax.[55]

Loriks had committed himself and the full resources of the SDFU to passage of the ore tax. During the petition drive, he put a public address system on his car, and in the fall of 1934, he and Fosheim traveled to every county in the state. They set up local committees to secure signatures on the initiatory petition. The slogan for the Farmers Union became "Tax gold, not Russian thistles." During the legislative session, Loriks was one of the most effective lobbyists in Pierre. Fosheim said Loriks could influence legislators in groups much more effectively than he could in individual conversations.[56]

Fosheim, now the vice president of the SDFU, was the ideal legislator to work for passage of the tax. He was a clever strategist who kept his temper, complimented his opponents, and won over the members of the Legislature. He became a real power and influence in the House. He was an articulate man, who carefully considered his priorities, and he committed his energies to the most important issues facing the Legislature.

Skilled leadership was essential as the opposition had the largest number of lobbyists in Pierre. Veteran legislators supported Homestake, and the Governor opposed the tax. Conversely, for family farmers, Union members, serving in the Legislature was an economic burden, and most of them could not afford to serve more than one or two terms.

Asking for a 10 percent tax on unrefined ore, Fosheim's bill was readily approved by the House on January 26, 1935, by a vote of 73 - 31. The original bill was amended to death in the Senate. The senators proposed a 5 percent ore tax with the tax imposed on the refined product rather than the ore itself. The House rejected the Senate's proposal. Governor Berry turned on Loriks and Fosheim. He questioned the credibility of their bill and said it would not raise needed revenues. He implied that the Democratic party should be purged of dissidents, and in the next election, candidates should be selected who would be loyal to the party.[57]

By this time, Homestake's conservative allies had launched a smear campaign against proponents of the ore tax. Loriks in his annual address to Union members reported that "The opposition in their drunken frenzy referred to us frequently with their insane desire to discredit us as Communist headquarters down in the flats!. They proclaimed that we were nothing but a 'chronic bellyache,' always against everything."[58] Loriks and the Union had not been referred to as socialist or communist until the possibility of an ore tax had first been introduced. The first time he heard the word communist, the SDFU president looked it up in the dictionary to determine its meaning.[59]

The Farmers Union president almost got into serious trouble at an evening meeting in Phillip, South Dakota, in early 1935. While a blizzard raged outside, Loriks had to contend with a group of belligerent miners, who had been given the day off and a stipend to attend the meeting. Many of these miners were former farmers who had found employment at Homestake. Hinting that they might lynch him, they let Loriks know that the ore tax would cost 600 of them their jobs. Their dire prediction was wrong. After the tax passed, not a miner lost his job.[60]

In a more restrained manner, the *Sioux Falls Argus Leader* stated that the ore tax was a bad bill. Union leaders quickly retorted that Homestake had taken $300 million from the state and never paid a cent of tax on production. They believed that the state should get some of that money and not world-wide stockholders.[61] The SDFU and the Democratic party encountered problems getting their message over the airwaves. WNAX radio was the only station permitted to broadcast at night, and the Gurneys, owners of the station, were partisan to the Republican cause.[62] D. B. Gurney, president of the House of Gurney, had earlier warned both Everson and Loriks that when using the station's facilities, they should confine themselves to the best interests of the Farmers Union. They should be very careful of making any libelous statements or charges that would tend to get the broadcast station into court.[63]

The ore bill was finally referred to a conference committee. Fosheim and Henry Mundt represented the proponents. They developed a strategy whereby Mundt argued vigorously for a high tax rate. Fosheim held back and observed the struggle between the low and high tax groups. Then, Fosheim, now the supposed noble neutral, stepped in with a compromise solution. He recommended a 4 percent gross tax on unrefined ore, and the committee readily accepted his suggestion.[64] The tax was expected to raise $640,000, a significant increase over the $160,000 currently being paid to the state by the mining companies.[65] A Union goal for eight years had finally been achieved.[66]

The SDFU had dedicated itself to the passage of other legislation as well. Union members wanted to see an end to the gross income tax that became law in 1933. In this case, an old rival became an ally. The Farm Bureau favored a net income tax with rates ranging from 2 to 10 percent on personal incomes and on corporate incomes exceeding $250,000. There would be deductions for heads of families and each dependent for those persons paying the personal income tax.

In his inaugural address, sensing that there was growing hostility to the gross income tax, Berry urged that it be modified in the 1935 legislative session. As the session progressed, he said he would support equitable tax legislation. The Senate Tax Committee proposed to continue the gross income tax. The farmer- controlled House, responding to their constituents' wishes, soundly rejected the proposal 73-28. As with the ore tax bill, a compromise favored the farmers. The gross income tax was out and a net income tax was in. The tax on higher incomes was limited to 8 percent, and a sales tax, much to the Union's displeasure, was enacted into law.[67]

Obtaining a two-year moratorium on real estate mortgages was another significant triumph for the Union during the biennial session. The moratorium law was based on the Minnesota model, but the South Dakota measure did not provide as extensive debtor relief legislation as in Minnesota or in other states.

Never was the SDFU more active and unified politically than during the biennial sessions of the state legislature in 1933 and 1935.[68] The efficient organization and leadership of the SDFU had proved crucial in the struggle with Governor Berry and the vested interests that had dominated the state for decades. Fosheim and Loriks proved to be masters of legislative maneuvering and persuasion. Credit also must be given to the other Union members, who were in the Legislature as well as the rank and file, who sought signatures for the initiatory petition and who carried out a statewide lobbying effort to secure passage of the ore tax.

The advance of the Farmers Union during the 1930s was not one dimensional. In the midst of privation and political struggle, the SDFU pioneered an educational program that became a model for other state Unions. Moreover, the cooperative movement finally secured the wholehearted support of the State Union's leadership and fully emerged as an economic force to benefit farmers in future decades.

CHAPTER 6

WATERSHED YEARS — CONFLICT, COMPROMISE, UNITY 1928 - 1937

"If a house be divided against itself, that house cannot stand."
— Gospel of Mark, III: 25

Hard times, hard feelings, ideological convictions, strong personalities, battered egos, factionalism and generational struggle are terms that have been used and will be used again to describe the conditions experienced by the Farmers Union membership in South Dakota during the 1930s. In this sea of seeming negativism, nevertheless, the Union made gigantic strides towards a brighter future. Pragmatic, younger men realized that some of the old policies would no longer work. With enthusiasm, ingenuity and competence, they accommodated to the New Deal farm program, closed old wounds and completed the Farmers Union triangle by pioneering a remarkable education program and initiating a large number of cooperative enterprises.

THE RELENTLESS HEAT

The changes in the Union's policies took place during years of relentless summer heat. In South Dakota, farmers considered the continuing drought as much of a disaster as the Great Depression itself. In 1935, it seemed as if the state would have its first crop in years. Blistering July heat quickly destroyed quality, yields and price. An outbreak of black stem rust in the spring wheat area reduced the prospective crop by more than 100 million bushels. Conditions were worse the following year. There were thirty-two days in June and July when the temperature soared above 100 degrees. On July 18, 1936, one reporting site in the state stated that the mercury reached 116 degrees.

The pattern of scorching July heat and the lack of late summer rains was repeated in 1937. In addition to the drought, grasshoppers did considerable damage to many fields, and there was an outbreak of anthrax. South Dakota began to resemble biblical Egypt with its several plagues. Periodically, throughout the thirties, farmers also had to contend with beetles, webworms, caterpillars and Mormon crickets.

On October 20, 1934, President Loriks wrote to the president of the Minnesota Farmers Union, "It surely would be a treat to South Dakota if we should accidentally raise a crop once more."[1] In his report to the annual convention, Loriks reported that 80 percent of the state's cash crop had been wiped out by drought. Drought relief had been cut. In 1935, South Dakota received $225 million in relief; the amount was $135 million in 1936; and in 1937, relief assistance had been reduced to $20 million.[2] South Dakota farmers were truly wandering the economic desert of scarcity.

MAINTAINING MEMBERSHIP

Under such circumstances, it was remarkable that the South Dakota Farmers Union was able to maintain a significant membership. President Everson, in 1930, constantly expressed concern about what the deteriorating physical environment would do to membership. The membership in Districts IV and V, west of the Missouri River, did decline significantly in the early 1930s. Statewide, membership fell from 7,175 in 1929 to 4,660 in 1933. There was a remarkable resurgence in 1934 despite South Dakota's suffering destitution and drought. The SDFU gained 1,486 members in that year, but then there occurred a "bottoming-out" with only 4,242 enrolled in 1939.

Membership
South Dakota Farmers Union
1927 - 1940

1927	7,092
1928	7,017
1929	7,175
1930	7,694
1931	7,073
1932	5,603
1933	4,660
1934	6,146
1935	6,763
1936	5,611
1937	4,462
1938	4,431
1939	4,242
1940	4,723

Adjusting to the hard times, membership dues were reduced from $4.00 in 1930 to $3.00 in 1932.[3] Union organizers worked diligently to maintain the membership rolls and to bring new people into the organization.

The pool of farmers from which the organizers could recruit new members declined significantly. Unfortunately, many South Dakota farmers and ranchers did not survive the natural disasters of the 1930s. Too proud to admit defeat, many of them ran just ahead of disaster until they lost the race. Once comfortable farm families, who experienced the loss of homes and the depletion of resources and credit, often went down to defeat in silence and isolation because of their pride. Many, who left the farm, either moved to town or left the state and headed west to find water and survival on the Pacific Coast. Approximately 50,000 people left South Dakota during the thirties.

FEDERAL AID

Others were able to remain in the state because the federal government introduced programs that enabled them to "hang-on" throughout this decade of despair. It was a time when strong rural individualism was infringed upon. Going on relief was a traumatic step for many and not easily forgotten. It was noted earlier that old-line Union radicals, such as Batcheller and Everson, bitterly opposed New Deal farm policies until 1937 when their political dominance of the Union ended. Everson believed that the NIRA and AAA "were conceived not in the secret chambers of the Roosevelt Brain Trust, but in the staid and conservative United States Chamber of Commerce."[4]

Progressives within the SDFU, early on, realized that the New Deal was one of the best hopes for keeping the small family farmer on the land. By the end of 1934, 39 percent of the population of South Dakota was on the relief rolls, the highest figure for any of the forty-eight states. Government payments represented a substantial portion of the farmers' income. For example, in 1934, two-fifths of the receipts of farmers in Hyde County came from production control payments, emergency livestock purchases or relief grants. Many farmers received government crop and feed loans. By April 1935, 89 percent of all farmers received emergency relief. In Aurora County, approximately forty miles to the southeast of Hyde County, another area that emphasized livestock production and cash-grain farming, 63.8 percent of all of its farmers were on relief in March 1935. It is no wonder that the practical, younger Union members made their peace with the New Deal, and the Farmers Union became a firm advocate for government farm programs.

BRUISED EGOS

When the mantle of leadership passed from John Batcheller to E. H. Everson in 1928, the pro-legislative outlook at the top remained the same. The two men championed the cost of production, and they opposed the work of the Northwest Organizing Committee. Yet, the relationship between them was strained. Batcheller was reluctant to let go of the presidency. At the 1929 state convention, he ran against Everson, losing by a vote of 205 to 67.[5] He then became the board member from District I, and he attempted to manipulate SDFU policy from that position. Using the *Union Farmer* as his forum, Batcheller wrote letters attacking the Northwest Committee. He charged it with being an independent corporation, determined to control the NFU, and ignoring the sovereignty of the South Dakota Farmers Union.

An angry President Everson resented Batcheller's letters "to a Few of Our Union Boosters who resent Outside Interference." In his reply, Everson said the policies of the State Union would be decided in state conventions. "Now in conclusion," he wrote, "I wish to say that I have no bitter feeling toward you, Brother Batcheller, or any other member of the Union, and yet I fear you just about as much as I would the effect of being hit on top of the head with a soap bubble.!"[6]

Everson moved the publication of the *Union Farmer* from Yankton, the site of the paper's publication for over a decade, to Pierre in early May 1929. The Board of Directors

Homer Ayres, wearing a bowtie, seated in the center of SDFU delegates to the 1968 NFU Convention. Also shown, seated L to R: Joe Quillin, Melvin Johnson, Mrs. Melvin Slater and Paul Hofer. Standing: President Ben Radcliffe and Delegate Ralph Sapp.

returned the paper to Yankton in mid-December of the same year, making it easier for Secretary E. H. Sauber, who lived in the area, to work with the editor of the *Yankton Public Opinion*, publisher of the *Union Farmer*.

The war of words between the two "old fuzzy tails," a term used for Batcheller and Everson, by Homer Ayres, the Farmers Union bard from Sheep Creek Local near Zeona, touched most aspects of Union legislation, including the State Legislature. Although Hans Helgerson had been the SDFU lobbyist at Pierre during much of the 1920s, President Everson decided to serve as lobbyist to the Legislature during the 1931 session.[7] When he recounted the Union's successes to the annual fall convention, Batcheller challenged him from the floor of the convention. Everson had singled-out the passage of a law which returned 30 percent more of the auto license tax to counties for the development of farm-to-market roads. He also noted that the FU had secured a reduction of 16.5 percent in land values from the amount fixed by the Tax Commission in 1930. Batcheller, a legislative committeeman, charged that Everson had usurped the jurisdiction of the FU Legislative Committee and that he [Everson] had recommended that the committee be abolished. A furious Everson urged the delegates to investigate the charges, asserting that they would find them untrue.[8]The acrimonious debate between these two firebrands of the old guard subsided with the death of John A. Simpson. In the midst of the mourning and flowery tributes to Simpson, Everson had assumed the national presidency.[9]

A NEW OUTLOOK

Vice President Emil Loriks became president of the State Union and rallied to his side the progressive element within the Union. Affirmation replaced negation, beneficial compromises replaced intractable positions and a balanced program of education, cooperation and legislation replaced the inordinate emphasis on legislation in the preceding years.

PROBLEM RESOLUTION: THE COUNTY AGENTS

During his first year in office, Loriks resolved the Union's differences with the county agent system. In 1919, the American Farm Bureau, by law, had become a part of the agricultural extension program. Because of the resulting tie between the Farm Bureau and the county agents, both Batcheller and Everson believed the Bureau had an unfair advantage over the Union. They accused the Farm Bureau of having subsidized local stool pigeons, the county agents and their tax-paid organizers. The *Union Farmer* said that county agents promoted the Farm Bureau and 4-H clubs; the latter being used to educate farm youth for Bureau membership.[10]

In blind, unreasoning opposition, former President Batcheller had campaigned across the state against county agents. For a period of time, he had all but six of them removed.

Loriks did not fault the county agents. He believed they were tools of an organization, and despite being on the federal payroll, they had to go out and get members for that organization, namely the Farm Bureau.[11] He also realized that the agents were rendering a valuable service in carrying out the New Deal farm program at the local levels.

Rather than being blindly opposed to county agents, Loriks worked to separate the Farm Bureau from the extension program at South Dakota State College (later renamed South Dakota State University), in Brookings, South Dakota. Because the head of the Extension Service at State College had belonged to the same fraternity as Loriks at the University of Nebraska, the SDFU president's task was made easier. In his conversations with A. E. Anderson, Loriks found that most extension directors did not like the current system. Their agents were required to go out and secure membership for a certain organization which invited the enmity of every other group. Current requirements mandated that county agents had to secure so many members per year to hold their jobs.[12]

With Anderson's help, and without overtly involving the Farmers Union, Loriks convinced the 1935 Legislature to separate the Farm Bureau from the Extension Service.[13] The tension between the Farmers Union and the Farm Bureau eased considerably, and the two organizations were to work together on certain policy issues in the future. Most important to the Union, its members would now work closely with county agents in improving the family farm.

PROBLEM RESOLUTION: TUBERCULIN TESTS

For years, the Farmers Union and officials at South Dakota State College had disagreed on compulsory tuberculin tests for cattle. This was one issue on which Batcheller and Everson were in total agreement. It was the topic of several of Everson's radio addresses over WNAX, Yankton. He questioned whether the state could establish a police power (compulsory TB testing) until it was definitely established that no testing was dangerous to public health.[14] He argued that there was no evidence that any person contracted tuberculosis from cow's milk.[15]

Loriks accused Everson of hypocrisy regarding tuberculin tests. The president of the NFU had condemned TB tests "until he was almost black in the face in Lincoln County" after he had voluntarily submitted his own herd to a tuberculin test.[16] Many farmers were convinced that the tests were faulty. A Conde farmer reported that two good, young milk cows were condemned in his township and slaughtered. After he received the test results, the distraught farmers discovered that the cows had not been infected.[17]

The Legislature took up the issue in January 1937. House Bill No. 26, mandating the compulsory testing of cattle, was introduced by a joint-committee of the House of Representatives and the Senate. Many Union members wanted to "fight to the last ditch" to defeat the measure.[18] S. H. Lassen, of Brookings, an administrator with the U. S. Department of Agriculture, appealed to Loriks to end the Farmers Union's resistance to the testing program. State accreditation was Lassen's concern.[19] Loriks, who would be the role model for other Union members, resisted tuberculin testing until the fall of 1937. It was time to compromise and move on. In a letter to Lawrence Staley, secretary of the Hyde County Farmers Union, he wrote, "My herd has never been tested until now, I feel it is useless to resist any longer. The Dr. was up here Sunday when I was home, and he said many of my neighbors were refusing to test until I do, so you see, we have members who stay in the boat to the bitter end."[20] Another longstanding, contentious issue was laid to rest.

THE UNION AND URBAN LABOR

Many Farmers Union members maintained the Populist vision of solidarity between the farmers and urban laborers against the vested interests of businessmen. The New Deal's growing support of organized labor in 1935, and its open display of traditional Progressivism, was warmly received by Union and labor forces in South Dakota. Since the 1920s, the Farmers Union had made a determined effort to bring about better relations between farmers and labor unions. The Farmers Union and labor scheduled annual conferences and invited spokesmen from both groups to speak. The American Federation of Labor (AFL) convention in Rapid City during the summer of 1937 had been a pro-farmer convention, and the State Union reciprocated by permitting members of the State Federation of Labor to speak from the platform at the FU convention.[21] The Union supported repeal of South Dakota's right to work law, and organized labor expressed support for the Holiday movement.[22]

The SDFU rendered assistance to the workers at the John Morrell meat packing plant, Sioux Falls, when they were on strike from March 1935 to November 1937. President Loriks went through the picket lines at the plant, met with the strikers and gave them moral support.[23] Oscar Fosheim brought provisions to the strikers in 1935.[24] When the strikers called for a nationwide boycott of Morrell products, the FU membership participated in the boycott. The regional Farmers Committee for Action called on farmers not only to boycott Morrell products but also to refuse to ship cattle to meat processing plants. One of the farm spokesmen stated, "Action is being taken to support the strike morally and financially, as well as on the picket line."[25] Rural progressives, including Fosheim and Loriks, denounced the use of police in the labor dispute, and various groups appeared at the plant site pledging their support of the strike and providing food to the pickets.[26]

Albert Maag, President of the S.D. Federation of Labor.

President Loriks kept in close contact with Albert Maag, Huron, the president of the South Dakota Federation of Labor, and with Irvine Levine of the Worker's Alliance.[27] Maag and Loriks joined several hundred farmers and laborers for a Progressive Conference in the Mitchell Corn Palace on November 12-13, 1937. It was hoped that the two-day meeting would launch a real progressive movement in South Dakota, leading to a pro-progressive state legislature.[28] One of the unrealized goals of the conference was building a Farmer-Labor party, similar to the party created in Minnesota.[29] It was at the meeting in Mitchell that the conferees denounced the use of police in labor disputes. Although a progressive political party did not emerge in the state, many of the ideas that were expressed became future objectives for the SDFU. Homer Ayres, always the political maverick and a longtime Union member, praised Emil Loriks for his help in putting on the conference. In his usual folksy style, he wrote to Loriks, "I found out that you are a pretty well-bred Morgan (comparing men to horses). They don't lay down on the trail."[30]

ADMINISTRATIVE MATTERS

For seventeen years, the administrative office of the SDFU remained at Yankton. E. H. Sauber had been elected secretary-treasurer in 1920. Prior to being named to the position, he had been a farmer and an elevator manager. Shortly after his election, he moved the state office to Yankton, where he could work with the *Yankton Public Opinion* in publishing the *Union Farmer*. The *Public Opinion* was owned by the Farmers Union of Yankton County. Sauber's salary in the late twenties was $2,100 per year plus payment for the expenses he incurred on the road doing Union work. President Batcheller's salary in 1926 was $3,600 per year, from which he had to pay for office help.[31]

While serving during the terms of two very controversial and outspoken presidents, Sauber wisely remained aloof from factional quarrels in the organization. He kept silent when the most controversial issues were being discussed from the platform or in the *Union Farmer*.[32] Because of failing health, Sauber resigned his position in July 1937. He was replaced by Paul G. Erickson, the director of the Junior Education Program.[33]

The shift in Union leadership from the southern part of the state to the east-central area, Loriks was from Arlington and Erickson, a native of Clearfield, led to a transfer of the state office to Sioux Falls. The Board of Directors, on July 13, 1937, formally approved moving the state Union office to Sioux Falls.[34] The Board of Directors had been holding sessions, almost regularly, since January 1935, in the Cataract Hotel in Sioux Falls.[35]

Loriks believed Sioux Falls to be the center of most of the Union's activities although he was somewhat apprehensive about how the move would be received by the membership.[36]

The change in headquarters also meant a change in the location of radio broadcasts. In 1929, President Everson had made arrangements with Gurney's Seed and Nursery Company

Cartoon from 16 August 1939, *South Dakota Union Farmer*. Theme is keeping farmers and worker disunited.

for regular radio broadcasts every two weeks over WNAX, Yankton. His broadcasts had been accompanied by musical numbers from the WNAX orchestra.[37] Because it was more convenient, Emil Loriks, in July 1937, began making fifteen minute broadcasts every other Sunday over KSOO, Sioux Falls.[38]

Harmony at the 1934 state convention extended beyond the melodies of the Ericson Brothers orchestra of Alcester, annual performers at state conventions, as Emil Loriks was unanimously elected president of the SDFU for 1935. Oscar Fosheim, of Howard, was elected vice president. The Howard farmer's bald head was the subject of friendly wit prior to the vote. The *Union Farmer* commented, "Fosheim ... didn't need a spotlight to bring out

A Farmers Union orchestra, Market Day 1954

his points. The top of his head reflected directly on the "tax gold!' sign."[39] Loriks was easily reelected in 1935 and 1936, but his campaign for the presidency in 1937 was a different matter.

THE TRIUMPH OF THE PROGRESSIVES

Since 1935, the relationship between Loriks and NFU President Everson had reached the breaking point. Everson, by associating with E. E. Kennedy and Father Coughlin, was an avowed opponent of most New Deal policies. In this regard, by 1937, he was far removed from the mainstream thinking of the FU membership.[40]

The majority of the Union supported President Roosevelt's proposal to reorganize the federal judiciary in February 1937. The Supreme Court had angered the President by declaring legislation in the broad field of economic control as unconstitutional, including the much heralded AAA and NRA legislation. In response, Roosevelt proposed that if *any* federal judge, having served ten years, refused to resign after his seventieth birthday, the President be empowered to appoint a new judge to the bench, including, if circumstances required, six new judges to the Supreme Court.

Everson appeared before the Senate Judiciary Committee on March 22, 1937, and read a statement attacking the so-called "Court-packing bill." His resolution read, "Resolved that we are opposing the present bill in the U. S. Senate to enlarge the United States Supreme Court, and we recommend that Congress pass a law making it mandatory that the Supreme

Farmers Union Orchestra of Clarno Local, Lake County. L to R: Aurther Anderson, Ernest Moose, Lewis Johnson, Carl Anderson, Gerald Johnson, Clarence Knutson, Maynard Gudahl.

Court, in order to rate a law unconstitutionl, must bring in a decision of 7 to 2."[41] Progressive Union members objected to Everson's use of "we," and one of their leaders, M. W. Thatcher, head of the Farmers National Grain Cooperative, had developed a scheme that "If it goes the way I have planned, it will mean the end of Mr. Everson."[42]

E. H. Bremer, secretary to Senator Herbert E. Hitchcock, of South Dakota, described the scene after Everson read his resolution. After the "great self-styled leader," Ed Everson, read his statement, Senator Key Pittman of Nevada asked him if they were his views or those of the Union. The NFU president hesitated, said they were *his* views, and that no vote had ever been taken on his prepared statement.[43]

President Loriks was critical of Everson for opposing reform of the Supreme Court. He said that the NFU president gave comfort to the Liberty League, a well financed group which made virulent attacks on the New Deal, and other allied reactionary interests, including the *Sioux Falls Daily Argus-Leader*.[44]

In early summer, Loriks knew that the national president of the Farmers Union had committed himself to ousting Loriks from the state presidency at the annual convention at Huron in October. The campaign involved innuendo, false rumors and scurrilous, untruthful propaganda. Everson seldom openly attacked Loriks.[45]

Loriks could not fully comprehend Everson's motives. He had supported both Everson and Kennedy during the 1936 NFU election. It was the support of the SDFU that salvaged Everson's reelection to the presidency. The ensuing rancor distracted from the work that needed to be done to build a stronger State Union. Promoting personal politics only created internal dissension and strife among the membership.[46]

Farmer Union leaders in neighboring states, and allies of Loriks, kept a close eye on what was going on in South Dakota. H. G. Keeney, president of the Nebraska Farmers Union, believed that Everson's allies were William Randolph Hearst, the mining interests, the Consumers Cooperative of America and state payrollers. Keeney hoped "Mr. Everson will calm down before he destroys himself along with as much as he is able to inflict upon the National Union."[47] Morris Erickson, secretary of the North Dakota Farmers Union, expressed sympathy for Loriks' difficult position. He wrote, "The man (Everson) is entirely lacking in any principles or scruples. Thank heavens, he will be permanently retired in November."[48]

In the two weeks before SDFU delegates met in Huron to elect their new president, there was an escalation of charges and countercharges. E. E. Kennedy published a letter in the *Union Farmer* accusing Loriks of being on the federal payroll, receiving a salary of twelve to thirteen dollars per day and expenses. The letter said the SDFU president was a collaborator with the Department of Agriculture. Loriks appealed directly to Secretary of Agriculture Wallace to refute Kennedy's falsehoods, stating that he had neither received nor accepted a salary from the federal government since serving in the military in World War I.[49]

A letter from an I. N. Thompson, of Flandreau, that appeared in the *Sioux Falls Daily Argus-Leader*, and other state newspapers, charged that Loriks had sold farmers down the river to labor, that he had distributed CIO literature at the gates of John Morrell meat packing plant. The letter concluded, "Thank God for Everson, who pointed out that farmers and labor have nothing in common...." After careful checking, it was determined that there was no I. N. Thompson. The letter had been planted by Everson's supporters.[50]

Loriks, in turn, pointed out that Everson's prestige in Washington, D.C. was below zero. The cost of production bill had not seen daylight since John Simpson was the NFU president. The men who authored the cost of production legislation had consistently voted for the AAA and soil conservation since 1934.[51] Loriks further stated that Everson had raised his own salary from $3,000 to $5,000 and that he had insisted that his back salary be paid with 6 percent interest added to the amount due him.[52]

In his letters to Union colleagues and friends, the SDFU president indicated that Everson's physical and mental health was poor. Loriks told Fosheim that when Henry Mundt asked what was the matter with Ed Everson, he had replied, "Well, I said, it is difficult to answer that because words fail me. The most charitable thing I can say is that he is mentally screwy!"[53]

Writing to Morris Erickson on October 4, 1937, President Loriks summarized his views of the man who would deny him reelection:

> Poor old Ed, climaxing his career, writing his final chapter in the venom of hatred, suspicion, vindictiveness and distrust of his fellow men and co-workers and thereby under-mining his own health and both mentally and physically I believe, and spending all his time sowing the seed of internal dissension and strife in the Union. On the way out grasping all he can though it means a reversal of [the] principles that he has preached from the platform! It is a sad picture — a man so highly honored, to traduce the Organization."[54]

As the time for the convention approached, young Union progressives mustered delegates who would support the reelection of Loriks. Fosheim made sure that Miner County delegates, including Gust Moseson, manager of the Union Brokerage House in Sioux Falls, would by vote and voice work for his friend's reelection.[55] Homer Ayres from the West River area, Oscar Brekke in northeast South Dakota and Paul Opsahl in the north central region marshalled their forces for the SDFU president.

At the convention, Opsahl made one of the speeches in support of Loriks. What had been expected to be a close race was no contest. Everson and his coalition were crushed. A watershed had been reached. The progressives were now firmly in control of the State Union, and they could move forward freely with their agenda.[56]

The struggle between the progressives and reactionaries had yet to be completed at the national level. Everson continued his attack on Secretary Wallace's "regimentation program with its penalties and processing taxes."[57] He had opposed a conference of state officers in Omaha called to formulate a program for the National Farmers Union.[58] He sensed a conspiracy against him. The assumption was correct. Union progressives believed the time was at hand to secure a solid front in the official family of the Farmers Union. A successor must be chosen who would work with the New Deal.[59]

Realizing that his defeat was inevitable, Ed Everson, in a radio address over NBC, announced he would retire as national president at the NFU convention in Oklahoma City, Oklahoma, beginning on November 3, 1937. He said he was retiring because of ill-health.[60] At the convention, John Vesecky, of Salina, Kansas, current vice president and a moderate, was selected to replace Everson. Loriks was also elected to the national board of the NFU.

The era of faction within faction in the Union came to an end. The internal struggle within the Farmers Union at the state and national levels illustrated how an organization can split over means and neutralize its energies. The effects of the new unity soon became evident. By 1944, the Union had reached its highest membership total since 1921. Irvine Levine, of the Worker's Alliance, in a letter to Loriks on November 26, 1937, described the events of 1937 very well when he wrote, "The defeat of the reactionary forces is not merely a personal victory but one which involved the life and death of your splendid militant farm organization."[61]

Given the divisions and tensions that existed in the SDFU during the 1930s, it is remarkable that the State Union created a sound education program for the Junior locals and a successful camping program for its youth (See Chapter 7). At the same time, the cooperative program began to reach fruition (See Chapter 8). From 1934 on, there was an ambitious program of building cooperative oil companies. In 1937, one of the world's largest grain marketing cooperatives, the Farmers Union Grain Terminal Association, was established. In the midst of recurring natural disasters, the Union membership displayed dedication, hard work, ingenuity and perseverance in completing the Farmers Union triangle.

Chapter 7

COMPLETING THE TRIANGLE — EDUCATION 1928 - 1938

"He Loves His Country Best, Who Strives to Make It Best"
— National Motto for Junior Partners

"It was no accident that the word "Educational" was placed first in the name of the organization.
It was significant that the Farmers Union was born in a schoolhouse."
— Mrs. David Lovinger, October 1930.

While direct-action tactics, legislative policies and intra- organization divisiveness captured the headlines, the South Dakota Farmers Union, with wisdom and ingenuity, moved towards fulfilling the goals of the Union as determined by its founders in 1902. Early Farmers Union insignias emphasized the tools of farming, the plow, hoe and rake, and the three basic elements of the farm organization — education, cooperation and legislation.

Education is at the base of the equilateral triangle, the symbol of the Farmers Union. The two sides, legislation and cooperation, are vital elements in the survival of family agriculture, but education is the base that is the key to success of all other activities. Structural engineers use triangular trusses to counter stress in their building projects. Likewise, the base and two sides of the Union triangle met the socio-economic stress placed upon the American farmer. During the agricultural depression of the 1920s and 1930s, the South Dakota Farmers Union completed its triangle, and in doing so, it provided the methods and organizations needed to ensure the survival of the family farm.

THE NEED FOR YOUTH EDUCATION

The foresight of the Union's founders laid the foundation for the youth program of the SDFU. The first constitution and by-laws of the NFU contained a section that stipulated that children of members, between the ages of sixteen and twenty-one, were to be counted as members of the Farmers Union.

Nothing was done concerning this farsighted provision until the late 1920s when the Union was nearly thirty years old. The realization dawned that the Union's first members were no longer young. What could be done to maintain the organization? It was imperative that the Farmers Union avoid the mistakes of past farm organizations. These now defunct farm bodies had all followed the same pattern. Founded in a time of economic distress, their adherents were initially filled with zeal, but as soon as the greatest distress had been allayed, their zest for action dissipated, and they returned to their normal farming routines. In turn, as the original members grew older and their ranks were thinned by death, there was no new enthusiastic group to replace them.

Recognizing what had happened to the Farmers' Alliance, Populists and other predecessors, and realizing that many of the members of the Union were aging, Farmers Union members in South Dakota began to think seriously about methods which might be introduced to bring a steady flow of new strength into the Union. There was a need for youth in the Union and a grave need for education. It was imperative to develop an educational program which would keep the Union, its aims, principles and achievements before the membership at all times.

SOUTH DAKOTA LEADS THE WAY

Much of the credit for pioneering in the field of education must go to the South Dakota Farmers Union. Out of its Junior program grew the pattern of education which eventually included both junior and adult education for the National Farmers Union. In October 1919, the editor of the *South Dakota Union Farmer*, John W. Batcheller, called for a junior column in future editions of the paper. Mrs. Minnie Lovinger, Ethan, South Dakota, the wife of David Lovinger, the first FU county president in South Dakota, answered the call in November. Already editor of the Women's Department, Mrs. Lovinger introduced a column called the Children's Nook. She, affectionately, headed the column "Dear Nookies" and signed it with her chosen pen name, Penelope.

Children wrote letters to Penelope, and they eagerly watched the Children's Nook for new topics on which they were to write letters. The topics assigned by Mrs. Lovinger were always designed to interest boys and girls residing on farms. Out of this first educational venture grew a system of credits and awards. One credit was given to a youth for each letter received by Penelope and five credits for each letter printed in the *Union Farmer*. Twenty-five credits were needed to be eligible for an award. Mrs. Lovinger also encouraged the submission of essays, awarding the winner a box of crayons.

The Junior program continued in this vein until the annual state convention in October 1928. Three years earlier, however, at the 1925 annual convention, a resolution was approved, recommending that locals organize their juniors into Junior locals to be conducted on the same basis as Senior locals.[1] The juniors would be educated in the principles of the Farmers Union.[2] By 1928, a group of interested citizens, led by Michael E. Stevens, Huron, asked that the program be changed to provide more learning experiences for the children. Young people wanted to do more than write letters and provide entertainment at local meetings. They wanted to learn about the Farmers Union and farm problems. Convention delegates responded to Steven's concerns by recommending 1) that each local organize a Junior local and 2) that a study be made of the issues he had raised.[3]

The SDFU Board of Directors authorized the organization of Junior locals. In July 1929, they met in Pierre and appointed a four-person committee to explore Junior locals and to draw up a course of study for juniors. The committee was given full authority to promote the interests of FU children. The Board appropriated $100 for the committee's work and instructed the committee to report back at the next annual convention.[4]

During the State Fair at Huron in September 1929, the committee came together in a fateful meeting. Gathering at the Beadle County courthouse in a hot, stuffy room, the four committee members went to work. Seated around the table were Mrs. Minnie Lovinger, Ethan; Michael E. Stevens, Huron; Mrs. O. H. (Elsie) Olson, Bijou Hills and F. C. Morgan, Chamberlain.[5]

Mrs. David (Minnie) Lovinger, First Educational Director for the SDFU and first National Farmers Union Educational Director.

Since the 1928 SDFU convention, each committee member had gathered material and ideas which could be used to formulate a Junior program. It was not an easy task as they were pioneering in an uncharted field. There was consensus in the group that farm youth must be made aware of the problems faced by their parents, that they should know how cooperatives helped people, and that they should be familiar with the causes of war.

It was imperative that farm youth be taught to speak for agriculture, and they should learn parliamentary procedure and cooperation. Other points of discussion included topics, activities, a constitution and by-laws. The historic meeting went on for hours, with the committee working late into the afternoon. So engrossed were they in their work, that they discovered upon concluding the meeting that they were locked in as the courthouse had closed for the day.[6]

The committee made its presentation at the state convention in Rapid City on October 8-9, 1929. F. C. Morgan, chairman of the group, was accompanied to the platform by other members of the committee. Morgan proceeded to list the committee's accomplishments and recommendations. The group had prepared a Junior constitution and by-laws that included a preamble, a membership section, the duties of Junior local officers and a senior guide (leader) and a definition of some activities.

The committee recommended that:

1. Juniors should be a part of the adult local and not a separate organization.
2. Meetings should be in conjunction with and precede the senior meeting.
3. Meetings should be both entertaining and educational with topics chosen for study.
4. Activities should include club work with a county exhibit.
5. A Junior Farmers Union County Day should be observed once each year with a program, contests and exhibits.
6. Juniors should establish a book or magazine exchange.
7. Provisions must be made by the SDFU Board of Directors for a Junior Department in the *Union Farmer* and pins or buttons be made for junior members.
8. A Farmers Union textbook needed to be prepared. Michael E. Stevens, who brought forth the idea, urged that it be a primer that dealt with big issues simply and briefly.[7]

For Minnie Lovinger, it was the realization of her long-held conviction that there was a place in the Farmers Union organization for young people. In her remarks to the delegates, she said:

> I want to talk with you for a few moments upon the most important subject in the world, "The Children," and also their connection with the Union. Many of us here today believe that the idea of Junior locals is the biggest thing that the South Dakota Farmers Union has undertaken to carry on. Somehow, we have overlooked or neglected this great opportunity that lay at our door.[8]

She indicated that the Junior educational program would emphasize the three components of the FU triangle — cooperation, education and legislation. Junior local meetings would teach young people the value and necessity of cooperation and prepare them to become better Union workers than their parents were before them.[9]

Elsie Olson and Mike Stevens followed Mrs. Lovinger with brief remarks. Stevens concluded by saying that "The best way to build the Farmers Union for the future is to train the children while [they are] young."[10] Stevens had brought twenty-seven young people from Beadle County to the convention, and they demonstrated a regular Junior local meeting. The delegates responded with loud and lengthy applause and enthusiastically endorsed the new Junior program.

On January 9, 1930, the state Board of Directors commended Mrs. Lovinger for her efforts toward creating the Junior local. They decided to replace the "Children's Nook," a feature in the *Union Farmer* for ten years, with a Junior Department, where the activities of the new Junior locals could be reported. The February 12th edition of the Union paper published the complete constitution of the Junior Farmers Educational and Cooperative Union of South Dakota and gave instructions for choosing a guide for a Junior local from senior members of the same local. The planning committee classified young people between 16 and 21 years of age as Juniors. Local dues were set at twenty-five cents per year. A new song had also be been composed for the juniors. Sung to the tune of "Stand Up For Jesus," the words were:

> Come on, come on, all juniors and join the Union song. There's always room for boosters To help the cause along. We're strong for education So we begin in youth To organize as Juniors We're bound to know the truth.[11]

The delegates who came together for the annual state convention in Yankton, in October 1930, formally established and arranged financing for the Junior local program. At that time, twenty-seven Junior locals were already active. A few traditionalists rejected the new education program, associating it with the Farm Bureau.[12]

While South Dakota did the pioneer work in Junior education, other state Unions were getting their youth involved in local meetings and programs. Union leaders determined that a national program for youth must be implemented. At the 1930 annual NFU convention in St. Paul, Minnesota, a concerned group of Union members cited the following three reasons for a youth education program:

1. The Farmers Union was an organization of farm families where the young learned to share the responsibilities of the farm; therefore, they should be an active part of a farm organization which represented their family interests.
2. In order to make membership meaningful, the young needed to learn about the Farmers Union in a way they could understand.
3. It was vital to train new leadership for the future welfare of the Farmers Union. The FU allowed each member, age sixteen and over, the right to vote, to hold office and to serve as delegates to conventions.[13]

One of the final acts of Charles E. Huff, as president of the NFU, was to appoint a committee to work out a national pattern of youth education. Two South Dakotans, Minnie Lovinger and Mrs. O. H. (Elsie) Olson were on the committee. The other members were Gladys Talbott Edwards and Mildred Kay. Mrs. Lovinger was named the first National Farmers Union educational chairman.

The committee quickly agreed upon three basic principles that would be applied nationally. The principles provided for:

1. A uniform study topic, ending in a national essay contest. Lovinger had introduced a statewide essay contest in South Dakota in 1930. Her first topic was the need to establish farmer-owned cooperatives.
2. An annual topic to be based upon a social problem which concerned the welfare of the people.
3. A membership stipulation that juniors must be a part of the chartered local and be sixteen to twenty-one years of age.[14]

The committee emphasized that cooperation was possible only through education. Courses should stress human relationships. Emphasis should be placed on the problems that confronted humanity. The committee's working document read, "By means of ... playing together, working together and studying together, we cherish the hope that our young people of the Farmers Union, with the young people of other nations, will get the idea that the inhabitants of the world can be lifted out of their miseries, and that is their job."[15]

Gladys Talbott Edwards, National Director of Education, 1930-1953.

At the 1932 state convention in Huron, the delegates voted to adopt the study program approved at the national convention, intended to lead up to a national essay contest. At least, five states were using the same study subject in 1933 with the expectation that at the next national FU convention, a study program would be adopted and used in all of the various state organizations. As of November 1933, there were eighty-eight Junior locals in South Dakota.[16]

THE JUNIOR LOCALS IN ACTION

The first topic chosen for the national essay contest was "World Peace." In the two Dakotas, Farmers Union juniors studied the issue and wrote essays for contests at the local, county and state Union levels. The winning essays were read at the national convention. Farmers Union members listened and praised the program that had provided the framework for this quality work. The Four Minute Speech contest was introduced in 1934. Its purpose was to teach young people to speak adequately and briefly upon a subject related to agriculture. The essay contests were eliminated from Junior work in 1937. They, along with the speech contests, were difficult to judge, and juniors were not interested in them. They preferred achievement work and awards programs.

Junior work grew rapidly during the early 1930s. Youth, unable to attend high school or to find jobs in cities or towns, turned to the Farmers Union and discovered, as the National Junior Committee had hoped, a place to study the problems of the world and to develop a feeling of responsibility for those problems.

Mrs. Russell (Florence) Tilberg, Ethan, South Dakota, led many Junior local meetings. Florence, a member of Dodd Local 16, was in the first Junior class led by Minnie Lovinger. Minnie was a neighbor and inspired Florence to become a leader in the FU educational program. Mrs. Tilberg became educational director of her local in 1939, director of education in District II in 1940, and in 1943, she became a member of the Union's State Education Committee. She served as an education leader for locals for forty-nine years.[17]

For many years, the fourth Friday of each month was "Union Night" for the members of Dodd Local 16 and adjacent locals. The adult members of Dodd Local 16 assembled at a nearby school while the juniors met in a home; quite often, it was the home of the senior guide. The sessions usually lasted three hours with lessons based on the FU state curriculum. Following roll call, students would study the assigned topic for the year. They had texts, prepared and provided by the SDFU Education Committee. In her experience, Mrs. Tilberg said that the sixteen to twenty youngsters who were present were usually prepared. Because other social options for them were limited, they looked forward to the sessions. Cooperation was constantly emphasized in the meetings. During the course of the evening, songs would be sung and either cake or sandwiches served to those in attendance.[18]

Mrs. Russell (Florence) Tilberg, A leader in SDFU education for 49 years.

Failing health and a debilitating stroke early in 1932, caused Minnie Lovinger to resign from her state and national Junior work. This grand and wise woman, who laid the foundations for the Junior local program died in 1939. Badger Clark, South Dakota's poet laureate, authored lines that appropriately described her achievement. She "snatched the thorny chance, broke the trail that others followed." Her "soul has never left them in their sowing and reaping."

Mrs. Elsie Olson was appointed to Mrs. Lovinger's position, and she immediately took over

The 1941 Torchbearers.

as editor of the Junior Department in the *Union Farmer.*[19] When E. H. Everson became president of the NFU in 1934, he made Mrs. Olson the national director of Farmers Union Junior and Educational activities.[20] During her tenure as national leader, a course of study for young people under sixteen years of age was implemented. They, too, began to learn about cooperation, the Rochdale pioneers and the pioneers in the Farmers Union.

The Torchbearer program was introduced in 1937. After two years as juniors, two people could be nominated by their peers in their local for the Torchbearer competition. The recognition was based on leadership in the local, class work, attendance at camps, achievement, writing of poems and preparation of study books. The study books were evaluated by the senior guide. The state office made the final decision on who would be the Torchbearers to be recognized at the annual state convention. Upon promising to work for the cooperative movement, to practice tolerance and brotherhood and to strive for a warless world, they received a gold badge in the form of a flaming torch.

Two juniors went to the stage of the Ritz Theatre in Winner, the site of the 1938 state convention (October 11-13), to receive their badges.[21] In front of the delegates the recipients pledged "... to bear my torch with clean hands and a courageous heart," and "to keep my torch ever one of those which shall light the way to a warless world."[22]

During the same convention, the first awards were made in the Minuteman Project. It was a program through which young men and women, who were capable and forceful speakers, were trained to present the program of the Farmers Union. Three junior Minutemen gave their qualifying speeches and were subjected to questions by the SDFU del-

The 1982 Torchbearers elected at the State Camp.

egates present at Winner. The topics of the three speeches were "Man Marches On," "Our Agricultural Problems and How Cooperation Can Solve Them" and "Racial Problems." Each of the young speakers was awarded a Minuteman badge and a camp scholarship for 1939.[23]

Within two years of its inception, the Junior local program involved more than 2,000 young people speaking for agriculture and learning about cooperatives. By 1940, the program included Farmers Union members in fifteen states and involved thousands of young members in study and camping programs.[24]

THE SDFU BEGINS A CAMPING PROGRAM

Another important reason for the loss of interest in essay and speaking contests was the introduction of the camping program in 1934. The camps placed emphasis on cooperation and not competition. Summer camps were to become the most signficant part of the Farmers Union youth program.

During the summer of 1934, Mrs. Elsie Olson, the state Junior leader, attended a joint Montana-North Dakota youth camp at Sidney, Montana. Inspired by what transpired, she returned home and made plans for the first SDFU state camp.

South Dakota's first camp was held in conjunction with the annual state convention held at Rapid City in October 1934. More than 100 young campers stayed in cabins in the Baken Park area of Rapid City and commuted to convention activities at the Alex Johnson hotel in downtown Rapid City, where they participated in educational activities, including speech-

A Kitchen Band at Camp (1956).

making. A planned excursion for the young campers to Mt. Rushmore took longer than expected when two cars stalled attempting to climb the first hill west of Rapid City.[25]

Teaching and learning at a camp site was enthusiastically embraced by the participants, and plans were made for a more ambitious program, separate from the convention, for 1935. Two camps were held in the early summer that involved more than 300 juniors. In order to accomodate farm parents, who could not afford to have their youngsters travel long distances, an East River FU camp was held on the Lake Madison Chautauqua grounds, and a West River camp was held at Methodist Assembly Park, near Pactola, in the Black Hills.[26] These same sites were used for the camping experience in 1936. A variety of locations, including Camp Watymca on the shore of Lake Kampeska, near Watertown, in 1937, were used until 1959 when Camp Placerville in the Black Hills became a permanent home for the Farmers Union state camp.

The SDFU, as early as 1933, became involved in locating a site where juniors from the several member states could come together for a camping experience. At the state convention in Winner, a site-selection committee reported that it would recommend to the national membership a camp site approximately eight miles from Rapid City.[27] The recommendation was enthusiastically received at the NFU meeting in Omaha, Nebraska, on November 21-22, 1933, but the first Farmers Union All-States Camp was not held until the summer of 1936.[28] The Unions of Montana, North Dakota, Minnesota, Wisconsin and South Dakota in conjunction with the FUGTA and FU Central Exchange, had set up the Farmers Union Cooperative Education Service. Sponsorship of the All-States Camp was one of the first

All-States Camp, Estes Park, Colorado, July 1938.

NFU All-States Camp at Camp Judson, Black Hills, 1940 or 1941.

projects of the Service. Twenty-three juniors from South Dakota met with juniors from four other states at Camp Judson, Pactola, South Dakota. In this recreational setting, they discussed "Peace and Patriotism" and made plans to study the cooperative movement at next year's All-States Camp, again at Camp Judson.[29]

In assessing the value of the first camps, Mrs. Olson wrote that "it was interesting to watch the splendid leadership developing and the young people themselves at these camps .... My own estimation of the camps is that they were among the most valuable undertakings of the Farmers Union in the state to date."[30] Following the East River and West River camps in 1935, a committee of five junior youth made their own evaluation of the camping program. They described the value of the camps as inestimable, "especially as regards leadership training, education and cooperation and exchange of ideas of different parts of the state."[31]

During the 1930s, campers arose early to the call of a whistle, the ringing of a bell or the sound of a trumpet. Before breakfast, the participants gathered for flag raising ceremonies and recited the pledge of allegiance. Following breakfast, work (K.P.) teams, organized on the opening day of camp, proceeded with their assigned tasks for properly maintaining the camp, before going to class. Class topics at the state level included farm issues, cooperatives, parliamentary procedures and other key issues of the era. In the mid and late 1930s, much attention was given to the peace movement because of increasing militarism in Europe and East Asia. In the fall of 1937, juniors were circulating petitions favoring the passage of the Ludlow Amendment. Congressman Louis Ludlow of Indiana had proposed a constitutional amendment that would have required that any declaration of war be ratified by a referendum vote of the people, except in the case of invasion.[32] Signers of the petitions were asked to contribute at least one cent to the cause, half of the money received would be used to promote peace and the other half to finance junior education, which really meant the same thing.[33]

District III, Richmond Reserve Co-op Band, Marvin Winter, Adviser.

Guest speakers, including the president of the SDFU, made presentations to general assemblies of the campers.[34] Topic periods were kept short and to the point to hold the interest of the campers. Discussion was encouraged to develop critical thinking among the participants.

Camp members were taught to use their hands as well as their heads. A period was set aside each day for handicraft, the study of arts and crafts. By the close of a camp, many juniors could proudly display innovative and beautiful products, the results of the youngsters' creativity. Classwork was interspersed with recreational opportunities, everything from group games to hiking. Evenings came to a conclusion with camp fires and singing.

The high winds, lack of rainfall and repeated crop failures of the 1930s affected the camping program. Camp registration fees were kept low, $2.00 per camper in 1935, to encourage juniors to attend summer camp. Many FU locals, nevertheless, had to hold special fund raising events to send their youth to camp. Box socials, coffee and pie events and one-act plays were among the means used to raise the needed funds. The meagre income of the juniors' parents and the limited access to entertainment contributed to the popularity of the camps. Camps provided an accessible and constructive diversion from living "on the margin" economically and socially. Within a few years, the Union also had county camps, district camps and Junior Reserve (ages 13-16) camps.

The topics and activities at one of the Davison County camps took place in the following order: (1) Traffic and Society, (2) Cooperatives, (3) Alaska and Hawaii, (4) Handicrafts, (5) Banquet, (6) Film, (7) Party and (8) Lights Out. The banquet was usually held on the last

Farmers Union State Camp at Chautauqua Center, Lake Madison, 1937.
Emil Loriks is seated in the second row on the right side. Standing two places to his right is Loyd Larson.
Larson served on the Board of Directors as District III chaiman from 1947-1964.

night of the camp and the young people were asked to dress in their "Sunday best" for the event.[35]

Florence Tilberg of rural Ethan said Davison County camps were held at a nearby schoolhouse, lasting two or three days. On one occasion, the girls slept in the school while the boys slept in the Tilberg barn, accompanied by Florence's husband, Russell. As late as the 1950s, forty or more youngsters attended the three to four day district camps.[36]

The juniors quickly made their presence known at the South Dakota State Fair in Huron. Shortly after the SDFU received its charter, the Farmers Union, beginning in 1919, annually erected a tent on the fairgrounds during the days the fair was in session.[37] By 1936, booths within the tent were made available for junior exhibits. In addition, juniors assisted in booths that displayed the various activities of the SDFU, including insurance sales, the livestock commission, cooperatives, and the wholesale FU Brokerage Company. It was estimated that over 6,000 people visited the FU exhibits in 1936.[38]

The foundations for SDFU's successful youth program were laid in the 1930s. Before the decade was over, the people responsible for its inception were departing from the scene. Because of ill-health, Minnie Lovinger had departed early. Elsie Olson resigned as national chairman for FU education and as SDFU Junior leader in November 1935. Paul G. Erickson, Clearfield, South Dakota, was appointed state Junior leader in 1936. Erickson left the position in mid-1937 when he was asked to replace long-time state secretary E. H. Sauber, who resigned because of ill-health. During Erickson's brief tenure, the Board of Directors merged the Junior and Education Departments with the state office, which was now located in Sioux Falls.[39] Mrs. Arnold (Marie) Styles would lead SDFU educational activities into the decade of the 40s.

In the years ahead, new education directors with new ideas would come to the fore, new programs would be introduced and new curriculums would be devised, but the aims of the juniors remained basically the same. Juniors aimed to:

1. Create a better world by becoming better citizens.
2. Develop tolerance and understanding to insure liberty and justice to all.
3. Develop an understanding of the problems of agriculture to better meet the challenges of farming.
4. Establish a strong, militant organization that will help solve the foregoing problems.
5. Build cooperatives to prevent the wealth of the farms from being centralized in the hands of a few.
6. Promote economic democracy, and thereby safeguard political democracy.
7. Love fellow human beings, whatever their race, creed or color.
8. Work toward a warless world.[40]

Chapter 8

COMPLETING THE TRIANGLE — COOPERATIVES 1928 - 1938

"We must build economic power through economic organization — through cooperation."
— Emil Loriks (1936)

"South Dakota Goes Cooperative"
— *South Dakota Union Farmer*, 15 April 1936.

"Love your neighbor and co-operate."
—President C. S. Barrett, October 1927.

A promise and a purpose were fulfilled between 1920 and 1940. It was during these two decades that cooperativism became a permanent part of the Union's ideology. The promise and purpose were in that part of the preamble to the Farmers Union's 1909 constitution which read:

> In the course of modern industrial development, we find it necessary that the farmer not only apply the principles of scientific agriculture, *but that he systematize his business by cooperation* and apply the principles of scientific commerce....

For the Union, the cooperative movement became more than a means for securing better prices for the consumer and producer, a means by which to omit the middleman. It helped unify the members who came together and pooled their financial resources to begin, to plan and to manage *their* cooperative. It was their responsibility to maintain the enterprise against private competition.

Cooperatives were the result of the farmer's individualism. They were the product of taking things into his own hands. The Farmers Union viewed cooperatives as "self-help institutions, a grassroots effort to reinstate man as master of his economic life."[1] Cooperation was more than an economic device. In its Junior education program and in summer camps, the ethical values of cooperation were emphasized. Cooperation led to unselfishness, peace, service to others and the unity of humankind.

Original store of the Rochdale pioneers.

THE ROCHDALE PIONEERS

A prevailing theme of the Farmers Union was that education must precede the effective operation of cooperative agencies. An important part of the Junior curriculum was the study of the Rochdale Movement. In 1844, twenty-eight weavers residing in Toad Lane, Rochdale, England, decided that they must take matters into their own hands to improve their lot in life. Victims of the Industrial Revolution, these textile workers worked long hours and received a pittance for wages, as low as forty-five cents per week. Living conditions were as bad as working conditions. Their miserable dwellings lacked furniture and bedding. Their food was contaminated with starch in the sugar, ground bone in the flour and iron filings in the tea.

Several of these weavers knew of the Chartist movement to bring about democratic social and political reform in England between 1838-1848. Having heard discussions of the value of cooperatives, they decided to form the Equitable Society of Rochdale Pioneers, whereby they would use a cooperative to save money and to purchase better food. Each member saved, in one year, a British pound (about five dollars) and used the money accumulated as investment capital to begin a store for the weavers in a rented, rundown building. The store opened on December 21, 1844. Overcoming financial obstacles and the opposition of their employers, the store, by the end of 1845, had increased its capital from the original 20 pounds to 181 pounds.[2]

The Rochdale pioneers established the essential characteristics of a true cooperative. These characteristics included:

1. The principle of one person, one vote
2. Open membership
3. Business would be conducted on a cash basis only

Thirteen of the original Rochdale members.

4. Dividends would be allocated on the basis of patronage
5. There would be no proxy voting for directors
6. The organization would be apolitical and nonsectarian in nature
7. Cooperative education would be a regular function of the broader organization
8. Reserve would be set aside for expansion and education
9. Share capital would be paid a moderate fixed return on investment.[3]

Theodore Saloutos, a noted mid-20th century agrarian historian, argued that too much importance was attached to the Rochdale consumer group. He argued that it was an urban movement created to improve conditions for poorly paid urban workers "rather than a rural movement seeking lower marketing costs and higher returns from farming."[4] Saloutos sought recognition for a group of Wisconsin cheese producers, who formed a cooperative to manufacture cheese in 1841. This cooperative had revived an old Swiss system of cooperative marketing to seek the highest possible returns for its produce.

THE FEDERAL GOVERNMENT AND COOPERATIVE BUILDING

The post-World War I agricultural depression caused farmers to look for more efficient methods of selling their produce. Cooperative marketing associations appeared to be the solution. The federal government also stimulated the cooperative movement. In the early 1920s, the farm bloc wielded considerable influence in the Congress. Senator Arthur Capper of Kansas, and Representative Andrew Volstead of Minnesota, authored a bill that on February 18, 1922, became known as "the Magna Carta of Cooperative Marketing." The Capper-Volstead Act legalized cooperative marketing associations and defined the terms

Jerauld County Farmers Union Building, the "Big Brick," Wessington Springs.

under which producers engaged in interstate commerce could organize. In the same year, Congress passed the Grain Futures Act. The act sought to protect cooperatives from discrimination by boards of trade and chambers of commerce.

President Herbert Hoover expressed his support for cooperative marketing in the presidential campaign of 1928. A campaign promise became a reality with the passage of the Agricultural Marketing Act on June 15, 1929. Considered by some agrarian historians to be the "most important single piece of legislation ever enacted in behalf of cooperative marketing," it called for the establishment of a Federal Farm Board and authorized a revolving fund of $500,000,000 for low interest loans to help cooperatives and to engage in price stabilization operations.[5] The Federal Farm Board, consisting of eight members and the Secretary of Agriculture, was to promote the marketing of farm commodities through agricultural cooperatives and stabilization corporations.

The cooperative marketing of grain was not as successful during the twenties as was the marketing of livestock. It was only in the late 1920s that the Union and non-affiliated cooperatives emphasized the cooperative marketing of small grains. The South Dakota Wheat Growers, headquartered in northern South Dakota, engaged in wheat pooling. Through these pools, growers hoped to strengthen their bargaining position. In forming a wheat pool, a manager was hired and given complete control of storing and selling the farmers' wheat. Farmers, normally, were bound to the pool for a period of five to seven years.[6] They hoped that pooling would eliminate some of the evils of competition and make it possible for grain producers to control the wheat market.

Redfield Farmers Union Creamery.

Beginning in the twenties, four types of cooperatives emerged in South Dakota. The Farmers Union associated with and promoted cooperatives in the general areas of insurance, grain elevators, livestock sales and wholesale and retail marketing. Soon after the creation of the first county Unions, purchasing agents secured needed food and supplies for Union members at wholesale prices. In the 1920s, the SDFU was responsible for the Farmers Union Mutual Insurance Company of South Dakota and the Farmers Union Livestock Commission at Sioux Falls (See Chapter 3). The Farmers Union Brokerage at Sioux City was also incorporated in South Dakota.

MERCANTILE COOPERATIVES

Convinced that the cooperative movement offered the only real promise of better livelihood for farmers through self-help agencies, Union locals, as early as 1915, organized a co-op elevator at Ethan, a cooperative supply at Mitchell and a mercantile co-op at Mount Vernon.

One of the Farmers Union food co-ops with the greatest longevity was organized at Frederick, South Dakota, on May 9, 1918. Twenty-three farmers met at the Frederick Equity Exchange to organize their own store, the Frederick Co-op Mercantile Company. They elected a board of directors and voted to capitalize the company at $50,000, divided into 500 shares at $100 each. A branch store was added at Hecla in the 1920s. The Great Depression took its toll as net savings of more than $7,000 in 1922 dipped to $650 by 1934. Although the

Farmers Union No. 253, Rumpus Ridge, Canistota. Local 253 operated a co-op store, manged for 49 years by Local Secretary Alfred Waechter. Pictured are: Mrs. Waechter; sons Jim, left and Tony, right and co-op employee Jack Kuyper.

branch store at Hecla had to be sold, the cooperative at Frederick survived and flourished during World War II and the post-war years. The co-op installed its own meat processing and locker facilities. By 1970, however, the board of directors was forced to convert that store into a Jack and Jill franchise because of the difficulty of getting supplies from Consumers Cooperative of Superior, Wisconsin.[7]

Another of the first general merchandise cooperatives in South Dakota was established at Wessington Spring in February 1920. The Jerauld County Farmers Union purchased a brick building and an inventory of farm implements to enter the retail business. Begun exclusively as a farm machinery store, the building was remodeled into a department store, adding groceries and complemented by a cream station and a hardware store. By June 1930, the Jerauld County Farmers Union Mercantile Company had 315 stockholders.[8]

Representative of a co-op mercantile in the southeastern part of the state was the store at Rumpus Ridge, near Canistota. Farmers Union Local No. 253 built the store and meeting hall in 1923. The co-op did such a good business that in March 1925 the local voted to increase the size of the building. Two years later, the growth of the co-op made it necessary

John Stekly with the Tyndall Farmers Union Oil Company delivery truck in 1940.

to construct another building. Alfred Waechter was the secretary-treasurer of the local from its inception on December 15, 1917, and manager of the store until his death in April 1967. The latter duty being assumed by his wife and sons at that time.[9]

THE FARMERS UNION CO-OPERATIVE BROKERAGE

The primary outlet for Union members to purchase merchandise had been the Farmers Union Brokerage and Produce Commission, organized at Sioux City, Iowa, in October 1924. Within eighteen months, it received a certificate of incorporation from South Dakota. In the spring of 1931, arrangements were made to open a branch of the Brokerage in Sioux Falls. A large warehouse was leased just north of the Great Northern depot and on the main road leading to the stockyards. Situated at 221 North Weber Avenue, the warehouse was conveniently located for unloading carloads of flour, feed, salt, fencing and other commodities. Paul Batcheller managed the facility.[10]

During the early thirties, the Brokerage suffered economic reverses because of the Great Depression. Crop failures and low prices reduced the purchasing power of farmers. The sale of binder twine and coal, the two primary sources of income for the Brokerage, plummeted. Poor crops in 1933 reduced the sale of binder twine by almost 80 percent. In order to move its stock, the Brokerage was forced to reduce its prices, regardless of cost.[11]

Headquarters of the Farmers Union Brokerage Company, Sioux Falls.

One of the cardinal principles of a successful cooperative was violated when too much merchandise was sold on credit. By 1933, the accounts receivable for the firm stood at $3,852.11.[12]

Accountants recommended that $3,000 of this amount be eliminated. Federal taxes became another issue. The board of directors for the Brokerage assumed that the firm was permanently exempt from federal taxes because it was a cooperative. The Brokerage lost its exempt status when non-Union members, beginning in 1929, made up a substantial number of the cooperative's customers. The management of the co-op did not become aware of this change of status until 1935, by which time it was liable for back taxes.

Even the best of intentions did not help the Brokerage's plight. The manager of the Brokerage at Sioux City bought a retail coal yard in 1932 to provide work for its employees when business slowed during the winter months. The bulk of the coal was purchased by non-Union members, compounding the Brokerage's tax difficulties. The Internal Revenue Service levied approximately $14,000 in taxes against the cooperative. The amount was later reduced to $12,800, and through a series of appeals to the Treasury Department, the Farmers Union was permitted to amortize the payments over a thirty-month period.[13] The matter had not been resolved as the decade came to a close.

In the meantime, significant changes had taken place at the management level. George Batcheller, the manager of the Sioux City Brokerage resigned in 1936, and soon purchased the coal yard. On March 1, 1936, the transfer of the head office of the Farmers Union Brokerage Company from Sioux City to Sioux Falls took place. The move was based on the

Branch of the Farmers Union Brokerage Company, Mitchell.

central location of Sioux Falls and the possible installation of a petroleum products plant beside the warehouse at 221 North Weber Avenue in northeast Sioux Falls. Paul Batcheller became the manager of the slightly restructured cooperative.[14] When Batcheller took charge of the Brokerage, Henry Helgerson was named the manager of the Farmers Union Oil Depot, a part of the Brokerage's operations. Helgerson had managed the Yankton County Farmers Union Cooperative Oil Company for four years.[15]

Branches of the Sioux Falls Brokerage were established at Mitchell and Watertown in 1935 and 1936, respectively. Much of the business of the Sioux City, Iowa, site was transferred to Sioux Falls when the head office was moved to South Dakota. President Loriks encouraged the FU membership to buy their needed supplies at one of the Brokerage locations. He encouraged Union members to think of cooperatives beyond a few dollar dividends. He envisioned a vertically integrated cooperative structure from the local producer to the marketing terminals and to the processing of and manufacturing of the Union's own products and supplies. If the Farmers Union set its mind to it, it could build a new economic order similar to what had been accomplished in Scandinavia.[16]

The Brokerage was renamed on January 20, 1937. Henceforth, it was to be called the Farmers Union Co-operative Brokerage. W. C. Hermann, of Rockham, was chairman of the Brokerage board, and Enoch Hofstad, of Elk Point, served as secretary.[17] In the fall of 1937, Loriks and the SDFU Board of Directors selected Gustav L. Moseson, of Howard, and a cousin of Oscar Fosheim, to be the new manager of the Brokerage.[18] Although his appointment became official on December 8, 1937, Moseson had spent his evenings the prior month

FARMERS' UNION STATE EXCHANGE

FARMERS EDUCATIONAL & CO-OPERATIVE UNION OF AMERICA S.D. DIV.

IS THE FARMER'S OWN STORE

AUTOMOBILE TIRES AND TUBES

30x3	Plain	$13.65
30x3	Non-Skid	15.50
30x3½	Plain	17.90
30x3½	Non-Skid	18.35
31x4	Plain	27.15
31x4	Non-Skid	28.50
32x4	Plain	27.95
32x4	Non-Skid	29.20
30x3	Tube	3.05
30x3½	Tube	3.55
31x4	Tube	4.75
32x3½	Tube	3.95

BLOWOUT PATCHES

3½ and 4 inch	.50
4½ inch	.75
5 inch	.90

CREAM SEPARATORS.

750 pound capacity	65.00

CARTS

Barrel Carts	9.25
Platform Carts	10.25

CORN PLANTERS

Case, Reg. Runner	73.00
Case, Single Disc	75.00
Hummer, Reg. Runner	58.00
Screen Corn Grader	1.25

CULTIVATORS

Case, 2 Row, 12 Shovels	100.00
Moline, D. D., 12 Shovels	100.00
Tongue Truck	10.00
Moline, Bal. Frame	52.50
Hummer, Leverless	43.00
Hammock Seat 6, S	45.00

CORN SHELLERS.

1 Hole Sheller	12.50
2 Hole, hand and power, with carrier	30.00

CORN SCOOPS

Scoop Shovels	1.70
Scoop Forks	1.75

DISC HARROWS

All Disc Harrows equipped with tongue truck, tongue and 4-horse evener, three lever and hard oil cups.

8 foot 16-16	60.00
9 foot 18-16	65.00
10 foot 20-16	75.00

ENGINES

Challenge 1½ H. P.	62.50
Fairbanks 1½ H. P.	60.00
Lindsay 1 H. P.	55.00
Fairfield 4 H. P.	145.00

FEEDMILLS

6 inch Burr, Grinder	18.00

FORKS

Hay forks, our best	1.15
Manure forks	1.10
Straw forks	1.35

GALVANIZED IRON BASKETS

Heavy, 1½ bushel	1.50
Rope Handle, 1 bushel	1.10
Double bottom, ½ bushel	.75
Elm Stave Baskets	.35

GRINDSTONES

Double Treadle	6.75
Power	9.50

HOG OILERS

Sipe, Non Waste	12.00
Sipe Hog Oil, per gal.	.50

HOG WATERERS

Ideal, Single	23.50
Ideal, Double	28.00

HAY SLINGS

Double rope sling	6.00
Sling lock	2.25

HARNESS

You ought to see them:

Plow Harness, 1½ in.	42.00
Team Harness, 1½ in.	62.50
Extra Heavy Harness	70.00

HORSE COLLARS

Prices from	$3.00 to $7.50

HORSE BLANKETS

Cotton, 80x84	2.50
Mixed, 80x84	4.50
Mixed, 84x90	4.50
Woolen, 84x90	5.00
Woolen, 90x96	5.50

HORSE BRUSHES

Fibre, "Bull's Eye"	.25
Leather, black bristles	1.00
Oval Block, leather	1.50

HALTERS

Leather	$1.25 to $2.50
Rope Halters	1.10

CURRY COMBS

6 Bar, black steel	.10
8 Bars, lacquered steel	.25
Reform, leather handle	.25
Arch Comb, spring steel	.25

OVERALLS AND JACKETS

Men's Overalls, striped blue white material, full cut, double stitched. Has two large front pockets, watch and pencil pocket on bib. Well made, every way	$1.65
Heavy Blue Denim Overalls, especially well made	$2.00
Extra heavy, Indigo Blue Denim Overalls	$2.60
Extra heavy Denim Jackets	$2.60

PLOWS

Hummer Gang, 14 in.	90.00
Moline Gang, 14 in.	105.00
Hummer Sulky, 16 in.	57.50
Moline Best Ever, 16 in.	67.50
Moline G. E. Sulky	55.00
14 inch walking plow	18.50

STACKERS

Acme No. 2, with Transp.	72.50

HARROWS

Steel Flex. 160 teeth	52.00
U Bar, Lever, 120 teeth	40.00
U Bar, Lever, 90 teeth	30.00
Plow Harrow, 35 in.	9.50
Gang Harrow, 47 in.	11.00

REGULAR RAKES

Moline 10 ft., self dump	46.00
Acme 10-ft., self dump	44.00
Acme trus frame Sweep	32.00
Acme 2 wheel Sweep	27.50

SCALES

Fairbanks, 1000-lb.	30.00
Pitless, 5-ton	135.00

SPREADERS

Independent, 70 bu.	165.00
Moline, 70 bu.	180.00

STOCK TANKS

2 x 6 Cypress	15.00
2 x 7 Cypress	18.00
2 x 8 Cypress	22.00

SALVET

10 pound package	.50
20 pound package	1.00
40 pound package	2.00

TANK HEATERS

Nelson	12.00
New Wrinkle	15.00

WAGONS

Mandt Special, complete with 36 inch box	130.00
Mandt Regular, complete with 26 inch box	135.00
Lafayette Truck	60.00
Calmar, Steel Wheel	50.00
Keller, Steel Wheel	52.50
36 inch wagon box	35.00
Wagon Reaches	1.25
Wagon, Dbl. T. & N. J.	3.65
Steel Wagon Poles	8.25

WASHING MACHINES.

White Way, twin tub	55.00
White Way, single tub	42.00

Above machines equipped for power.

White Way lever tub	15.00

WIRE GATES

Dreadnaught, 14-ft.	14.00

FARMERS' UNION STATE EXCHANGE

Price List, Farmers Union State Exchange, Mitchell.

at the Brokerage office acquainting himself with the business. Following a visit with Moseson, Loriks wrote to Fosheim, "I feel very optimistic about the whole thing and I like his [Moseson's] attitude. I feel that under his management the whole set-up will undergo a wholesome reorganization and will be put on its feet and will be run in a business-like manner."[19]

Moseson had helped to organize the Farmers Union in Miner County, and he had served as president of the county organization. He was also president of the State Association of Mutual Insurance Companies before coming to the Brokerage. Moseson graduated from Augustana College in 1915 with a degree in academic business. He taught school for one year prior to managing a large plantation at Bluefield, Nicaragua, for two years. While residing on this Central American plantation, he saw how the workers were exploited at the company store. It developed in him a deep concern for working people.[20] Approximately four months after Moseson's appointment, in April 1938, the Minnehaha County Farmers Union Co-op Oil Association opened a service station adjacent to the Co-op Brokerage on Weber Avenue.

During the summer and fall of 1938, the overall financial situation of the SDFU deteriorated. At the annual convention in October, at Winner, it was recommended that the Union's cooperative enterprises sell only for cash and that the managers of co-ops be held responsible for credit granted. It was even suggested that the new SDFU president, John Skage, serve half-time as president and half-time as the field man for the Brokerage.[21]

Within a year, the Union's financial status had improved. Farmers on the Prairie Plains were once again favored with good weather, and the national economy began to recover with the threat of a general war hanging over Europe. Responding to an increased demand for supplies, several improvements were made at the Co-operative Brokerage. New features in the structure, as of August 1939, included a model display room, a new storage and supply room, a spacious sales room and new office facilities. Added to the inventory were more tires and auto accessories, electrical goods, high-grade tools, pump and well fittings, feed mills and water systems.[22]

The property on North Weber Avenue, Sioux Falls, had become more than a supply center for Union members. During the summer of 1931, the SDFU Co-operative Grain Company had set up an agency just north of the Brokerage to sell carloads of grain and hay at reduced prices to needy members. There were sections of South Dakota that had a surplus of grain and feed that could not be disposed of for a profit. This surplus would be distributed to those areas of the state where farmers contended with drought conditions.[23]

Gustav "Gus" Moseson, Manager of the Farmers Union Brokerage Company.

During the annual state convention at Mitchell in October 1931, an Agricultural Credit Corporation was created to meet demands for money to finance feeder loans. The new corporation was also located at Sioux Falls, in connection with the Sioux Falls Livestock Commission. Facilitating the program was C. H. O'Hern, the manager of the Livestock Commission.[24] Within a year, the corporation had

Ernest Rusch, manager, Lake Preston Farmers Union Oil Co-op, 1948.

made ninety-nine loans, totalling approximately $103,000. Six of the loans had already been paid in full, and coupled with several partial payments, about $20,000, had been returned to the corporation. With a capital stock of $10,000, it was necessary to secure additional capital. This was provided by the Livestock Board. By increasing capital by $50,000, the corporation gained a rediscount privilege of another $50,000.[25]

COOPERATIVE OIL COMPANIES

Among all of the SDFU cooperative enterprises during the 1930s, it was the phenomenal increase in the number of cooperative bulk oil and gasoline stations that gained the most attention. The first Farmers Union station of this kind had been established at Yankton. As of 1930, the Union had three bulk oil stations, located in Yankton, Woonsocket and Redfield.

One of these stations, the Sanborn County Farmers Union Oil Company of Woonsocket, began operation on January 12, 1928. It opened at a site that had been a part of the "gas war" during the early 1920s. The growing popularity of the automobile and increased road construction caused national oil companies to raise the price of gasoline to 26.6 cents per gallon. Governor William H. McMaster, believing that the oil companies were making an unconscionable profit, ordered the State Highway Department to sell gasoline at its district headquarters for sixteen cents per gallon. The 1925 Legislature reaffirmed McMaster's action and authorized the State to sell gasoline, oil and lubricants if retail prices were unreasonable or excessive. The State had built a facility to dispense petroleum-based products

Farmers Union Oil Company, Wessington, 1948.

when the U. S. Supreme Court, on October 28, 1925, ruled that states should not engage in the retail gasoline business.

Farmers Union local members from Woonsocket, Harmony Hall and Letcher, led by John W. Murtha, president of the Woonsocket local, negotiated with the State of South Dakota to purchase the facility at Woonsocket. The station and two fifty-foot lots were bought for $2,500. A loan was secured by signature of each of the temporary board members, who served for the first three months until their successors were elected at the first annual meeting on January 15, 1929. The first $240 needed for business operations was borrowed from twenty-four members who gave ten dollars each.

Articles for incorporation of the Sanborn County Farmers Union Oil Company were signed on June 26, 1928. During its early years of operation, the co-op borrowed money from the Sanborn County Bank, Woonsocket, to build its inventory and to pay expenses; in March 1930, the bank loaned $2,000 to purchase a carload of lubricating oils and greases. The oil station had few difficulties during the thirties because of the wise handling of accounts receivable. Much of the credit for good management has been given to B. J. Dankey, who capably managed the station from 1935 to 1968.[26]

S. Douglas Murphy, who helped build Harmony Hall, Local No. 447, and a Union member for seventy years, served on the Sanborn County Farmers Union Oil Company board of directors from 1937-1974. He was chairman of the board for twenty-five years. He said that Consumers Cooperative Association (CCA) of North Kansas City, Missouri, sold its first railroad tank car of gasoline to the Sanborn co-op. CCA was the Sanborn co-op's wholesale

S. Douglas Murphy, Woonsocket and Clement Kloucek, Scotland, Co-op Hall of Fame Awards, 1993.

supplier until 1943 when Farmers Union Central Exchange, St. Paul, Minnesota, assumed that role.[27]

The Sanborn County oil co-op continually expanded its facilities and increased its services throughout the 1930s. A cream-buying station was added in 1932 and continued to function until the mid-1940s. A service station and small bulk plant were built at Letcher in 1933. A service station was started at Alpena in 1934.[28]

All considerations for expansion of the Sanborn County co-op were first discussed by the board of directors and presented to the stockholders at their annual meeting. During the co-op's first fifty years of operation, credit policies were an issue at board and annual meetings. Despite the many years of poor crops and low prices, the Sanborn County cooperative continued to grow and prosper. There were years when the co-op paid a dividend, and there were years when payments were deferred. In its first fifty years, nevertheless, retired share capital stock paid to patrons totalled $674,766.16.[29]

Murphy, as chairman of the Sanborn County co-op's board of directors, always reminded the stockholders that the Farmers Union should be their banker. He impressed upon them the purposes of the FU cooperative movement — to serve as a balance wheel to keep prices in line, to provide a source of supply, to assure a saving to members, to control quality and to provide better services. For many years, the stockholders completed the annual meeting with a pancake supper or sandwiches and good conversation.[30]

Approximately one year after the establishment of the Sanborn County Farmers Union Oil Company, the Redfield Farmers Union Oil Company came into being. Recognizing the

Farmers Union Oil Company, Redfield.

need for a cooperative oil company in Redfield, area farmers held an organizational meeting and on June 10, 1929, scheduled their first official board meeting. W. H. French was elected chairman of the board. The oil board worked closely with the board of the Spink County Farmers Union. At the time, the Spink County FU operated a cream station and produce business. Richard Turbandt, who managed the cream station, became the first manager of the oil company. The Redfield FU Oil Company opened for business on July 19, 1929.

The Redfield cooperative did not escape the problems associated with the Great Depression. Looking for ways to improve the co-op's economic status, groceries were added to the products available for sale. Despite the economic hard times, a new building site was purchased in 1935 and a new building constructed which housed the service station, produce and grocery business. In 1941, the co-op eliminated the grocery business and on January 22, 1942, affiliated with the Farmers Union Central Exchange. After that time, the cooperative began to prosper because of loyal patrons, better crops and prices. A new service station was completed in 1949.[31]

By 1931, there were eight Farmers Union bulk oil stations in South Dakota. An association of FU oil stations was formed to purchase supplies of gas and oil in large quantities and to secure reduced prices based on volume purchased.[32] The momentum for more and more FU cooperative bulk oil stations continued to grow throughout the 1930s. Members were continually reminded that "in Union there is strength, power and economy." The question confronting them was, "What am I doing to help my organization (the Farmers Union) in building a cooperative commonwealth in South Dakota?"[33]

When President Emil Loriks made his report to the annual SDFU convention in the Sioux Falls Coliseum in October 1936, he proclaimed that 1936 marked the dawn of a new era in the cooperative field.[34] He had a right to be optimistic. Six months earlier, the *Union Farmer's* banner headline proclaimed "SOUTH DAKOTA GOES COOPERATIVE." The paper announced that new cooperatives had been organized in Beadle, Brown, Davison, Day, Deuel, Grant, Hanson, Hyde, Jones, Kingsbury, Marshall, McCook, Miner, Moody, Roberts and Tripp counties.[35]

As with the Sanborn County Cooperative Oil Company, many of the new bulk oil co-ops had committed and strong leadership. Henry A. Vinz served as president of the Farmers Union Co-op Association of McCook County from its inception in 1937 to 1962. Once again, a group of determined farmers set out to get signers on notes of ten dollars each. Sixty-three persons signed on, and with a loan of $6,000 from the Farm Security Administration, a vacant lot was purchased and a small building, equipment and tanks were placed on the site in Salem.[36]

John Skage was the moving spirit in the organization and building of the Deuel County FU co-op. Beginning in 1931, he sold $10 shares of stock, telling potential stockholders that "There is no limit to what we can accomplish if we work together." By 1936, over $1,770 had been raised, and the Deuel County co-op opened for business at Brandt. The co-op later established stations at Astoria, Clear Lake and Gary with the main bulk plant being located at Toronto. For many years it was the largest South Dakota affiliate of the Farmers Union Central Exchange. Fifteen years after its beginning, the cooperative had a net worth of $215,000. The Deuel County Farmers Union Oil Company, managed by Skage between 1936 and 1960, returned almost $5.2 million in net savings to patrons on total sales of more than $81 million in the first fifty years of operation.[37]

Martin C. Nelson was president of both the Roberts County Farmers Union and the Roberts County Farmers Union Cooperative Oil Association when the co-op was started in the spring of 1936. Several hundred farmers subscribed $2,500 at the organizational meeting and many volunteered to double their subscriptions. The new enterprise also had the support of the City of Sisseton. Another fast growing cooperative was the Kingsbury County Farmers Union Cooperative. Managed by Edward Alberts, the Kingsbury County co-op emphasized that it handled only Farmers Union products and brands.[38]

ARE WE FOR STRICTLY FARMERS UNION CO-OPS OR WHAT?

The emphasis upon using only Farmers Union products indicated the Union's growing concern with the operations of the Consumers Cooperative Association in South Dakota. The CCA, formerly the Union Oil Company, had been selling gasoline in the state since the late 1920s. In 1934, the firm was given permission to make a presentation on its policies to delegates to the annual SDFU convention in Rapid City. CCA wanted an established outlet in South Dakota to sell its products. The North Kansas City based cooperative believed the SDFU Brokerage in Sioux Falls offered the best opportunity for selling CCA products.[39] During the spring of 1935, the SDFU board of directors studied the consequences of associating with the CCA.

The board journeyed to North Kansas City and met with Consumer Cooperative Association officials. The CCA's executives made two proposals for the SDFU board to consider and to bring to the annual convention in October. The first proposal called for the CCA to take over control of the Brokerage, exchanging shares of stock of the Union Oil Company for stock of the SDFU brokerage. An attempt would be made to alter the by-laws, allowing two members of the South Dakota Farmers Union to serve as directors on the Union Oil board.

CCA's alternate proposal would have the Brokerage withdrawing from the sale of all petroleum products but serving as a bulk oil station for Union Oil. CCA would reciprocate by not competing with the Brokerage in any of other items that the Brokerage sold in its trade area.[40] Union officials did not believe that the CCA was interested in the educational program of the Farmers Union, and they made their own proposal to the 1935 convention. Their plan to operate a compounding plant at the site of the Brokerage in Sioux Falls received approval.[41] Moving quickly, the board of directors moved the headquarters of the Brokerage from Sioux City to Sioux Falls and made plans for the new petroleum compounding plant.

After further discussion, the board contracted with the Farmers Union Central Exchange to supply the Brokerage with petroleum products rather than the SDFU operating its own blending plant.[42] The Brokerage received from Central Exchange products on its own specification at a fair price. The agreement proved satisfactory, and after one year, the South Dakota Farmers Union label appeared on all containers of petroleum products.

The decision having been made to affiliate with the Central Exchange, the *Union Farmer*, throughout much of 1936, carried on a campaign for the purchase of Farmers Union products. Union members were continually reminded that the Farmers Union Central Exchange was a "Farmers Union Cooperative Institution" and that it supplied gasoline, kerosene, lubricating oils, tires and batteries. In addition, the Central Exchange had its own compounding plant. Farmers Union co-ops that were already established and purchasing through old-line concerns were urged to "lose no time in making ... arrangements to purchase future petroleum supplies through their own organization — THE FARMERS UNION!"[43] One of the co-ops that changed affiliation was the Moody County Farmers Union co-op at Flandreau. For several years, the Moody County co-op purchased its gasoline from the CCA, which helped finance its operations. Gust Moseson, representing the Brokerage and the Central Exchange, used his persuasive powers to convince the co-op to buy from the Central Exchange, assuring both an adequate gasline supply and financing.[44]

Articles in the *Union Farmer* indicated that the CCA could not be trusted. The paper proclaimed that the Kansas FU Co-Operative Association had severed its relations with the CCA because it refused to make brokerage payments.[45] President Emil Loriks carried on a lively correspondence with Farmers Union officials in other states regarding the CCA and its president, Howard Cowden. In a letter to H. G. Keeney, president of the Nebraska Farmers Union, Loriks accused Cowden of spreading subversive propaganda and of trying to join the National Farmers Union board of directors for his own gain.[46] James G. Patton, a member of the FU national board, said that in building a farm program, one must know the difference between Farmers Union cooperatives and Cowden's program. The Farmers Union needed a unified organization to protects its rights in dealing with other organized groups.[47] M. R. Miller, secretary of the Missouri Farmers Union, apologized to Loriks when the Missouri FU identified with the CCA in co-op promotion. He facetiously suggested that the program of the CCA could benefit the SDFU by joining them in the slogan — "Join a cooperative" and add "Be sure it is a Farmers Union cooperative."[48]

By early 1938, Loriks was working closely with Emil Syftestad, the general manager of the FU Central Exchange. A conference of all of the SDFU co-ops was held at Sioux Falls on September 20, 1937, at which Syftestad was a presenter. Later on, the SDFU president requested Central Exchange promotional materials, including films of the Central Exchange. Of particular interest to Loriks was the Exchange's compounding plant. Gleefully, he wrote Syftestad that at a co-op institute at South Dakota State College at Brookings, the program "left the Bureau, Grange, and CCA out in the cold."[49] The competitive battle lines between the Farmers Union and the CCA had been drawn. The final resolution of this struggle in favor of the SDFU did not occur until 1953.

THE OTHER COOPERATIVE VENTURES

Early in 1930, the Sioux Falls Livestock Commission was made a joint-marketing agency for three states. Cattle sales from farmers in southwest Minnesota and northwest Iowa were encouraged by admitting a representative from the Unions in each of these states to the governing board. Each Union would share in the profits, proportionately to the volume of business furnished.[50]

The success of the livestock commissions, the livestock marketed by all FU agencies totaled $110,000,000 in 1929, caused the SDFU to have a vested interest in six different commissions by 1932. The South Dakota Union had some degree of ownership in the Farmers Union firms in Chicago, St. Paul, Omaha, Sioux City, Huron and Sioux Falls.[51] The commission houses in Sioux City and Sioux Falls were able to return to members 40 to 50 percent of commission charges in 1931 because of the large volume of business and good management.[52] Despite the adverse economic conditions, the Sioux Falls Livestock Commission managed to pay dividends. For example, the firm paid a dividend of 15 percent because of savings in 1936.[53]

There was one failure. The Farmers Union Livestock Commission that began operating in Huron in January 1930, under the management of R. V. Haedt, did not last four years.[54]The commission house closed permanently on September 1, 1933, because of the severe drought, the Great Depression and a closed market at Huron.[55]

On January 9, 1934, the SDFU board of directors placed the directorship of the livestock commissions under the supervision of selected board members. The assignments included: E. N. Hammerquist to the Chicago commission, William Sinkular to Omaha, W. C. Hermann and John W. Batcheller to Sioux City, and the foregoing men plus Jesse Kistler to the Sioux Falls Livestock Commission.[56]

After seven years in existence, the Farmers Union Mutual Insurance Company of South Dakota had $10 million in policies in force.[57] Prior to May 1, 1931, the firm wrote fire and tornado insurance on farm property. Beginning on May 1, the FU Mutual Insurance Company began to write insurance policies for automobiles because the Union had decided that other insurance companies charged excessive rates.

All of the insurance agents were farmers who understood the value of farm property. One of the requirement of Farmers Union Mutual was that all policy holders must be members of the Farmers Union.[58] Aside from the addition of automobile insurance, the state Union did not make any significant changes in its insurance cooperative until the 1940s.

After a year of planning, a Farmers Union produce company was organized at Mitchell in 1930.[59] With its general office in the old FU Exchange building in the Corn Palace city, the produce company became a marketing agency for poultry and eggs for Union members in twenty counties in eastern South Dakota. At the Mitchell site, eggs were graded and poultry fattened and dressed for market.

As of 1932, farmers in Tripp and Gregory counties also marketed their own poultry and dairy products. Union members owned and operated a creamery at Burke, where cream was churned into butter, and the proceeds returned to members.[60]

Many of the cooperatives established in the thirties later checked off membership dues through their earnings and set aside 5 percent of the net earnings for educational purposes.[61] Education, of course, was the main requisite for cooperative building, and in the Union plan, it had to precede the effective operation of cooperative agencies.[62]

As of June 1937, the Classified Directory of Cooperative Associations in South Dakota listed the following number of cooperatives:

52 cooperative oil companies
40 cooperative creameries
18 general merchandise cooperatives.[63]

Emil Loriks, president of the SDFU from 1934-1938, said that building cooperatives, such as marketing and supply cooperatives, was probably the biggest achievement of his administration.[64] In a 1937 speech, he affirmed his belief in the Farmers Union and the cooperative movement. He said:

> Thirty-five years ago, Newt Gresham and his intrepid little band, ten hard-hit, destitute, poverty stricken farmers of Rains County, Texas, organized the first Farmers Union local. They wrote the Farmers Union Constitution, the Magna Charta, the Declaration of Independence from economic tyranny and despotism, that will ring through the ages.
>
> It stands today like the Rock of Gibraltar, untarnished and unsullied by the vicious attacks of demagogues and charlatans, imposters and fake medicine men with their quack cure-alls and nostrums. In the raging storm on the rock-bound coast of economic dictatorship, it stands as a beacon light to guide the ship of economic democracy to a safe haven.[65]

REGIONAL COOPERATIVES: THE FARMERS UNION GRAIN TERMINAL ASSOCIATION

Cooperative grain elevators became identified with the Farmers Union soon after the first locals were organized in South Dakota. During the 1920s, many of them did not use the FU name, preferring to use a name identified with the particular community. By the late twenties, the Farmers Union Cooperative Grain Company was organized for the purpose of furnishing co-op elevators and members of the FU with terminal marketing facilities for grain on a cooperative basis similar to the system of livestock commission agencies. The SDFU encouraged members of the cooperative to use the facilities of the FU Terminal Association at St. Paul, Minnesota.

The FU Terminal Association was built upon the ruins of the old Equity Cooperative Exchange. The Exchange, an offspring of the American Society of Equity, had been established in the spring-wheat area in 1911, and following the outbreak of World War I, had developed a chain of country elevators in Minnesota and North Dakota as well as a terminal marketing agency in St. Paul. Poor business management, and opposition from the private grain trade, forced the Equity Union Marketing Association into receivership in 1923. It was revived as an agency of the Farmers Union in 1926 and took the name of the Farmers Union Terminal Association.

In 1930, the FU Terminal Association handled more than 15,000,000 bushels of grain annually, claiming to be the largest cooperative grain-marketing association in the United States. It announced the distribution of $35,000 in patronage dividends, the first grain cooperative on the Prairie Plains to distribute dividends to its members.[66] The times favored the Terminal Association. Grain production was up, business was plentiful and the national president of the Farmers Union, Charles E. Huff of Kansas, favored the Association's efforts.

The Agricultural Marketing Act of June 15, 1929, established the Federal Farm Board to promote the marketing of farm commodities through agricultural cooperatives. President Charles Huff, a staunch supporter of the Federal Farm Board, the Northwest Organizing Committee and other leaders of the Union helped to launch the Farmers National Grain Cooperative (FNGC) in the fall of 1929. It was the first national marketing cooperative incorporated under the Farm Board. When he was replaced by John A. Simpson as president of the NFU, Huff became president of the Farmers National Grain Corporation.

In mid-November 1929, the South Dakota Wheat Growers, a wheat pool, and the FU Terminal Association turned over their selling facilities to the FNGC. Ed Everson, president of the SDFU, urged all county cooperative elevators to market their grain through the federal cooperative.[67] The SDFU Cooperative Grain Company was also associated with the FNGC. The FU Terminal Association became the largest stockholder in the Farmers National. With the onslaught of the Great Depression, however, the purchase of grain by the FNGC failed to stop a drop in prices.

The progressive and pro-Farm Board faction of the Farmers Union withdrew from the six year old Corn Belt Committee in May 1931 and established the National Committee of Farm Organizations as a lobbying group. This action, which ended the Corn Belt Committee, infuriated President Simpson, who was already very outspoken in his opposition to the Farm Board.

The SDFU grain cooperative, shortly thereafter, severed its relationship with the Farmers National Grain Corporation. The delegates to the annual state convention, meeting in Mitchell's Corn Palace in October 1931, voted to withdraw from the Farm Board sponsored federal cooperative. The decision, motivated by the radicals in the SDFU, was directed against the leaders of the Northwest Organizing Committee, because they played a leading role in the National Grain Corporation.[68]

The following year, 1932, the leading farm organizations, including the Farmers Union, established the National Farm Conference. Following the election of Franklin D. Roosevelt as president in

M.W. Thatcher, Grain Terminal Association founder and its General Manager until May 1968.

November, the National Farm Conference, in December 1932, proposed a price parity, production control and processing tax bill similar to the New Deal's AAA legislation in 1933.

Since 1927, M. W. Thatcher had been a major role player in the NFU. He was chairman of the NFU's Northwest Organizing Committee from 1927-1930 and helped develop the Union's strength in the spring wheat area. After the Democrats came to power in 1932, he spent considerable time in Washington, D.C., serving on several important committees dealing with agricultural affairs. Thatcher, in 1936, lobbied Congress and the administration for the Commodity Exchange Bill. The legislation would regulate the practices, often unfair, of the large grain exchanges in Chicago, Minneapolis and New York. The Commodity Exchange Bill amended the Grain Futures Act of 1922 to require that cooperatives and other associations be admitted to membership on the exchanges. The concept of open marketing had long been a goal of farm cooperatives.

Thatcher returned to St. Paul from the nation's capital in June 1937 to manage the Northwest Farmers National Grain corporation. As a "Washington Insider," he knew that the Farmers National Grain Corporation would be dissolved a year hence. Believing a grain marketing cooperative on the Prairie Plains to be essential, he returned to Washington, D.C., to convince President Roosevelt and Secretary Henry A. Wallace that such a cooperative was needed. Thatcher proposed to get operating capital by securing a loan from the Farm Credit Association for $300,000. He asked that the Bank of Cooperatives provide a line of credit of $900,000 and that the Farm Security Administration (FSA) loan individual farmers twenty-five dollars to buy stock in the proposed cooperative.[69]

Thatcher brought his plan to Roosevelt who referred it to Wallace. Secretary Wallace agreed to the proposal, and with the formal dissolution of the National Grain Corporation, the Farmers Union Grain Terminal Association (FUGTA) began marketing grain on June 1, 1938, after a seven year absence from the market. The GTA combined the FU Terminal Association, a purely paper organization, with the Northwest National Grain Corporation. The Northwest staff joined with the board of the Terminal Association to manage the newborn cooperative. Dan O'Connor was president of the FUGTA and Thatcher became its general manager.[70]

Thatcher sent out salesmen to sell stock in the GTA. Farmers responded by borrowing money from the Farm Security Administration to purchase their stock. When conservatives in the Congress objected to the FSA's cooperative loan program, it was pointed out that the GTA had agreed to set aside one-half of its net earnings to protect the government on its loans to farmers.[71]

The new cooperative had eighteen months in which to pay back the $300,000 capital loan to the Farm Credit Administration. The deadline proved to be no problem. By 1939, the GTA had 220 subsidiaries. More than 250 locals marketed their grain through the FUGTA. In 1940, the FU Grain Terminal Association was pronounced the largest grain marketing cooperative in the world.[72]

REGIONAL COOPERATIVES: THE FARMERS UNION CENTRAL EXCHANGE

The FU Terminal Marketing Association became a holding company in 1927 when it organized a Delaware corporation known as the Farmers Union Exchange. The Exchange would be used to sell needed products to Union members residing on the Prairie Plains.

Profits from these sales would be directed to the educational and organizational activities of the Union or placed in the treasury of the Terminal Association. The Exchange marketed staple items such as coal, feeds, fertilizer, fencing, flour, gasoline, grain loaders, grease, oil, tires and twine. Items such as coal and binder twine would be handled on a brokerage basis. The financial statement for the Union's Exchange in December 1928 indicated that the Exchange started with a cash capital of $1,000. On that meagre amount of capital, it had handled, in approximately eighteen months, over $500,000 of business, primarily in twine.[73]

The previously discussed Union bulk-oil stations began to appear on the Prairie Plains in 1928-1929. In the latter year, the FU Exchange of South St. Paul began its oil program. Ironically, the previously discussed Consumers Cooperative Association, then known as the Union Oil Company, was organized in the same year. Initially, marketing only petroleum products, the CCA began to market other consumer goods.

When the oil business of the Central Exchange reached new heights in the summer of 1930, plans were made to establish a new Farmers Union Central Exchange, under the cooperative laws of Minnesota, and to disband the old Farmers Union Exchange. The new Central Exchange began operations on January 15, 1931, and the managerial reins were soon in the hands of the capable Emil A. Syftestad.[74] The financing of the cooperative came from a handful of believers, who invested $525, and an advance of $75,000 from the FU Terminal Association. The Central Exchange became a very successful enterprise, beginning business operations with over fifty local oil cooperatives. By 1936, the Central Exchange had built its own compounding plant for lubricating oils.[75]

Syftestad began courting the various state Unions in order to secure the business of their bulk-oil cooperatives. President Loriks and the heads of the various petroleum cooperatives in South Dakota were invited to attend the Central Exchange's annual business meetings in St. Paul.[76] Loriks had already allied himself with the Central Exchange, and he worked diligently to direct business away from H. A. Cowden's CCA to the Central Exchange.

The cooperative movement was a valuable tool for recruiting new members into the South Dakota Farmers Union. Longtime Union members indicated that the economic advantages offered by cooperatives caused them to join the Union. For several people it was the purchase of binder twine and rape seed at reduced prices that attracted them, with others it was the dividends obtained through their oil co-op, and others cited the excellent dividends gained by selling their livestock through a Farmers Union livestock commission.[77]

Emil Syftestad, General Manager, FU Central Exchange 1932-1957.

Nationally, cooperative building declined in the 1940s, but the large and established cooperatives continued to grow. The cooperatives affiliated with the South Dakota Farmers Union, although they received less publicity than the legislative activity of the Union, became vital to the economic and social health of the organization. The ties between the Farmers Union and cooperatives had been historical and voluntary, not legal. The SDFU leader-

A 1936 Co-op tractor.

ship, nevertheless, encouraged its members to participate in those co-ops that bore the Union's mark of approval or name.

L. S. Herron, a Farmers Union leader from Omaha, Nebraska, attended the annual convention of the SDFU in 1937. Following the convention, he wrote to Loriks that the real issue at the convention was not who would be elected state president but whether the South Dakota Farmers Union would vigorously push a program of economic cooperation. Cooperation had won, and Herron wrote an article for the *Nebraska Union Farmer*, heading it "South Dakota Goes Co-operative."[78]

The Farmers Union triangle was completed at a time when fascist states in Europe and East Asia were attempting to conquer neighboring countries. By the late 1930s, an arms race was underway, and the world stood on the threshold of a general war. Throughout the 1920s and 1930s, the Union had campaigned for peace. When the United States was finally drawn into World War II in December 1941, the Farmers Union mobilized its members to produce foodstuffs for the Allied cause and saw many of its young men enter the armed forces.

Chapter 9

ON ACTIVE DUTY FOR PEACE AND COUNTRY 1938 - 1947

The Farmers Union stands ready to help preserve our Democracy and our form of government. There will be no bottlenecks in agriculture where the production of food is concerned.
— John Skage, *Union Farmer*, November 5, 1941.

Within the short span of fifteen years, Americans experienced two traumatic events, the Great Depression and World War II. By 1938, the first of these events was nearing its conclusion. Conditions were improving for South Dakota farmers. Rainfall increased, commodity prices were somewhat higher, and total cash farm income was on the rise. Up until the end of the decade, nevertheless, the AAA benefits program provided a substantial portion of farm income.The total cash farm income, including government payments, for South Dakota farmers in 1939 was $188,100,000; in the following year, this income totaled $140,000,000. Government benefit payments in the foregoing totals amounted to $23,100,000 in 1939 and $20,000,000 in 1940.[1]

New Deal agrarian legislation had helped the state's farmers in numerous other ways. The Farm Credit Act of June 16, 1933, and the Farm Mortgage Refinancing Act, which became law on January 31, 1934, provided much needed farm-mortgage relief to thousands of South Dakota farmers who were in danger of losing their equity through foreclosure proceedings. The federal government's refinancing program made it possible for farmers to liquidate old mortgages and to renew loans at lower interest rates. The Farm Credit Administration (FCA) also developed a system for long and short term loans. From May 1933 to July 1938, FCA loans to South Dakota Farmers totaled $83,378,000.[2]

As the pall of the Great Depression began to lift from the land, the shadow of the next traumatic event, another general war, gave people little cause for rejoicing. The forces of aggression were at work overseas. In response to the growing fascist menace, President Roosevelt asked Congress, in 1938, for start-up appropriations of $300,000,000 for defense. British and French orders for munitions, planes, guns and tanks also began to descend on American industry. While these demands for armaments stimulated the American economy, questions were raised about the future foreign policy of the United States.

THE UNION STANDS FOR PEACE

Prior to 1914, the Farmers Union had championed pacifism but was moderately internationalist. At the end of the First World War, the Union supported Woodrow Wilson's proposal for a League of Nations. President C. S. Barrett went to the Paris Peace Conference as an official spokesman for American agriculture. Disillusioned with the failure of the major powers to adhere to the League's principles, Union members became isolationists in the early 1920s. The spirit of isolationism was most pronounced on the Prairie Plains, and most rural people believed that isolationism would bring peace in the inter-war years.

John A. Simpson advocated the "Devil Theory" of history, which argued that bankers, big business and the armaments makers had led Congress to declare war against the Central Powers on April 6, 1917. He accused Eastern financiers of making loans to the Allies and of wanting war to protect these loans. Union members were not alone in their isolationist stance during the twenties and thirties. Their views reflected American public opinion throughout the forty-eight states.

Early on, the SDFU proclaimed its opposition to any form of compulsory military training. At the annual convention at Chamberlain in November 1924, the delegates passed a resolution expressing their disapproval of compulsory military training at the University of South Dakota. The Reserve Officer Training Corps program should be discouraged rather than encouraged.[3]

The state's union leadership viewed with alarm the growing threat of war during the 1930s. Responding to the conflict that had broken out in Ethiopia in 1935, President Loriks said that the Union must strive to keep America out of war. The SDFU had been born in the throes of war (World War I). "The job is ours," he said, "to build up and fortify public sentiment against war!" Repeating Simpson's view on who were the war mongers, Loriks argued that World War I had been fought for the House of Morgan.[4]

Two years later, the Union launched an all-out effort on behalf of Joint Resolution No. 199, introduced by Congressman Louis Ludlow of Indiana. The resolution, first introduced in 1935, called for a national referendum on the declaration of war. Ludlow's proposed constitutional amendment stipulated that except in the event of an invasion of the United States, the authority of Congress to declare war should not become effective until confirmed by a majority vote in a nationwide referendum. Supporters of the amendment believed that those who fight the wars, and pay for them, should have something to say about declaring wars.[5]

In September and early October 1937, South Dakota's Juniors circulated petitions favoring the Ludlow Amendment. Nationwide, the Farmers Union hoped to secure a million signatures by the time Congress convened. Signers of the petitions were asked to contribute at least one cent to the cause of peace. Half of the money received was used to promote the peace movement and the other half designated for Junior education, which meant it also was used to promote peace.[6] A national poll revealed that 73 percent of those polled favored this type of referendum. Only the intervention of President Roosevelt, who said such an amendment to the Constitution would cripple any president in the conduct of foreign relations and lead other nations to believe that they could violate American rights with impunity, caused the House of Representatives to return the resolution to committee.

From 1937 to 1941 anti-war cartoons appeared frequently in the *Union Farmer*. A late 1937 edition contained a cartoon with the caption "NOW is the time to protest against war! After we are at war, it will be *too late*!"[7] A song written by Irving H. Palmer of Aberdeen,

"Just Stay This Side of the Pond," was the hit of the 1938 SDFU convention. Palmer's nine-year old daughter was asked to sing it several times for the delegates, and NFU President John Vesecky invited her to sing the song at the national convention at Madison, Wisconsin. The lyrics described the wars being fought in foreign lands and urged the United States not to become involved in foreign entanglements. The song was published by the SDFU Juniors and sold for thirty-five cents.

World War II began in the first days of September 1939, and the delegates to the 1939 convention at Rapid City rallied against involvement in the European conflict. The Junior Committee, in its report to the convention:

1. Reaffirmed a belief in the principles of the Ludlow Amendment.
2. Favored an increase in the arms embargo.
3. Proposed that the manufacture of armaments should be a national monopoly.
4. Urged the withdrawal of American citizens from war zones.[8]

In response, SDFU President John Skage urged the convention to pass resolutions memorializing Congress to nationalize the entire munitions and arms industry and make the private manufacture of weapons and ammunition illegal.[9]

As the Nazi juggernaut rolled across Western Europe in the spring and summer of 1940, the President and the Congress introduced additional defense measures. The Union took a stance against the Burke-Wadsworth Compulsory Military Training bill, the first peacetime program of compulsory military service in American history. The Union declared there was no reason for passing such a bill, and there was no likelihood of an invasion of the United States. The bill, which made every male between eighteen and sixty-five liable for the draft, despite major opposition, became law on September 16, 1940.[10] Several South Dakota locals passed resolutions opposing any form of draft. They argued that World War I had resulted in the loss of 100,000 American boys, and the real winners in the war were creditors and contractors.[11]

The Union's opposition to the conflicts raging in Europe and East Asia peaked in the summer of 1941. When President Roosevelt ordered ships of the U. S. Navy to patrol areas of the North Atlantic far beyond the nation's territorial waters, the SDFU board of directors sent a resolution to the President and the state's congressional delegation expressing its opposition to any form of convoy activity because it could lead to American involvement in the European war.[12]

By this time, however, the Union's national leadership, although its official sentiments were isolationist, had become more sympathetic towards defense measures and assistance to Hitler's European opponents. NFU President John Vesecky wrote: "Because no nation, no matter how peaceful and self-contained, is safe from the attacks of the madmen, it behooves us Americans to be prepared to meet the menace of power-seeking maniacs."[13] After James G. Patton assumed the national presidency in 1941, he advocated substantial material aid to those nations bearing the brunt of the fighting "against the menace to our civilization."[14] When President Skage made his report to the state Union's annual convention in October 1941, he said the Farmers Union stood ready to help preserve America's democracy and form of government. There would be no obstructions to the production of food.[15]

THE UNION AND THE WAR EFFORT

In early December 1941, the United States became directly involved in World War II. President Patton immediately wired President Roosevelt that he had the wholehearted support and cooperation of the Farmers Union in defending the United States. Serving as a role model for the Union membership, Patton gave direct service to the government. During the war, he was a member of James Byrnes' Economic Stabilization Board and of the War Mobilization and Reconversion Administration. He was also a consultant to the American delegation at the founding conference of the United Nations.[16]

The increased demand for food put farmers in a favorable bargaining position. The Farmers Union, nevertheless, was the only major farm organization not to object to price controls and to pegging farm prices at 100 rather than 110 percent of parity. Even though they knew that price controls would level off rising farm prices, Union leaders put the national interest before the interests of the organization.[17] Because of its position on price controls, the FU gained new stature and respectability in the eyes of the Roosevelt administration. As the Union became a strong supporter of the Administration's policies, the Farm Bureau took an opposing position. In addition to opposing price controls, the Farm Bureau criticized the Farm Security Administration, the food stamp and other surplus disposal programs. As the Farm Bureau grew progressively disenchanted with price-support measures, the Union's enthusiasm grew. Patton later argued that World War II illustrated how the federal government must play a leading role in the nation's economic system. The government, Patton stated, should provide economic stability and use its power to make for a more equitable distribution of economic wealth among all the nation's citizens.[18]

Farm production in South Dakota reached a new high during the war years. Fewer acres were cultivated than in the 1920s, but through the use of intensive farming and the seeding of cropland that had lain idle because of the AAA crop reduction program, grain production increased dramatically. Farmers had to overcome a variety of handicaps. Much of their machinery was old or worn out. It was difficult to find repair parts, and because of the demand for weaponry, the production of new farm machinery was limited. Ingeneous ways were discovered to keep machinery operating, and mechanics and blacksmiths were in considerable demand.

There was also a shortage of people to operate the machines and to harvest the crops. Thousands of young men were called into the military, and many farmers, hearing of wages being paid in defense factories, left South Dakota for jobs in Michigan and California. The Farmers Union helped to change Selective Service requirements. Exemptions from the draft were based on essential war units. In 1943, the esssential units were reduced from sixteen (the 16-cow rule) to eight. Essential farm producers were classified according to the following representative war units if the farmer demonstrated a diligent effort to increase production:

One war unit = 1 milk cow
= 20 feedlot cattle
= 20 acres in wheat
= 5 acres in carrots, etc.

A percentage was given to each special type of production. Thus, one acre of wheat was 0.5 percent of a war unit.[19]

Attempting to counter the growing shortage of farm labor, the FU participated in Farm Mobilization Day in early January 1943. In mustering rural support, James Thatcher, a

Central Exchange employees Drum and Bugle Corps (1940s).

brother of M. W. Thatcher, had composed a song in 1943 entitled "Soldiers of the Soil."[20] Mobilization councils were created at the local, county, state and federal levels to assist farmers in securing and protecting much needed experienced farm manpower. The councils would represent farmers before draft and rationing boards to insure maximum production from the agricultural sector of the American economy.[21]

In deference to the war, the state Union did away with colorful floats, automobiles and trucks in the parade that was a part of the annual convention at Aberdeen in October 1942. Featured in the "foot parade" was the forty-piece drum and bugle corps from the FU Central Exchange in St. Paul, made up of youthful employees from the Exchange.[22] Due to the shortage of gasoline and tires, the Board of Directors chose Mitchell for the 1944 annual convention because of its central location. The SDFU also supported the war effort through the purchase of war bonds. These bonds, totaling $50,000, contributed to the financial health of the organization in 1944.[23] The war even affected the state Union's leadership when Melvin Thomsen, the state secretary-treasurer, was drafted into the army on January 11, 1945.[24]

South Dakota's agricultural prosperity continued after World War II. There was a mild recession for a few months in 1946-1947, but the American economy quickly recovered and began to expand. A high demand for American farm products, because of the millions of starving people in Europe and Asia and favorable growing conditions at home, favored both production and marketing. The cash income from crops and livestock, exclusive of federal payments, neared the half-million mark in 1946 for the first time in the state's history. The high point was reached in 1947, when total cash farm income reached $673,000,000, an average income of $7,600 per farm.[25] Farmers could now move to eliminate the indebtedness they had incurred during the Great Depression.

With 17 million bushels of wheat in their bins, South Dakota's Union farmers cooperated with the Mercy Wheat Campaign in 1946. M. W. Thatcher, the general manager of FUGTA, developed the plan to feed starving people in Asia and Europe. Farmers lent wheat to the government, which shipped the grain. At the elevator, the farmer received a U. S. government receipt that indicated the type of wheat, bushels and grade. He also named the date he chose for selecting the price of his grain. Finally, the farmer notified the designated federal agent what day he wished to cash in his Wheat Loan Certificate. The government wheat price support program backed the wheat certificates as if the wheat were retained on the farm.[26]

The Farmers Union did not return to isolationism after World War II. The Union strongly supported the United Nations. The NFU believed that a free, democratic nation had the responsibility to protect itself and weaker nations against aggression through the United Nations.[27] Programs were favored that would eradicate the presumed causes of war such as starvation, human exploitation, dictatorship and colonialism. The Union affirmed support of foreign economic aid and the Food and Agricultural Organization. President Patton served as president of the International Federation of Agricultural Producers, an international organization of farm groups.

The FU did not abandon its traditional anti-military position. When House Bill No, 4278, calling for Universal Military Conscription, came before Congress in early 1948, the SDFU drafted a resolution at a January workshop in Mitchell expressing unanimous opposition to the conscription bill.[28] The Union believed that the United States should not follow in the paths of Prussia, Russia, Austria and France in adopting peacetime compulsory military training.[29] Despite public opposition, Congress approved a new Selective Service Act on June 24, 1948, which called for the induction of qualified men between 19 and 25 years of age for twenty-one months of service.

The beginning of the Cold War caused the Farmers Union to question parts of President Harry Truman's foreign policy. The Union objected to the Greek-Turkish aid program (March 1947) and the formation of the North Atlantic Treaty Organization (April 1949). This opposition was not based on the resurgence of the old isolationism, but the Farmers Union's new concept of internationalism. In establishing the Truman Doctrine and NATO, the United States had not worked through the United Nations. During the 1950s and 1960s, the Farmers Union would make the United Nations a priority in its educational program.

THE SDFU AND "HOME FRONT" POLITICS

While the Farmers Union emphasized non-intervention in international politics during the 1930s, it played a much different role in South Dakota politics. The passage of the 4 percent ore tax by the 1935 State Legislature elevated Oscar Fosheim and Emil Loriks into leadership roles in the progressive wing of the Democratic party. The two Union stalwarts, who had advanced the slogan that "it was better to tax gold than Russian thistles," became known as the "Gold Dust Twins" throughout South Dakota.

The two men benefitted from the factional split within Democratic ranks. The defeat of Tom Berry in the 1936 gubernatorial race had been a major setback for conservative Democrats and cleared the way for the farm progressives to exert influence in the party.

Prior to the state primary election in 1938, the leader of the Democratic progressives, Congressman Fred Hildebrandt, announced that he would not run for reelection. His withdrawal from politics resulted in a closely contested Democratic primary with six candidates vying to succeed him.

Emil Loriks became one of the six nominees for the position. He deliberated long and hard before accepting the nomination, consulting with close friends within the Union's ranks. In a letter to Roy Brazell of Volga, the chairman of the SDFU Legislative Committee, he wrote:

> [I] was mighty glad to get your reactions for it was the most difficult decision ever confronting me whether to accept the nomination if drafted.... [I] Am not in position to participate personally in any way in a primary campaign so it will just have to take its course, and if the nomination comes my way and I enter the campaign, I will want to be relieved of the work I am doing, which at present keeps me going night and day.[30]

The SDFU board of directors unanimously supported his candidacy. They said the House of Representatives was in dire need of another spokesman to work and vote for justice and equality for the farmers of the nation.[31]

Loriks accepted the nomination and went on to receive 39 percent of the vote in the spring primary election, without having actively campaigned. He needed only 35 percent of the Democratic vote, and, thus, he became the Party's candidate for the First District congressional seat. Meanwhile, Oscar Fosheim won the Democratic nomination for governor by an almost two-to-one margin.[32]

Immediately after his victory, Loriks prepared to submit his resignation as SDFU president, believing that the organization needed a full-time president.[33] Union executives throughout the region, such as Morris Erickson, James Patton and A. W. Ricker, offered their congratulations but also expressed their regrets at the loss of one of the Union's top leaders.[34] In his letter to Loriks, Ricker said he was disappointed that Loriks had become a candidate for Congress. Ricker wrote that "you can pick up a Congressman most any place, but real farm leaders are one in a million."[35] Loriks' resignation became effective on July 10, 1938. In his last radio broadcast as president of the SDFU, he praised his successor, Vice President John Skage, relating Skage's deep roots in the Farmers Union and his skills as a builder of cooperative enterprises.[36]

During the fall campaign, Loriks discovered that he was somewhat of an amateur in politics. He did not realize that campaigning takes money, and he never asked for contributions. The National Farmers Union sent him $200, which he reluctantly accepted. The State Union contributed no money to his campaign. Before the voters went to the polls, he had spent about $4,000 in his contest with Karl Mundt. Loriks did not openly seek urban labor's support although he knew this group was sympathetic to his cause.[37]

The tactics used by his Republican opponent, Karl Mundt, proved to be the worst part of the campaign. Loriks became the victim of disinformation. He was accused of being a leftist and a socialist because of his unceasing efforts to build cooperatives. Fictitious letters were planted in several state newspapers. A certain Esther Erickson, supposedly from Willow Lake, had written a letter to the *Watertown Public Opinion* in which she said: "Businessmen beware, Emil Loriks is running for Congress. He has been active in organizing elevators, oil companies and creameries, but next will be socialized medicine."[38] A careful investigation proved that Esther Erickson did not exist.

His recent Union opponent, E. H. Everson, appeared on the same platform with Mundt and accused Loriks of selling out to the New Deal. Even an attorney for the Farmers Union, Louis Smith of Sioux Falls, said that the choice was "the New Deal versus Americanism" in the election.[39] The bogus charge that Loriks was a paid collaborator of the Department of Agriculture was raised again. Mundt admitted later that he had researched Loriks' loyalty to the United States, but it was usually Mundt's supporters who falsely charged that the former SDFU president was soft on communism.[40]

The divison within the Democratic party was detrimental to the progressive candidates, and the improved economy tended to help Mundt and Fosheim's opponent for governor, Harlan J. Bushfield. Despite these adverse factors, as well as the disinformation spread by Mundt's supporters, the House race was surprisingly close. Mundt's winning margin was 16,443 votes. Fosheim did not fare as well in the conservative resurgence. He lost to Bushfield by 21,877 votes, 149,362 to 127,485.

Loriks resumed Union activities with great zeal. He was elected to the board of the NFU and assisted Thatcher in founding the Farmers Union Grain Terminal Association.[41] He was also named the state and regional administrator for the Farm Security Administration in 1946. As administrator, Loriks was in a position to help in the founding of additional cooperatives in South Dakota. At the 40th anniversary convention at Oklahoma City, Oklahoma, in November 1942, he was elected secretary-treasurer of the NFU. Loriks defeated the incumbent, James M. Graves, by a vote of 67.5 to 42. Loriks was elected because of his vigor and experience on the national board. As the nation moved into a period of total war, Loriks was believed to be the person who possessed the energy to be available to work in any phase of the national program, including traveling anywhere, day and night.[42] He played a role in moving the national headquarters from Oklahoma City to Denver, Colorado. He resigned as secretary-treasurer of the NFU on December 31, 1945.[43]

Oscar Fosheim returned to the political arena in 1940. He ran for Congress against Karl Mundt and lost by more than 40,000 votes. In addition to the continued Republican resurgence in South Dakota, Fosheim's support of Roosevelt's foreign policy contributed to his defeat. Favoring assistance for the beleaguered British was not a vote-getter on the Prairie Plains in the fall of 1940. Mundt resorted to the same tactics he used against Loriks in 1938, intimating that Fosheim was soft on communism. Thirty years after the election, Fosheim remarked that "Mundt was a great Red baiter .... He Red-baited himself into office."[44] During the 1940s, Fosheim returned to Union activities. After conducting a successful drive to bring new members into the SDFU, he was elected president of the organization in 1944.

CONFRONTING A CONSERVATIVE RESURGENCE IN PIERRE

Emil Loriks described his successor, John Skage as a protege of John A. Simpson. Skage, he said, believed that the Farmers Union must stand on its two legs, legislative and cooperative, but that those two legs rested on a program of education which provided the foundation for the entire structure.[45]

Skage had committed his life to the Farmers Union. He was born on April 2, 1904, in Brandt. He grew up in Brandt and farmed near that community until 1936 when he became the manager of the Deuel County Farmers Union Oil Company. Skage played a key role in organizing the Farmers Union in Deuel County, and as noted in Chapter 8, he was the driving force in the establishment of the Deuel County co-op. While president of the SDFU, he still found time to manage the cooperative. At the time he became president, he was serving as a field man for the FU Co-operative Brokerage. For many years, Skage helped select the legislative tickets that supported the Union's program in the State Legislature. He moved to Sioux Falls in 1960 where he joined the staff of GTA Feeds until his retirement in 1969.[46] Aside from Loriks, John Skage was the foremost founder of farmer-owned cooperatives throughout the state. Because of his work as a cooperative pioneer, the Union honored him with the prestigious Award for Meritorious Service at its 71st annual convention in Aberdeen in October 1986. Two years later, on September 15, 1988, Skage was inducted into the South Dakota Co-op Hall of Fame.

The new president's position was confirmed at the convention in October 1938. The delegates elected Roy Brazell of Nunda as vice president and Paul Erickson as secretary-treasurer. The new administrative team and the Board of Directors quickly turned to legislative matters as South Dakotans had elected a very conservative governor, Harlan J. Bushfield, less than a month after Skage's election. Bushfield's Republican predecessor, Leslie Jensen, had been very sympathetic to the farmers' interests. He unenthusiastically backed Bushfield while telling Oscar Fosheim he personally preferred him during the 1938 gubernatorial campaign.

During the 1939 and 1941 legislative sessions, the Farmers Union ardently supported a graduated land tax law. The bill, designed to discourage large landholdings, would have increased taxes on landholdings held by a person or corporation in excess of $75,000 valuation, with a graduated raise in the taxes as the holdings increased above this amount. The Union believed the bill would encourage the home ownership of family-size farms.[47]

When the House of Representatives rejected the bill 38-34, in February 1939, the SDFU kept the issue alive through a petition drive in the fall of 1940. As the legislators assembled for their biennial session in January 1941, Roy Glover presented them with petitions containing the names of 6,000 farmers and businessmen from throughout South Dakota, requesting that the Legislature approve a graduated land tax amendment to the constitution.[48] Gilbert Larson, the FU lobbyist, operating from his headquarters at the Waverly Hotel in Pierre, urged farmers to come to the capital city and speak with their elected representatives. The Senate approved the bill 26 to 9, but, once again, the House refused to refer the SDFU's foremost legislative plank to a vote of the people by a vote of 40 to 33.[49] President Skage referred to the bill's opponents as dictators.[50]

John Skage, President, SDFU, 1938-1944.

The Union's position was upheld on two measures during the 1939 legislative session. The SDFU successfully opposed a measure to raise the required number of signatures from 5 to 15 percent for initiative and referendum petitions. A proposal to have the Legislature begin in late January and run through March was blocked. Union-affiliated legislators pointed out that in a normal crop-growing year, farmers were in the fields by late March.[51]

The SDFU became caught up in several controversial measures during the 1941 session. The ore tax had been raised from 4 to 6 percent during Jensen's administration. The Farmers Union argued that a further increase in the ore tax could mean the elimination of the three cent sales tax. To the contrary, Governor Bushfield supported Homestake's request for a reduction in the ore tax from 6 to 5 percent. If passed, the state would lose $175,000 to $200,000 per annum. The proposal was approved in the Senate, but the SDFU was able to kill the measure in a House committee.[52]

The FU and the Bushfield administration also clashed over the Union's attempt to repeal the sales tax on farm machinery. The measure was passed by the Senate, but the House Committee on Assessment and Taxation, chaired by Frank L. Messner of Brookings, refused to report it out of committee. Messner, suspected of being the primary opponent of the tax reduction proposal, strongly favored an appropriation of approximately $100,000 for an Agricultural Hall at State College at Brookings.[53] Many pro-Union legislators favored the Agricultural Hall but refused to support the appropriation unless Messner reported the sales tax repeal on the farm machinery measure to the House floor.[54]

Faced with a stalemate, the Governor invited several Union members to a conference to advance his solution to the dilemma. He said he would support the sales tax repeal and the Agricultural Hall if the FU dropped its opposition to the reduction in the ore tax. Bushfield threatened to use his veto power if any of the measures were passed independently. The "Bushfield Package" was presented to a House caucus in the closing hours of the session. Three times during the caucus, a FU member, who opposed the Bushfield proposal asked to speak but was not recognized by the chairman. The frustrated legislator and several of his colleagues left the room, causing the caucus to adjourn.[55] Commenting on the failure of the three measures, Skage told delegates to the 1941 SDFU convention that "... the farmers are still paying the sales tax on their farm machinery, Homestake is still paying a 6% ore tax, and State College will have to wait at least another two years for the agricultural hall."[56]

At the previous convention at Yankton, the SDFU went on record to require that the state Secretary of Agriculture be elected by the people, believing it to be a more democratic way of filling the office. The proposal was also a reaction to Governor Bushfield's appointment of E. H. Everson as State Secretary of Agriculture in January 1939. The *Union Farmer* commented that Everson was "working with a friendly administration and THAT ADMINISTRATION IS IN COMPLETE CONTROL."[57] The Union-motivated bill passed both houses of the Legislature during the 1941 session. True to form, Governor Bushfield vetoed the measure. He argued that the governor should not be shorn of his appointive power.[58]

During the acrimonious session, steps were taken to provide some tax relief. A two-mill state property tax was eliminated, and the sales tax was reduced to two cents. Bushfield's recommendation for repeal of the state income tax went unheeded. Incomes $3,000 and over were subject to the tax. The net income tax law had brought in nearly a million dollars for the fiscal year that ended June 30, 1941. Bushfield was a U. S. Senator when the Legislature repealed the income tax in 1943, the last time an income tax was imposed in the state.[59]

The Farmers Union fought off yet another attempt to reduce the ore tax in the 1943 legislative session. The House Assessment and Taxation Committee proposed to reduce the tax to 2 percent on all mined ore in excess of 100,000 tons annually. At the time, Homestake paid into the state treasury more than $1 million a year.[60] Opposition to the bill was led by Representative Charles Bruett of Hanson County. Using data prepared by President Skage,

Bruett pointed out that Homestake paid dividends totaling over $9 million per year from 1935 through 1941, an excellent pay-out because the true value of Homestake's properties totaled about $10 million.[61] Lobbying by the Farmers Union contributed to the House's rejection of the bill by a vote of 41-32.

The SDFU, by way of the *Union Farmer*, informed the membership of the vote of each House member on the ore tax. A precedent was established, and in the ensuing years, the paper published each member's voting record on a variety of issues. Reiterating the Union's position, Vice President Roy Brazell read the name of each House member and his vote on the issue to the convention delegates assembled in the Huron College auditorium the following October. He told the audience to support legislators on the basis of their vote on the ore tax and their friendliness toward Union positions.[62]

Homestake's proponents did not give up. A 2 percent ore tax bill was presented to the legislators in January 1945. In the bargaining that followed, a proposal to lower the tax to 4 percent was approved. The Union fell three votes short in the House of keeping the ore tax at 6 percent.[63] Fosheim, the Union's leading advocate for higher ore taxes and president of the SDFU in 1945, attributed the setback to the defeat of "Union legislators" in the primary and general elections of 1944. If the Union continued to be insufficiently represented in the Legislature, he said, the ore tax might be eliminated entirely in the next legislative session.[64] His fear was unwarranted as the 4 percent ore tax remained in place following the 1947 legislative session.

The Farmers Union did not fare well with its tax proposals during the 1940s. The same could be said for its allies in organized labor. They, too, fell victim to the conservative mood of the electorate. The Republicans dominated the election of 1946 and only four legislative seats were won by the Democrats. The conservative mood was expressed in the overwhelming support voters gave to a constitutional amendment that inserted a "right-to-work" provision in the bill of rights to the State constitution. The 1947 Legislature responded to the amendment by enacting three anti-labor measures. These measures made picketing unlawful, outlawed the closed shop and permitted labor unions to sue or to be sued.[65] It seemed that the best hope for legislative relief for farmers resided not in Pierre but in Washington, D. C.

THE UNION AND THE DEVELOPMENT OF THE MISSOURI RIVER BASIN

President Franklin D. Roosevelt visited South Dakota during the dark days of the Great Depression. When his special train stopped at Pierre, several of the state's farm and labor leaders, including Farmers Union President Emil Loriks, were invited into Roosevelt's private car for an hour's briefing on federal assistance. During the course of the conversation, the President "pointed to the Missouri River and said, 'This is the greatest resource of your state. It's got to be developed.'" He told his small audience how Sweden had developed its power resources for the benefit of thepeople. In a similar manner, Roosevelt stated, the Missouri River should be developed for flood control, irrigation, navigation, hydroelectric power and soil conservation.[66]

The Farmers Union was one of the first organizations in the state to work toward the fulfillment of the President's vision for the Missouri River Basin. At the 1934 state convention, the SDFU went on record for the development of hydroelectric power from the Missouri's waters. The Union's resolution read: "We realize that the Missouri River is South Dakota's greatest underdeveloped natural resource, and we know full well that the future welfare of our people is going to be inseparably linked with the availability of cheap sources of electrical energy."[67] The following year, the SDFU conducted a campaign advocating development of the Missouri River; President Loriks took to the airwaves over WNAX to promote the Union's position.[68]

The National Farmers Union became interested in the development of the Missouri River Basin in the early 1940s. President James G. Patton cited five major reasons for the Union's interest: 1) the scarcity of power for REA cooperatives and the prevailing high electric rates, 2) floods in the lower basin, 3) the need for irrigation water, 4) a need for western phosphate development, and 5) a need for industrialization in the Basin to boost the level of income.[69] The Union created its own regional resources development committee, composed of state presidents and national board members. The committee spent time in the Tennessee River Valley, studying all phases of the Tennessee Valley Authority (TVA). They returned to the Prairie Plains thoroughly convinced that a single, unified development agency was needed in the Missouri Basin. The TVA became the role model, and the great preponderance of FU locals in every state in the Missouri Basin voted 100 percent for a Missouri Valley Authority (MVA).[70]

Federal agencies, regrettably, disagreed on priorities in developing the river. General Lewis Pick, speaking for the Army Corps of Engineers, emphasized flood control. He wanted dams located for flood control and navigation. W. Glenn Sloan, the Assistant Director of the Bureau of Reclamation, favored a number of reservoirs on the headwaters of the Missouri River basically for irrigation and hydroelectric power. He wanted dams where they would do the most good.

In 1943, the SDFU told Governor Merrill Q. Sharpe that they would back him in any postwar effort to bring flood control, irrigation, navigation and power development to the Missouri Basin.[71] Sharpe, governor between 1943-1947, played a prominent role in the preliminary planning for the Basin project. He organized the governors of the river basin into the Missouri River States Committee. When a compromise was worked out between the Pick and Sloan factions, the Missouri River States Committee acted in liaison with the Inter-Agency Committee that represented the federal government.

James G. Patton, President, NFU, 1940-1966.

The Pick-Sloan Plan, which led to the passage of the Flood Control Act of December 1944, angered the FU leadership. NFU President Patton said the Bureau of Reclamation and the Army Engineers entered into "their shameless, loveless shotgun wedding contract to create the basis for some claim to unity in the Basin."[72] Loriks described the Pick-Sloan Plan as a hydra-headed monster, being administered from the nation's capital by a number of government departments and agencies.[73]

If the Missouri Valley Authority were implemented, the Union believed that the administration for the development of the Missouri should be where the program existed. The SDFU envisioned a governance system similar to cooperatives; the manager and the board of directors of the co-op always resided in the area where the co-op operated. The MVA was patterned after the TVA, a unified, integrated program for overall development. Loriks suggested that the governors of the states in the Missouri Basin could help in the appointment of the MVA board along with the President of the United States. At the time, the President appointed the board of the TVA.

The Flood Control Act vested administrative authority in three federal agencies and in the governors of the ten states in the Missouri Basin, ten governors, who at the time, came and went every two years. The Union condemned what it considered a hodge-podge compromise that did not take into consideration "the total needs of the entire River Basin and the best interests of the people."[74] Throughout the spring of 1945, the *Union Farmer* published article after article championing the MVA and Senate Bill No. 555. The bill called for the creation of a board of directors located in the Missouri Valley that would be empowered to give direct answers to direct questions without red tape, bureaucratic interference and political jockeying.[75] The FU linked the MVA with the maximum production and transmission of electric power from the Basin at a wholesale price. An idea opposed by the private utilities lobby.

Conflicts of interest between federal agencies as well as between different regions in the Missouri Valley led to the rejection of the Farmers Union's proposal for a centralized agency similar to the Tennessee Valley Authority. Union protests against the "mud dams" contemplated under the Pick-Sloan Plan were also ignored. Again, using the TVA model, the SDFU wanted indestructible, concrete, multipurpose dams that would provide great quantities of low cost electric power and enhance the possibilities for irrigation and navigation. The movement of spring wheat between Pierre and Sioux City on the Missouri was of great interest to the Union.[76] In accordance with the Flood Control Act, construction of the first of four rolled-earth, multipurpose dams on the main channel of the Missouri River within South Dakota began at Fort Randall in the final days of Governor Sharpe's administration.

One other proposed federal waterways project had attracted the attention of the Farmers Union in the months immediately following World War II, the St. Lawrence Seaway. The FU had a longstanding interest in supporting a twenty-five foot navigable channel from salt water to the Great Lakes, and the maximum development of the hydroelectric potentialities of the St. Lawrence River. The waterway's cheaper transportation costs would give farmers at least five cents more per bushel of wheat. With the support of Secretary of State George Marshall and former President Herbert Hoover, it appeared that this new avenue of trade would become a reality in the near future.[77]

Between 1941 and 1945, the SDFU made a major contribution to the Allied triumph over the forces of fascism. Many of its sons and daughters served in the military forces, and South Dakota's farmers joined with other tillers of the soil to provide the foodstuffs necessary to feed not only our own citizens but those of our allies as well. In this same period of time, the Farmers Union developed a close working relationship with the Roosevelt and Truman administrations.

World War II and plentiful rainfall had brought unaccustomed prosperity to South Dakota' farmers and success to their cooperatives. No longer beset by surpluses, farmers could produce as much as hard work and their ingenuity would allow with the assurance there would be a favorable price for what they produced. An improved, rural economy led to a conservative resurgence in South Dakota. As a result, only parts of the Union's legislative agenda were enacted. There were setbacks at the polls, and some of the decisions concerning taxation went against the farmer's interests. Although the Union's objective of a TVA-type of management of the Missouri River Basin was not fulfilled, the development of the Basin held great possibilities for improving the economic and social life of South Dakota's residents. While the SDFU had served the nation and pursued legislative matters, significant developments were taking place within the organization as it pursued growth and economic stability.

Chapter 10

PROBLEM SOLVING AND GROWING PAINS 1938 - 1949

Never throw rocks at those who go down,
for those who go up will also come down.
— Oscar Fosheim, 1944.

A brief period of good feeling pervaded the South Dakota Farmers Union prior to America's involvement in World War II. The administrative shift from Loriks to Skage in 1938 went smoothly as both men pursued similar agendas. Their views were also in harmony with the national leadership following the election of John Vesecky in 1937. The discordant elements, led by Everson and Kennedy, were no longer effective. Vesecky sought to cooperate with the Roosevelt administration in finding the best possible solution for the farm problem. His successor in 1940, James G. Patton, at age forty, the youngest man to lead the Union, also worked closely with New Deal administrators.

CREATING A CENTRALIZED NATIONAL ADMINISTRATION

Despite his comparative youth, Patton had served several years in administrative positions for the Farmers Union, as national secretary of the NFU, and most recently, as president of the Colorado Farmers Union. He brought strength to the national organization. He was politically oriented, dramatic and incredibly energetic. Patton had discerned that the NFU was really a loose federation of state organizations, possessing little national character. His primary goals in the 1940s included the creation of a full-time national education department, a permanent legislative office in the nation's capital and a solvent *National Union Farmer* that would reach members on a regular basis. The NFU president, secretary-treasurer and education director would operate out of a national headquarters building at Denver, Colorado.[1]

The National Farmers Union enacted a significant constitutional change in 1946. The board of directors was enlarged to include all of the state presidents, instead of the five state Union heads previously elected by the national convention. According to John A. Crampton, in *The National Farmers Union, Ideology of a Pressure Group*, this action had the greatest nationalizing effect of all of the changes initiated by the Patton administration.[2] Another constitutional change made the position of secretary-treasurer appointive instead of elective.

President Patton delivered an inspirational address at the 1941 convention of the SDFU in the Corn Palace at Mitchell. He discussed the preservation of the family-sized farm.

Modernization had begun to affect the status of the family farm during the 1920s. A 1940 document prepared by the Bureau of Agricultural Economics, "Technology on the Farm," stated that "Machines alone are expected to displace 350,000 to 500,000 additional farmers during the next ten years."[3] Skage echoed Patton's thoughts, stating that the SDFU was threatened from within by the disappearance of small farms. With the resources of the nation dedicated to winning World War II, however, a significant decline in the number of family farms was postponed until the 1950s.

WORKING TOWARD A SOLVENT AND UNIFIED SDFU

The first half of Skage's seven-year tenure as president (1938-1944) was given over to solving the SDFU's financial problems and to giving further unity to the organization. In the latter case, the media was used extensively. The Union sponsored radio broadcasts on a regular basis over KSOO, in Sioux Falls, and WNAX, in Yankton, providing news on Union activities and policies and livestock reports. The *Union Farmer* continued to be the primary outlet for Union views and information. It was a costly enterprise, and for a short period of time, the SDFU was heavily in debt to the *Yankton Public Opinion* for printing the *Union Farmer*.[4] The Fosheim administration, 1944-1947, solved this problem by transferring the paper's publication to the Inter-County Cooperative Publishing Association of Frederic, Wisconsin. The transfer meant the Union saved almost $3,000 per year in publication costs.[5]

When the Union met in Rapid City for its annual convention in 1939, the Board of Directors had to borrow $500 to pay the expenses of the convention.[6] Two years later, the debts of the SDFU had been paid, and a cash balance of approximately $3,500 remained.[7] By 1943, the state FU held $40,000 in government bonds and had $8,263.20 in the bank.[8] The financial gain was due to the dissolution of the Farmers Union Co-operative Brokerage and its sale to the Farmers Union Central Exchange.

One of Skage's closest allies during his administration was Roy Brazell. He was the vice president for five of the seven years that Skage was president. In addition, Brazell served as the Union's legislative lobbyist. The SDFU had its first female state secretary-treasurer when Helen Shepherd of Sioux Falls served in that position from October 1941 to October 1942. Otherwise, Paul G. Erickson served in that capacity during the Skage administration.

Oscar Fosheim, SDFU president from 1944-1947, continued to maintain the Union's debt-free status with a comfortable reserve. He announced with great satisfaction to the delegates at the 1945 convention that the Union possessed $47,000 in bonds and $13,863 in cash. In addition, the regional cooperatives, the GTA and Central Exchange, and the Sioux Falls Livestock Commission had contributed almost $17,500 to the cooperative educational fund.

The SDFU also regained its tax exempt status, in October 1945, when the Union received a letter of tax exemption from the Internal Revenue

Roy Brazell, Volga. Vice President of the SDFU for three terms.

Department. The FU had lost its income tax exemption status twelve years earlier, on August 3, 1933. The exemption was restored because the federal government decided to recognize the FU as an educational organization. Fosheim cautioned that the Farmers Union must dispose of its educational funds properly. He also noted that the Union was now exempt from the capital stock tax.[9]

He also followed through on President Patton's call for the Union to cultivate urban allies. It was necessary, Patton said, for the FU to "join hands with city consumers who belong to organized labor, to win legislative battles."[10] In November 1947, the SDFU and organized labor in South Dakota established the South Dakota Joint-Legislative Council. The Council, a political lobby, maintained an office at Pierre during legislative sessions. Urban labor and the SDFU pooled resources, financial and human, to oppose or promote legislation beneficial to both groups.[11]

In response to President Fosheim's request to move the state office to a more centralized location, the state Board of Directors took action on January 16, 1945. After considering Mitchell and Huron for the new site, the board decided by a four-to-one vote to move the office from the Strand Building in Sioux Falls to a six- room complex on the fourth floor of the Western Building in Mitchell. The SDFU began operating from its new quarters on February 1, 1945.[12]

Fosheim's administrative team included Elmer Beck as vice president and Ed Backlund as secretary-treasurer. Backlund succeeded Melvin Thomsen, who shortly after his election in 1944, was drafted into the U. S. Army. Fosheim, a former legislator, and Sam Ulrickson represented the Farmers Union at the biennial sessions of the state Legislature until the appointment of Roy Glover as legislative council.

Approximately thirty years after the first local had been established in South Dakota, the dues for a member were $4.00 per year. Seventy-five cents of this amount went to the local, seventy-five cents to the county and $2.50 to the state. National and district dues and the publication and mailing costs for the *Union Farmer* were taken from the latter amount.[13] Delegates to the annual state convention were allotted according to the following scheme: 1) Each chartered local in good standing with five or more annual dues paying members was allotted one delegate. 2) County Unions having three or more locals in good standing were assigned one delegate and one legislative committeeman.[14]

In an effort to do a better job of getting the Union's program before the membership, an annual state winter workshop was introduced. The workshops were first held in the basement of the Congregational church at Mitchell in mid-January. Topics considered were legislation, cooperatives, credit unions, health and recreation.[15]

Policy-making was also democratized. Formerly, the state president named a Resolution Committee. He could, if he wished, select people with his views, and 90 percent of resolutions formulated in this manner were passed at conventions. The SDFU Board of Directors proposed a change in the by-laws whereby the president would name a County Councillors Committee, consisting of all legislative committeemen and all county delegates elected by the membership to develop resolutions. The recommended change in the by-laws was approved at the 1946 state convention.[16]

The growing financial strength of the SDFU and the constructive and unifying policies of Skage and Fosheim during the 1940s were sometimes overshadowed by the sharp differences of opinion over the fate of the Brokerage and the success of a membership campaign. Strong personalities reacted in different ways to these two issues. Divisions in the State

Union first occurred when deciding on the future of the Farmers Union Co-operative Brokerage.

THE DISSOLUTION OF THE CO-OPERATIVE BROKERAGE

Despite a steady growth in the business of the Brokerage after 1937, the Sioux Falls-based cooperative continued to experience financial problems.[17] The firm still owed the Internal Revenue Service back taxes. There were increased costs associated with its retail business, and financial losses mounted because of the continued extension of credit to co-op members.[18] Concerns were expressed about the work of the Brokerage's fieldman and its auditor. Questions were raised as to their effectiveness in the field. Some South Dakota cooperatives were considering using a non-Farmers Union auditor to evaluate their accounts.[19] In an effort to build-up cooperative enterprises across the state, Gustav L. Moseson became the supervisor of field work for the Brokerage on January 1, 1941. His replacement, as manager of the Co-operative Brokerage, was C. H. Halonen of Valley City, North Dakota.[20]

A more aggressive approach to aligning cooperatives with the Farmers Union was needed because of the increasing competition from the Consumers Cooperative Association. The CCA said it had pioneered the cooperative distribution of petroleum products in South Dakota and should have first consideration in selling its products in the state. Enoch Hofstad of Elk Point, the secretary of the Brokerage board, retorted that the Farmers Union should have first consideration because it had pioneered cooperative education in South Dakota. He pointed out that even CCA fieldmen admitted that they found it easy to organize cooperatives where the Farmers Union had sown the seeds of cooperative education.[21] The CCA proposed to offer refined fuels, lubricating greases and a complete line of paints to the FU Co-operative Brokerage on a cooperative basis with the first cost of these commodities being competitive with the current source of supply, that being the Farmers Union Central Exchange.[22] No deal was possible because by mid-1939 the leadership of the Brokerage and the SDFU were determined to oust the CCA from South Dakota.[23]

The future of the Brokerage was decided at a fateful meeting of the Brokerage board on April 19, 1941. The Bank of Cooperatives at Omaha, Nebraska, had agreed to loan $45,000 to the Brokerage. The manager of the Brokerage used the money to pay-off all accounts payable, including the money due the Internal Revenue Service. Repayment of the taxes was to be made in monthly installments. The loan was made possible because of Emil Loriks, Emil Syftestad of the Central Exchange, Joe Fortner and others. W. C. Hermann, president of the Brokerage board said, "If it had not been the willingness of the St. Paul boys to merge us with their Association, I do not believe we would have received the loan."[24] Hermann believed that the best thing for the cooperative movement in

Oscar Johnson, Oldham.
Member of the Board of Directors, Farmers Union Central Exchange.

South Dakota was to merge with the Farmers Union Central Exchange. It became his goal to work toward that end during the summer of 1941. It would solve the question of the fight with Howard Cowden's CCA, and it would consolidate the cooperative movement on the northern Prairie Plains.

By this time, Gust Moseson was actively involved in the merger issue. Soon after he assumed management of the Brokerage, he had realized that the survival of the cooperative depended on a merger with the Farmers Union Central Exchange. Driving an old Buick, which had a small body and a big engine, he was on the road six and one-half days a week, persuading the state's cooperative leaders to support a merger. W. C. Hermann, Oscar Johnson of Oldham and William Sinkular were also deeply involved in the cause.[25]

Plans to secure the support of the Union's membership for the merger proceeded smoothly until July 9, 1941, when the SDFU Board of Directors met. A motion by Enoch Hofstad opposing a merger with Central Exchange was approved by a vote of four to three. W. C. Hermann, Paul Erickson and John Skage had opposed his motion. Hofstad reasoned that the merger would put employees out of work, that the SDFU would give up control of the Brokerage and of the business that it had created.[26] Proponents of the merger held a conference at Redfield on August 28. Representatives from the Brokerage and Central Exchange were present. The conferees passed a resolution deploring Hofstad's remarks, accusing him of creating suspicion, division, dissension and distrust.[27]

During the weeks before the annual convention, there was a bitter exchange of views through letters and on the pages of the *Union Farmer*. Hofstad said the struggle over the Brokerage was between those favoring "Enterprise Control" and those who wanted "Membership Control." He associated "Enterprise Control" with the pro-merger group, including in their numbers the leaders of the Northwest Organizing Committe, M. W. Thatcher, C. C. Talbott and A. W. Ricker, and SDFU progressives John Skage and Emil Loriks. Hofstad urged Union members to follow in the footsteps of John Batcheller, who had earlier taken a stand against the Northwest Committee.[28]

Despite Hofstad's negative campaign, the pro-merger forces had their way at the annual convention in Mitchell on October 14, 1941. W. C. Hermann provided a status report on the financial condition of the Brokerage. The Brokerage board, adhering to instructions they had received from the delegates to the 1940 convention, closed the three branches of the firm at Mitchell, Watertown and Sioux City. The sale of the inventory from the closed branches enabled the Brokerage board to repay $15,000 of the $45,000 loan obtained from the Bank of Cooperatives. The payable amount of the Brokerage as of October 1941 totaled a little over $30,000 compared to the firm's $70,000 obligation in October 1940.[29]

President Skage then announced that the stockholders of the Farmers Union Co-operative Brokerage agreed to invite the Farmers Union Central Exchange to extend its business into South Dakota. Formal affiliation with the Central Exchange would eliminate the duplication of expense and service. It would give South Dakota's Union members the advantage of the combined purchasing power of about 300 local cooperative oil companies representing the approximately 100,000 farm families affiliated with Central Exchange in 1941.[30]

The delegates then narrowly approved the following resolution:

> Be it further resolved, that this corporation hereby recommends said Farmers Union Central Exchange, Inc., to the members of this corporation and urges their full cooperation with and patronage of said Farmers Union Central Exchange, Inc., for the achievement of the common purposes for

> which both said Farmers Union Central Exchange, Inc. and Farmers Union Cooperative Exchange were organized.[31]

An ensuing motion by the Hofstad forces to instruct delegates to vote against such a merger at a forthcoming stockholders' meeting of the Brokerage was defeated by one vote.[32]

At a special meeting at Huron, ballots favoring the sale of Brokerage stock were counted. The vote favored the sale 136 to 64, but under state law it was necessary for 75 percent of the stockholders to vote in favor of such an action. The negative vote of 32 percent of the stockholders forced any decision on the fate of the Brokerage into the next year.[33] Meetings of the divided Board of Directors were filled with charges and counter-charges. When Hofstad called for a special meeting of the Board on December 19, 1941, President Skage refused to accede to the request. Vice President David Tiede of Parkston, an ally of Hofstad, called the group together. Hofstad was still seeking information on the "Redfield Resolution" of August 28, 1941.[34] He wanted to know the names of all of the individuals who were present at the conference.

W. C. Hermann, the president of the Brokerage, asked the stockholders of the disputed cooperative to meet at the Strand Building in Sioux Falls on February 17, 1942. While making his request, he said the annual business of the Brokerage had seldom exceeded $600,000 in gross sales. The lack of volume meant the lack of purchasing power to take advantage of jobbers' discounts and the other advantages of large purchases available to the Central Exchange. Negotiations had been completed with the Exchange for the distribution on a wholesale basis of a complete line of farm supplies. The Central Exchange, having leased the warehouse occupied by the Brokerage would begin effective distribution on February 1, 1942, with Victor Johnson, son of Oscar Johnson, in charge. In arguing the case for the Central Exchange, Hermann noted that net savings for local affiliated cooperatives in 1941 were over $300,000.[35]

On February 17, the stockholders voted to liquidate the Brokerage by a vote of 183 to 42. Ironically, the decision came at a time when the firm was debt free. The loan from the Bank of Cooperatives had been repaid, and the federal income tax, which ultimately amounted to $21,367.22 had also been paid. Only one of the twenty-eight cooperatives voted against dissolution of the Brokerage. Most of the co-ops had already made arrangements to buy their requirements directly from FU Central Exchange, taking advantage of the Exchange's greater purchasing power.[36]

The opposition did not retreat silently into the night. A Circuit Court judge ordered dissolution of the Brokerage, naming W. C. Hermann and Enoch Hofstad as trustees for the liquidation, indicating that they could decide on the other trustee. Hofstad balked at this procedure so the court appointed John H. Rausch, a professional liquidator, as the third party.[37] The dissolution was contested by three stockholders. They were locals from Union and Turner counties, and Palisades Township in Minnehaha County. After three days of trial in April 1942, Judge Lucius J. Wall ordered formal dissolution of the Sioux Falls firm.[38]

The legal and financial parts of the bitter fight finally ended on March 6, 1943. A Circuit Court judge signed a judgment dissolving the Brokerage and ordered its board of directors to be trustees for the liquidation. About 6,000 checks were mailed to the stockholders and patrons pursuant to the order of the court. As the major stockholder, the SDFU received $41,620 for its share in the transaction. The Board of Directors, shortly thereafter, invested $40,000 of the amount in U. S. bonds.[39]

The property at 231 North Weber Avenue in Sioux Falls was sold to Central Exchange, and the St. Paul firm continued to operate the former Brokerage as one of its own branches. Within months, the Exchange was providing wholesale supplies in increasing amounts to most of the Farmers Union cooperatives in South Dakota. G. L. Moseson was named senior field representative for Central Exchange in the state. His co-worker in the field was Joe Fortner of Mitchell.[40]

In addition to the savings Union members experienced through their local cooperative's association with Central Exchange, there was a special dividend in educational funds. The Central Exchange set aside 5 percent of its annual net income for education. The share for a State Union was based on the volume of patronage of the FU Central Exchange in each state in relation to the total volume of business in the five states in which the Exchange operated. In 1945, South Dakota's share was $4,121.43.[41] The gains achieved through formal affiliation with Central Exchange had their personal cost. The bitterness generated over the dissolution of the Brokerage did not go away and contributed to the divisions within the State Union during the 1944 membership drive and the failure of John Skage's reelection bid in the same year.

DOUBLING THE MEMBERSHIP

Membership in the SDFU had not increased significantly with the revival of the rural economy at the beginning of the 1940s. There were 4,723 Union members in the state in 1940 and 5,880 in 1943, a gain of only 1,157. The Board of Directors in January 1944 decided to undertake an aggressive campaign to increase the Union's numbers. In a meeting at Mitchell, they hired Oscar Fosheim as director of organization. Fosheim was asked to assist in organizational and educational work, to help set up new locals and to educate new members.[42] His term would end at the time of the state convention in October.

Before reporting to the Farmers Union, Fosheim held an auction sale and rented his farm in Miner County. A minimal amount of preparation for a membership drive had taken place before he assumed his duties on February 1, 1944. Key individuals throughout the state had received letters from the state office asking them to attend a membership meeting. Fosheim had the presidents of the county Unions provide him with the names of influential men in their respective counties. The ideal was to have four such individuals in each township. Several of the men were not members of the Farmers Union, but they were placed on the mailing list of the Farmers Union Herald in March and had acquired some background on the Farmers Union.

The membership drive gained momentum as water-saturated roads dried out in mid-spring, and the recruiters could contact more farmers each day. In Day County, recruiters signed up 544 paid members in one day. Locals and county organizations staged contests in the quest for new members. Contests culminated with an oyster supper followed by a guest speaker. During the eight months of Fosheim's employment, the statewide membership increased from 3,718 on February 1 to 11,637 on September 15, 1944.[43]

The amazing feat accomplished by Fosheim and his dedicated recruiters was vigorously applauded at the annual convention in October. There remained a few disgruntled malcontents. During the course of the membership drive, one member of the Farmers Union told Fosheim that he "was a cheap, no good, washed-up politician." Fosheim agreed with him, telling the mean-spirited individual that he was okay and that they should "work together in building the Union."[44] Ironically, the 1944 convention program had not included a report by Fosheim, the director of organization. The Board of Directors arranged that he be called upon immediately to give a report of his activities.[45] He concluded his remarks to the delegates with optimism, stating that the Union's ultimate goal should be a paid-up membership of 20,000.[46]

While the membership campaign went on, a bitter rivalry played out at the leadership level within the state Union. The relationship between President Skage and David J. Tiede, his vice president for two years, had never been good. They had disagreed on the fate of the Brokerage, and Tiede argued that Skage did not permit him to participate in the decision-making process. Tiede likened himself to the spare tire on an automobile, "leaning against the body of a car and like on the new models sometimes ... pushed in the trunk and covered by a lid so as not to see where the other four tires are going, to be used on rare occasions."[47] His report as vice president in October 1942 reeked of sarcasm. Tiede said his report would be brief, because he had not been called on for assistance by Skage, only when his personal presence was specifically required. He accused Skage of hiring others who were not as responsible to the membership as were the elective officers.[48] He asked not to be reelected if the delegates approved of Skage's management style.[49] In the ensuing election, Tiede was replaced by Roy Brazell of Nunda.

The two men had major differences over the membership program. Tiede, now a member of the Board of Directors, accused Skage of not cooperating in the 1943 membership drive as it had been established and carried out by the Board of Directors.[50] He indicated that the Board had very little cooperation from the state officers in the early stages of the 1944 campaign. Only when the success of the organization drive exceeded all expectations, said Tiede, did Skage ask that the direction of the membership campaign be placed directly in his hands. The Board refused his request.[51] In his report to the 1944 convention, President Skage said that he had recommended the membership drive, and that some of the techniques used were similar to those used in a successful campaign in North Dakota.[52]

Rumors of a conspiracy against President Skage were page one news in the *Union Farmer* prior to the 1944 convention. The president of the SDFU generally oversaw the content of each edition of the paper. It was reported that two secret meetings were held in Huron on August 28 and September 22, 1944. Among those said to be present were several people affiliated with the Central Exchange and Board of Director's members W. C. Hermann and D. J. Tiede. The purported purpose of the meetings, according to the Union Farmer, was "to discredit, undermine and smear certain officers of the Farmers Union, especially the state president."[53] The paper went on to say that those present at these meetings had decided to support Oscar Fosheim for president, believing that Skage was spending too much time on private matters.[54]

Tiede presented the Board of Directors' annual report to the assembled convention delegates in the Corn Palace. He repeated the charge that Skage had not supported the Board in the membership campaign. The Board, he said, wanted to amend the by-laws and the constitution to prevent any individual from holding the presidency for more than four years in succession.[55] The proposal, obviously directed at John Skage, was not approved.

Before the balloting for president on October 18, an anonymous letter was circulated, containing charges against Skage. A motion was approved that the president answer the charges in the letter. Skage said he would respond only if the author or authors of the letter first signed it.[56] Although Skage led after the first ballot, he did not receive the necessary majority to gain his eighth year as president of the SDFU. Oscar Fosheim was selected to be the next head of the state Union on the second ballot. He won by a narrow margin, receiving 135 votes to Skage's 130.[57] The increased membership meant more delegates, and prior to the balloting, there had been some discussion on the eligibility of some of the delegates. State Senator Elmer Beck was elected vice president and Ed Backlund became the secretary-treasurer. The SDFU's Board of Directors for 1945 included David J. Tiede, Henry Vinz, Roy Glover, Harry Caslin and Ronald Veren. Glover and Caslin began their first terms on the Board.

When Roy Brazell announced Fosheim as president, the new SDFU head and Skage exchanged friendly greetings. Hoping to restore unity, Fosheim asked the delegates to rise and applaud Skage for his many years of service to the Union. "Never throw rocks at those who go down," said Fosheim, "for those who go up will also come down."[58] John Skage was later given the gavel which he had used to preside over seven conventions.

The seventh president of the South Dakota Farmers Union was a native of South Dakota. Born on the family homestead on January 14, 1895, he was the son of an early pioneer who helped to build the railroad through Miner County. After serving in the armed services during World War I, Fosheim spent a brief period of time selling real estate in California during the early 1920s. From 1923 until he headed up the membership drive in February 1944, he engaged in farming and livestock feeding on the family farm near Howard. Fosheim was persuaded to run as a Democrat for a seat in the state House of Representatives in 1930. He would go on to serve eight years as a legislator, introducing much legislation, including the celebrated ore tax. Water development became a key issue for Fosheim in the late forties. He favored a Missouri Valley Authority (MVA) modeled on the Tennessee Valley Authority. Through the MVA, he believed that construction of dams on the Missouri would open the way for water transportation on the upper part of the river.[59]

He was not a member of the Farmers Union during the 1920s. His interest in the organization was aroused when he discovered the financial advantages of selling hogs through the FU Livestock Commission. He said, "If the Farmers Union is trying to get more money for farmers, probably there's something good about them. So I joined the Farmers Union."[60] He played a major role in organizing the Miner County Farmers Union in the early 1930s.[61] President Fosheim had a great interest in history and read widely on that subject. In his speeches and conversation, he might comment on the world travels of the ancient Greek historian Herodotus or refer to the rules of the French Chamber of Deputies during the French Revolution.[62]

Farmers Union cooperatives in South Dakota gained in numbers and strength during Fosheim's three years as president. He concurred with Skage's plea that every cooperative write into its by-laws the check-off of dues and that every co-op contribute 5 percent of its net earnings to the state office to help finance education.[63] As part of the education program, the SDFU financed and directed a Junior trip to St. Paul to tour the Grain Terminal Association and the Central Exchange.

A major disappointment for the premier membership organizer was that the State Union's numbers remained almost stationary during his three year tenure. John Eide, the organizer for 1945, was handicapped by a limited gasoline allotment as World War II came to a climax in August of that year.[64] August Dahme, in delivering the state organizer's report for 1946, indicated that the Union's membership had increased by only 106 new farm families. He believed that at least 50 percent of South Dakota's approximately 60,000 farm families should be members of the Union. The Union's membership was 11,500 in 1946.

Setting a goal of 15,000 members for 1947, Dahme asked the 1946 convention delegates to approve his request that a specific sum of money be set aside in each county to be used for organization work. The county Unions would be rewarded. When the 1947 membership in the county equaled that of 1946, fifty cents per member would be paid to the county by the state Union. For each member over the 1946 total, two dollars would be paid to the county secretary-treasurer.[65] The delegates, many of whom had driven to Rapid City in an early October blinding snowstorm, favored the proposal, but it proved ineffective as the Union showed a loss of 262 members for 1947.[66] Fosheim attributed the lack of gain to prosperous farmers who had forgotten the needs of an organization and the growing conservative mood in South Dakota.[67]

Taxes and electric power had been high on the Union's legislative agenda in 1945. A Farmers Union request for the elimination of a two percent tax on fuel for farm power was passed by the Legislature.[68] A legislative resolution requesting that Congress prepare a constitutional amendment limiting income tax rates to 25 percent was strongly opposed by the Farmers Union, which accepted credit for the resolution's rejection. President Fosheim argued that such a limitation would only shift a large share of the tax obligation from the upper to the lower income brackets.[69] Sam Ulrickson, the Union's lobbyist, took on the private electric power monopoly, as he vigorously supported Governor Sharpe's Power District Bill and legislation that would facilitate REA construction in South Dakota.[70]

One of Fosheim's most significant accomplishments was restoring harmonious relations between the state officers and the Board of Directors. He also supervised the publication of a *Union Farmer* that was free from attack on its own members.[71] One area of the Union's activities where harmony prevailed was in the education program. It had the hearty support of both the leadership and the rank and file. The Junior and camp programs continued to thrive during the 1940s.

NEW PROGRAMS FOR A NEW GENERATION

The Junior program was strengthened in curriculum and instruction during the decade of the forties. Juniors were provided with new opportunities to engage in "hands-on" experiences. They provided the program for a Union-sponsored radio presentation over KSOO-Sioux Falls every Saturday morning. The program included interviews with farmers and truckers and advertising by the Co-operative Brokerage.

With the assistance of John V. Carlson, the manager of the Sioux Falls Livestock Commission, livestock loans were made to Juniors to encourage their participation in feeder and breeding projects. In the fall of 1942, Carlson reported that the program was successful beyond expectations.[72] Five years later, at a meeting in the Farm-Labor Hall at Mitchell, on November 18, 1947, the Union leadership took steps to form a SDFU Federal Credit Union. Upon receiving a charter from the Federal Deposit Insurance Corporation, Union youth would be able to borrow money from the Credit Union to finance livestock feeding and breeder projects in all parts of the state. The Farmers Union's Education Department also viewed the Credit Union as a means to further instruct the Juniors in practicing thrift and handling savings.[73] John Sieh of Groton was appointed as new Junior leader for the SDFU. One of his assignments was to promote the relationship between the Credit

1952 Torchbearers. Back Row, L to R: Ervin E. Winter, Marvin G. Winter, Alvin Geidel, Paul W. Opsahl, Maynard Winter, Jerald Stormo and James Sommer. Front Rox, L to R: Lois Murtha, Lillian Winter, Annetta Semmler, Darlene Hotchkiss, Gloria Hille, Elaine Winter and Leona Ravenberg, Director of Education, 1951-1957.

Union and young livestock breeders. Sieh had formerly been a representative of the FU Livestock Commission House at St. Paul.[74]

Mrs. Arnold (Marie) Styles of Brentford, the Junior leader in District III for several years, had been appointed as SDFU Junior leader by the Board of Directors on January 14, 1941. A person with formal teaching experience, she had headed-up the Junior program at the prior state convention in Yankton. Mrs. Styles would contribute much to curriculum development and the education of Junior leaders at the district, county and local levels in the next five years.[75] Her tenure as Junior leader was interrupted for several months when she was severely injured in an automobile accident on November 8, 1941. During Mrs. Styles' absence, Mrs. Olga Radke of Stratford served as acting Educational Director.[76] Junior and educational material, as well as Junior projects, continued to be displayed in the Farmers Union booth at the State Fair which was held in the late summer every year. Marie Styles returned to work on November 1, 1942, in time to assist with the evaluation of the writers' projects of 1942, which consisted of verse, editorials and feature stories.[77]

Leadership training for Junior leaders became structured during the 1940s. The training sessions, much like public school workshops, were held at least once a year in each of the five districts of the SDFU. These sessions, which involved the district education directors, usually lasted three or four days. In addition, a fieldworker was employed for the summer months. Elvera Schley, the fieldworker for 1944, working with Juliet Fleischl, a national fieldworker, conducted one day schools in some of the districts.[78] Leona Ravenberg of Canova, who became state Education Director in the early 1950s, served as fieldworker in 1945. Working in one county at a time, she visited locals, helped set up Junior meetings and displayed and explained education material. A primary function of the fieldworker was to promote a better understanding of the Union's educational program among parents.[79]

Age classifications for Junior work were changed as study topics were created for Union youth younger than sixteen. Young people whose ages ranged from eight to twelve were classified as Juveniles; youngsters who were age twelve to sixteen were called the Junior Reserves. A Junior committee finally decided that all Union youth under age sixteen would be called Junior Reserves.[80] A change in requirements at the higher end of the age range was also made. The Board of Directors decided on October 11, 1947, that any Junior who had performed any Junior work before age twenty-one was eligible to become a Torchbearer until he or she was twenty-six years old, providing that person was an achiever each year after age twenty-one.[81]

Marie Styles, after five years at the helm of the Junior program, retired at the end of 1945 because of ill health. Her successor was Mrs. Lena Husman, a former rural school teacher, and the current education director for Pennington County in District V.[82] Her tenure was rather brief as she resigned at the end of September 1947. She complained that disruptive factions were creeping into the organization. In her final report, she said that "any position not gained honestly, with all fairness, is not a thing to be proud of."[83] Husman provided no documentation to support her charges, but she may have been referring to a division of responsibility within the education program with the appointment of John Sieh as the new Junior leader of the SDFU. Hazel Brekke of Albee was Husman's replacement as Education Director in November 1947. In the following year, she also became Junior leader as Sieh was reassigned to fieldwork and supervision of the Junior camps.

Camp Judson was the site of the All-States Camp during the summers of 1939-1941. Beginning in 1939, four South Dakota Juniors were given scholarships to the camp based on merit. To qualify for a scholarship, the young person had to be at least eighteen years old, with an "A" grade in classes and conduct at a state camp and be recommended by locals officers and the state Junior leader. The FU Livestock Commission and the FU Co-operative Brokerage also offered scholarships to both the All-States Camp and the state camps. Competitors had to prepare an essay on the strengths of cooperative marketing.[84] The standard fee for the ten day camp session was $15.00. The number of state Unions participating in the camp had doubled by 1939. Juniors participated from Colorado, Idaho, Kansas, Minnesota, Montana, Nebraska, Ohio, Wisconsin and the two Dakotas. Since its inception in 1936, the All-States Camp had been held in the Black Hills with the exception of 1938, when the campers met at Estes Park, Colorado. By the late 1940s, the All-States Camp had found a permanent home in the "Centennial State."[85]

The state junior camps continued to meet at a variety of sites east of the Missouri River and at Camp Judson in the Black Hills. Following World War II, the camp fee was set at $10.00. The Union also arranged adult summer camps in the Black Hills for family units in the late 1940s. The Junior camps continued to emphasize themes that were of importance to the Farmers Union and the general citizenry at the time. During World War II, a regular camp feature included the showing of patriotic and educational films. In the immediate postwar world, the role of the United Nations in preventing further conflicts was stressed. When rural electrification became a major concern of the Farmers Union in the late 1940s, Mrs. Dora Haines, an official from the REA in Washington, D.C., warned campers to disregard the propaganda of private utility companies. These companies claimed that the REA

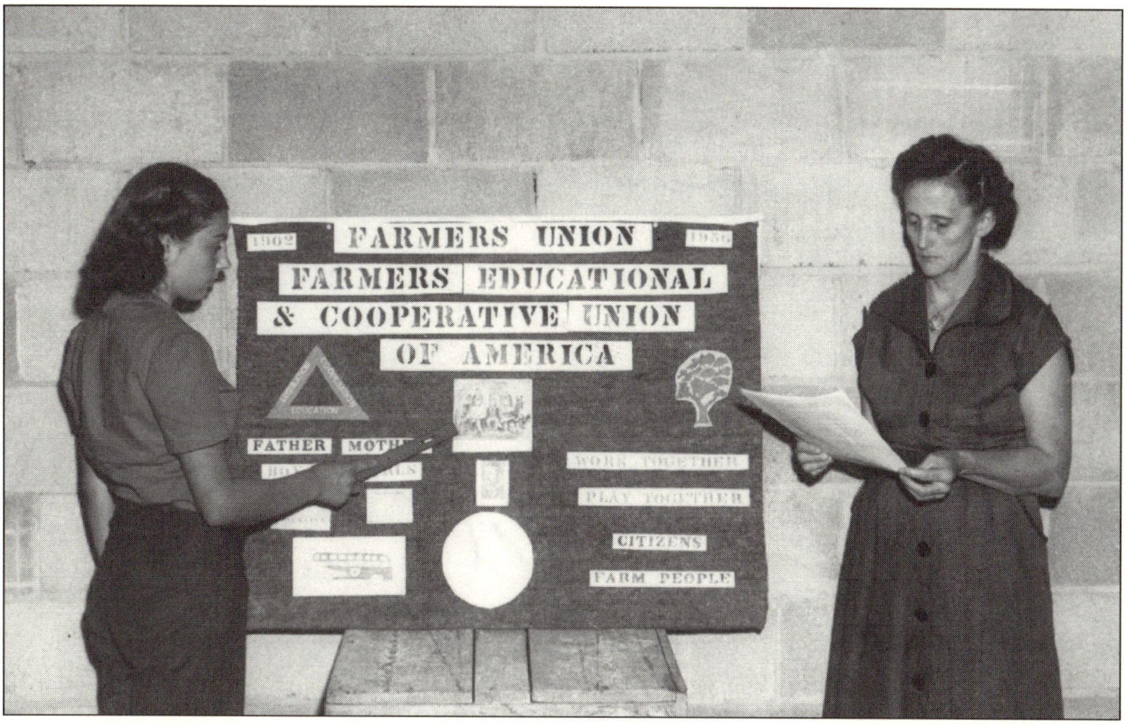

Farmers Union "Chart Talk," Hillard Local, Haakon County (1956). Mrs. Jewel Ferguson reads script.

would never be successful and that it would never provide farmers and ranchers with electricity.[86]

The *Union Farmer* often printed assessments of camp life by several of the participants. The following essay was published after the 11th annual Farmers Union State Camp, held at Camp Watymca, near Watertown, from June 24 to June 30, 1945.

> My Admiration
>
> I had never been to camp before and I didn't have the slightest idea what it would be like. Although I was slightly disappointed at first, I certainly learned to like it in a hurry. For a week the novelty of rising at 6:30 [A.M.] is interesting, but for a steady diet I don't think I would like it.
>
> Our teachers and speakers are the best.... They all gave very good talks that were interesting to everyone. At this camp no one had time to be bored. The classes that were held are [sic] "What Is This Farmers Union?", "The Cooperative Movement," journalism and speech. Many of the Juniors present at camp were appointed to give speeches in speech class and to write articles in journalism class. Community singing was common at camp.
>
> One of the first days student body officers were elected. We held a Student Body meeting nearly every day. A list of rules were made up. One of the most important ones was that all must be in their cabin by 10:45 every night except Friday, which was to be 11:45.

> We have plenty of free time to write, play, or do whatever we wish. Our evenings were spent in recreation. Raford Styles had charge of moving pictures. We had educational and humorous pictures. Three nights, programs were presented by the Juniors at camp.... They catered mostly to Farmers Union things, humor and music.
>
> .... Several nights there was dancing in assembly. On Thursday night the boys serenaded the girls.
>
> Many Juniors enjoyed frequent swimming periods although part of them couldn't swim, myself included. But wading around in the water was plenty of fun. The food is good and they have to serve meals in two shifts. On Thursday they served us big bowls of black raspberry ice cream. Was it good! What do you think!
>
> Of course there is always K.P. and A.P. (kitchen and assembly police) .... You meet new people and learn to cooperate....
>
> Camp Watymca was very crowded this year with about 170 people in a camp provided for 120. But I think our officers have taken care of that very nicely. They have managed to get extra beds for all of the Juniors. They have made room for these beds by pushing others together. All in all this is a pretty wonderful camp.
>
> Marilyn Dvorak
> Spink County[87]

Cooperative camp stores were introduced in 1942. Capital for the camp store was secured by selling shares to the campers. Organizers had to work quickly to get the camp store operative, because time was limited. A board of directors was elected, a manager hired and supplies purchased. The store was stocked with pop, candy, gum, Farmers Union T-shirts, emblems for FU jackets, etc. Campers learned to operate their store by the cooperative principles which governed all true cooperatives. A camp store audit was conducted. An inventory of the store's stock took place at the beginning of the session. When the camp ended, the store's receipts were apportioned out as would a Union in terms of the percentage of dues going to the local, district, state and national Unions.[88]

The parents of these campers continued to create new cooperatives during the war. New oil companies and co-op creameries, sponsored by the Farmers Union and financed by the Farm Security Administration, sprang up throughout South Dakota. The organization of cooperatives in the state reached its height in 1943.

THE SDFU'S COOPERATIVE ACTIVITIES

America's entry into World War II did not detract from cooperative building in South Dakota. President John Skage, a fervent believer in cooperatives, continued to use the resources of the SDFU to encourage and to support new and old cooperatives.[89] Addressing the delegates to the 1946 annual convention, Oscar Fosheim remarked, with great satisfaction, that money from the co-ops was not in the "pockets of fat, sleek men who lavishly wine and dine beneath glittering chandeliers." These same individuals he said, "Are now squealing like little pigs at weaning time because intelligent farmers turned to co-operative marketing, processing and procurement of farm supplies."[90] Former President Loriks played a pivotal role in creating new cooperatives when he was state and regional administrator for

State Camp Co-op Store Board (1953).
L to R: Mavis Stormo, Wayne Luening, Darles Hotchkiss, Harlyn Hellen, Richard Bong.

the Farm Security Administration (FSA) between 1940 and 1942. A goal of the FSA's Cooperative Division in South Dakota was to assist eligible farmers to participate in sound cooperative enterprises that would contribute to the farmers' rehabilitation. The FSA loaned money to farmers with which to buy stock in a cooperative.

Gust Moseson, who became the senior fieldman for the Central Exchange after the merger, along with his associate, Joe Fortner, helped start several new cooperatives in South Dakota. Moseson, a business-oriented pragmatist, knew that if you put a co-op in a community, the price of gasoline would drop. When traveling about the state, Moseson was asked — "What do you sell?" He replied, "I sell the economic advantage of the farmer."[91] One of G. L. Moseson's duties was to initiate efforts for FSA assistance for producer cooperatives in South Dakota communities. He recognized that many of the cooperatives would be unable to operate without the assistance of the Farm Security Administration. In a letter to Loriks, Moseson wrote, "Needless to say, the cooperative movement has gained momentum during the past few months primarily due to the publicity given to the finance program of the Farm Security Administration as it pertains to producer cooperatives."[92] In addition to the growth of Farmers Union oil companies, there was a considerable increase in the number of cooperative creameries sponsored by the FU and financed by the FSA in the early 1940s.

The state office retained half of the money cooperatives paid into the educational fund. The remainder of the amount was returned to the county Union to help finance the county's educational work, including paying expenses of the county director of education.[93]

Farmers Union Co-op Association, Mitchell (1948).

In the fall of 1944, more than 300 cooperative business enterprises, ranging from wool growers to burial associations, represented an estimated 50,000 stockholders in South Dakota. Meeting in Huron on October 2, 400 delegates from these cooperatives formed the South Dakota Association of Cooperatives. The delegates elected a nine person board of directors, drew up articles of incorporation, by-laws and approved a minimum budget of $12,000 annually to support a full time executive secretary. They proposed to establish a permanent office at a site yet to be determined.[94]

The cooperative movement, nationally, had to contend with a new opponent immediately after World War II. The National Tax Equality Association (NTEA) began a long-running campaign, including political lobbying and spreading false propaganda, to eliminate cooperatives in the United States. The NTEA claimed that cooperatives paid no taxes, an obvious falsehood because cooperatives paid the same taxes as any comparable business enterprise with the exception of an income tax. A true cooperative paid no income tax because it had no income; the profits of the cooperative were distributed to its members who paid an income tax.[95]

The Sioux Falls Livestock Commission had a substantial increase in volume during the war years. John Carlson, the Commission's manager, proudly reported at the end of 1943 that the Sioux Falls based operation was the only FU Livestock Commission in the United States that had paid a dividend of 30 percent for the past two years. Having built up a large reserve fund, the Livestock Commission saved its shippers large sums of interest by making feeder loans available through the firm's credit corporation at a low rate of interest.[96]

Farmers Union property, Sioux Falls Stockyards, Early 1950s.

The credit corporation also assisted Juniors in obtaining feeder calves which they fed for several months. The Livestock Commission purchased the stock for the Juniors with credit provided by the Farmers Union Credit Corporation. By purchasing well-bred Hereford calves for the youngsters, the Commission assured a top quality price at market rates. The program protected Juniors against financial loss if the animals died before they were marketed.[97]

South Dakota's Union members also profited through their affiliation with the Sioux City Livestock Commission. Refunds to SDFU membrs in 1946 totaled $172,828.46. In addition, the SDFU received the savings of non-members from South Dakota, who sold at Sioux City, and its share of the reserve account, the two sources of income totaled $58,848.78.[98] On October 12, 1945, the SDFU Board of Directors decided to formally withdraw from the Omaha FU Livestock Commission upon the payment in full of South Dakota's accumulated equity in the Omaha firm.[99] Despite this action, the SDFU, for several years following, continued to be represented on the board of directors of the Omaha FU Livestock Commission.

Fortunately, the financial success of the Sioux Falls Livestock Commision was not affected by a struggle within the SDFU Board of Directors as to who should sit on the Commission's board of directors. It was customary for the state Board of Directors. to serve on the Commission board. J. E. Broline of Winner, in a disputed election in late December 1941, had been elected to the SDFU Board from District IV. Some members of the state Board, led by Enoch Hofstad, ruled, without a hearing, that Broline could not serve on the Board. A series of law suits followed that acknowledged Broline's right to serve on both the SDFU Board of Directors and the Livestock Commission board. From April to the first part of October 1942, the pages of the Union Farmer were filled with charges and counter-charges by the spokesmen for the opposing sides in the Broline dispute. Skage, backed by Board members W. C. Hermann and Henry Vinz, indicated his strong support for Broline, in an exchange of letters with Hofstad published in the Union paper.[100] The dispute over Broline's right to serve was associated with a larger issue, the dissolution of the Brokerage. Broline supported dissolution, and he was a strong backer of the Central Exchange. Hofstad, as discussed earlier, viewed the Central Exchange as an enemy of what the SDFU had established in its three and one-half decades of existence.

The relationship between the Farmers Union Mutual Insurance Company of South Dakota and the South Dakota Farmers Union began to deteriorate during the 1940s. The Union's leadership began to question whether FU Mutual was truly a Farmers Union company. Although the firm made annual reports to the state convention, it no longer required membership in the SDFU, and it did not contribute to the Union's educational fund.[101]

In a meeting with FU Mutual's stockholders at Burke, in 1948, SDFU President Paul Opsahl demanded that the company either justify the use of the Union's name or use some other name for the firm. The company changed its name to the Missouri Valley Mutual Insurance Company on September 21, 1949.[102] During the prior year, the SDFU had affiliated with the National Farmers Union Life Insurance Company. The NFU Life Association had been chartered as a fraternal insurance company in Colorado on December 30, 1937, and started business on April 20, 1938.

The Farmers Union saw several advantages in entering an insurance program. NFU President James G. Patton said insurance companies were not businesses but "engines of democracy — engines for building the Farmers Union...." They met the needs of effective organizing. "You need money," according to Patton, "you need manpower, and you need something that will render a service to farmers to make it worthwhile for them to stay in the Union."[103] The appeal of insurance cut across commodity lines, and all farmers needed it. By making membership renewal automatic, through checking off the Union's dues from premium payments, state Unions could maintain a stable membership.[104]

In late June 1948, the NFU Life Insurance Company, through the SDFU's Service Association, was authorized to write life and hospital insurance policies in South Dakota. A month prior, the NFU Auto and Casualty Company had been licensed to do business in the state.[105] The Auto and Casualty firm wrote policies to cover casualty and liability insurance on motor vehicles, farm operations and fire insurance on property.[106]

Both of the insurance companies were cooperatives in all states in which they sold policies, with savings being returned to their policy holders. Each state unit controlled its own state agencies. The SDFU incorporated the Farmers Union Service Association in 1948, which served as the Union's insurance and adjustment agency. Paul G. Erickson was hired to manage the Service Association, a position he held until retirement in September 1961.[107] The SDFU Board of Directors became the board of the FU Service Association. President Opsahl served as executive secretary of the new agency. On a very warm day in July 1948, a meeting was called in Mitchell to help recruit Farmers Union leaders as part time agents for the fledgling insurance venture.[108]

Emil Loriks was seeking information on an entirely new cooperative venture in late 1946. He had become very interested in incorporating a cooperative telephone association in South Dakota. He sought advice from legal experts in Minnesota and North Dakota on possible articles and by-laws for such a cooperative.[109]

South Dakota Service Association, NFU Insurance Staff (1952).
From L to R: Helen Scheidegger, Clifford Bailey, State Manager, Paul Erickson and Marcella DeNure.

THE REGIONAL COOPERATIVES

SDFU elevators began doing business with the Farmers Union Grain Terminal Association in the late 1930s. At the annual state convention in 1940, the issue of the State Union's representation on the board of the FUGTA was considered. There were forty GTA elevators in South Dakota, and a large volume of grain from South Dakota was being handled by the grain marketing cooperative. At the GTA's annual meeting in late 1940, Emil Loriks was elected a director of the GTA for a three year term. He received 221 votes, the largest number of votes cast for any one director.[110]

By 1940, GTA was handling 29 million bushels a year for FU members.[111] It had become the world's largest grain receiving agency. Timing was right for the GTA and new local cooperative elevators. The long drought was truly broken, and the increase in grain prices plus stepped-up production because of the war in Europe benefited farmers. Conditions favored the growth of the GTA, and it did grow under the management of M. W. Thatcher. The young cooperative had a net worth of $2,400,000 in 1942 compared with its starting capital of $30,000 in 1938. On May 1, 1943, GTA acquired the properties of the St. Anthony and Dakota Elevator Company. This acquisition increased the number of elevators affiliated with GTA from 160 to 236. Other purchases of the Cooperative included the Brown Grain Company of Minneapolis and several lumber yards. The lumber yards became known as the Great Plains Supply Company.[112]

The acquisition of the elevator line meant that the newly acquired elevators had no experience as cooperatives and sometimes were in competition with Farmers Union elevators in the same community. Educational meetings were held to inform their farmer patrons how they could become owners of these elevators. The goal, according to Thatcher, was to have local co-ops buy the old-line elevators from the GTA. At the GTA's annual meeting in December 1943, M. W. Thatcher told stockholders that $80 million worth of grain had been handled during the year. The Cooperative's net worth now stood at $4.3 million.[113]

Tri-County Cooperative Association, Oldham.

There was resistance to Thatcher's managerial style by the former president of the FUGTA, Dan O'Connor. The differences between the two men were accentuated by a conflict between the strong personalities of Thatcher and Emil Syftestad, the manager of the Central Exchange. Upon his return from the nation's capital, Thatcher had expected Syftestad to surrender his role as manager of the Exchange, which Syftestad refused to do. The quarrel between the two men began in 1940, and it escalated in 1941 when the GTA board, with Loriks playing a major role, elected Ole L. Olson as its president over O'Connor. Thatcher wanted O'Connor removed from his positions with both the GTA and Central Exchange. The differences between Thatcher and Syftestad led to a rift between the two regional cooperatives. A visible distinction between them occurred in 1943 when the Board of Directors of GTA and CENEX became distinct. In the past, the Boards had shared members. The split strained friendships between loyal Farmers Union members. In 1944, the Central Exchange began holding its meetings in February rather than in December. Prior to 1944, the two regionals had met at the same time in December for the convenience of stockholders. The rift between the two cooperatives did not last long and the division, once viewed with alarm, actually strengthened both firms as they continued to grow in the ensuing decades.

At the first postwar meeting in December 1945, Thatcher reported that the GTA had a net worth of $9 million. Earlier in the year, the cooperative had paid $250,000 in cash for patronage benefits earned in 1940. Payments had been authorized when net worth exceeded $5 million.[114] Stockholders approved the construction of an office building in St. Paul at an estimated cost of $450,000. They also voted to build a terminal at Grand Forks, North Dakota, and an addition to the terminal elevator at Superior, Wisconsin. An action which had a direct impact on the SDFU was a vote to place the payment of educational funds in the GTA bylaws. The amendment said that a sum not to exceed 5 percent of the annual net income should be set aside to be used for the purpose of promoting and encouraging cooperative education. A similar action was taken by the FU Central Exchange. In prior years, since 1940, educational funds had been approved by resolution at the annual convention. The South Dakota Union was now assured of a permanent income in proportion to its business volume with the regional cooperatives.[115]

In 1947, the FUGTA exercised leadership in the National Federation of Grain Cooperatives. It edited that organization's *Co-op Grain Quarterly*. GTA had become a dominant force in the rural economy. It owned terminal and subterminal elevators with more than 10 million bushels storage capacity, a line of 137 local elevators and a durum wheat mill. GTA directed and financed three subsidiaries, including a grain and supply company, a feed manufacturing plant, and a firm that included forty-four branch lumberyards and twenty-nine coal yards. GTA was merchandising grain at terminal and interior markets in most of the forty-eight states, financing county elevator associations during peak seasonal shipping operations, plus accounting and auditing services for these cooperatives.[116]

Although limited by what it could sell during the war years, the Farmers Union Central Exchange also experienced phenomenal growth in the 1940s. By 1943 there was a shortage of batteries for farm radios and tires for automobiles and trucks. Gasoline was in short supply, even at harvest time. The bulk of the batteries, tires, gas transports and tank cars were assigned to the armed forces. The Farmers Union Transportation Company was created to deliver products to local oil companies. Yet, the cooperative which had begun with a subscribed capital of $525 had a net worth of $1.5 million in 1942.

The Exchange's dream of producing and distributing its own petroleum products became a reality late in the winter of 1943. On March 1, 1943, the Independent Refinery Company of Laurel, Montana, became the property of the refinery division of FU Central Exchange. Located fifteen miles from Billings, Montana, the Laurel site refined crude oil from Wyoming and Montana. Purchased for $1.25 million, the refinery produced over 25 million gallons of gasoline in 1942.[117] In late June 1943, Central Exchange joined with four other regional cooperatives to purchase the Globe Refinery of McPherson, Kansas, for $5 million.[118] The five cooperatives organized the National Cooperative Refinery Association. The McPherson facility, built in 1932, could produce 175 million gallons of refined fuels annually. A pipeline from the refinery at McPherson to Council Bluffs, Iowa, was already operative. The line was extended to Minneapolis and St. Paul.[119] The Exchange's total savings for 1944 and 1945 stood at $2,931,969. As of 1947, the Central Exchange planned to double its net worth of $5 million within three to five years.[120]

The National Farmers Union operated eight livestock commission firms in 1947. Compared with the GTA and Central Exchange, they had experienced little growth since 1920. The increasing tendency of growers to send stock directly to market by truck instead of rail in the postwar era, did not bode well for the future of terminal livestock cooperatives.[121]

One of South Dakota's own, James E. Wells, Jr., was named commissioner of the Bank for Cooperatives on July 14, 1947, becoming the top supervisory official of the cooperative bank system. He began as a grain buyer in the state. Later, as a county agent, he helped organize several livestock cooperatives. Wells gained banking experience in Mitchell before

Directors and officers of the South Dakota Rural Electric Association, elected at a 1950 meeting in Huron, were: Left to right: V.T. Hanlon, director; Oscar P. Johnson, director; A.J. Pew, secretary; Sam J. Ulrikson, vice president; A.C. Hauffe, president; Leo P. Flynn, attorney; and William Raabe, treasurer.

starting a government career in 1927. He was employed by the Federal Reserve Bank and by the Department of Agriculture before his assignment with the Bank for Cooperatives.[122]

The growth of the South Dakota Farmers Union and its cooperative affiliates parallelled an expanding economy generated by wartime needs. Prices of farm commodities soared between 1940 and 1950. No. 1 wheat, which sold for ninety cents per bushel in 1940, reached a high of $2.88 in 1948. Beef cattle prices per 100 pounds increased from $7.46 in 1940 to $23.20 in 1950. With the good came the bad. The increase in the index of prices of all farm products received by farmers, however, was misleading because prices paid by farmers for all commodities used for maintenance and production rose sharply during the 1950s. The sharpest increases being recorded in interest payable per acre, taxes and wages for farm labor. Continuing to defend the position of the small farmer, the Union had to battle against the policies of a hostile Secretary of Agriculture , Ezra Taft Benson. Through it all, the South Dakota Farmers Union truly came of age during the decade of the fifties.

CHAPTER 11

ATTAINING NEW HEIGHTS 1948 - 1961

It [the Farmers Union] is the only hope for agriculture. We must build our State and National Organization stronger if the family farm is to survive in America.
— Emil Loriks, August 10, 1956.

In the first decade following World War II, South Dakota's farmers experienced considerable prosperity. Cash income from crops and livestock, excluding government benefit payments, neared the $500,000 mark in 1946 for the first time in the state's history. Cash income would exceed the half-million dollar mark each succeeding year until 1955.[1] The South Dakota Farmers Union achieved a high membership of 21,600 in 1949. Throughout the fifties, the membership remained above 17,000. The Union moved into a new home with the dedication of a state office building in Huron in 1952. The SDFU launched a travel program, inaugurated Farmer-Worker Conferences and introduced a Livestock Cooperative Marketing Day. Better living conditions accompanied an improved economy. A veritable technological revolution followed the war. The switch from animal power to gasoline engines was completed. With improved roads and motorized transport, farmers brought their products to market with greater ease, putting an end to isolated farm life. There were 8,277 trucks traversing South Dakota's roads in 1921. That number had increased to 87,538 in 1955. Automobiles registered in the state in 1958 totaled 244,000.[2] By 1951, farmers harvested 95 percent of the state's corn crop with mechanical pickers, and the number of combines used by small-grain farmers in northern South Dakota increased each year.[3]

Electric power truly came to the countryside in the late 1940s. Electricity, more than anything else, reduced the differences between country and city living, and brought farmers into the modern age. It had many uses in agricultural operations, as well as in the farm home. Life became easier for rural residents with the use of electric motors to pump water, grind feed and power milking machines. Farm wives had electric lights, refrigerators, washing machines, vacuum cleaners and other electric household appliances. By 1953, the television set was becoming a fixture in farm homes, further linking rural areas with urban centers.

The production of farm commodities increased dramatically with new technology and modern science. With the introduction of the four-wheel drive tractor in 1957, labor costs per acre dropped to new lows. The invention of new harvesting machines eliminated most hand labor needed to harvest potatoes, tomatoes and sugar beets. Farmers increased production per acre by using artificial fertilizers and applied new chemicals to control weeds

Combines doing work formerly done by a threshing crew and teams of horses.

and the ravages of insect pests and plant diseases. Geneticists developed more productive varieties of food plants. Better feeds made for meatier cattle and hogs. New antibiotics checked animal diseases.

The foregoing contributions to production had their negative side. After 1951, overproduction and declining foreign markets caught agriculturalists in a price squeeze, and their relative share of national income declined. Most alarming was the considerable decrease in the number of family farms. The pursuit of efficiency via advanced technology created a surplus of farm labor. South Dakota lost 26,518 family farms in twenty years. The 1935 census indicated there were 83,303 farms in the state, most of them cultivated by families residing on the land. The 1954 census listed 62,520 farms, and of these, only 56,785 resided on the farm for which they reported. Between 1950-1954, the average number of farms decreased about 1,000 per year.[4] By 1960, 53,000 farms remained in South Dakota.[5] Nationally, the farm population declined by approximately 10 million people between 1933 and 1953 with the most radical drop, 8 million, occurring between 1940 and 1953.[6] Union leaders expressed alarm over this decline because of its negative impact upon the Farmers Union's basic belief in the family farm, membership and the organization's influence in the State Legislature and Congress.

Members of the Farmers Union were pleased when President Truman, during the summer of 1948, nominated Charles F. Brannan, a staunch advocate of many FU principles, to be Secretary of Agriculture. Brannan succeeded Clinton P. Anderson, a South Dakota native.[7] Many political pundits believed Brannan's tenure would be short as national polls predicted that the Republican candidate, Thomas Dewey, would defeat Truman in the

November election. Truman won the election, and the farmers' vote made the difference. The Republican dominated 80th Congress had responded to Truman's request for continued price supports for farmers with weakened price supports. With a bumper crop in 1948, the price of corn dropped from $1.78 per bushel in September to $1.38 in mid-October. Farmers expressed their feelings at the polls.

Truman believed that a permanent system of price supports for agricultural commodities was needed to avoid extreme variations in farm prices. He asked Brannan, a former administrator of New Deal agricultural programs, to come up with a plan that would insure farmers a stable income. The Brannan Plan emphasized production and abundance. It made it worthwhile for the farmer to produce commodities in a rational manner but kept consumer prices at levels where the average person could afford to buy. If the price of a given farm product fell below support levels, the government would make a direct payment to the farmer for the difference between the price he received and the support price. The Brannan Plan benefited small farmers as it would have excluded from production payments the yields of any one farm over and above a certain limit. Brannan argued that "Price supports are the farmer's equivalent of the laboring man's minimum wage, social security, and collective bargaining arrangements."

Union leaders at the state and national level "took to the stump" to support the plan when it came before Congress. Brannan's proposal was bitterly resisted by the American Farm Bureau Federation, led by Allen B. Kline. It is estimated that the "anti-Brannan Plan" campaign cost Farm Bureau members more than a half-million dollars in one year. Large-scale farmers objected to the upper limit on guaranteed income. Conservatives said the plan was too costly and socialistic.

Congress refused to enact the Brannan Plan into law. Truman said its opponents, who argued that one should cut down producing units until the fittest survived, lacked humanity. They would surrender the family farm to the absentee landlord or the corporate owner. The President was convinced that the farmer's sense of security was a vital part of the foundation of American life. The fate of the Brannan Plan was a portend of what would happen to pro-small farm legislation during the 1950s.

In addition to confronting a conservative dominated Congress, the Farmers Union came under attack from ideologists of the far right. A "Red Scare" and political witch hunts were promoted by spokesmen on the extreme right of the political spectrum. McCarthyism made its impact on Congress and the mass media. A spokesman for this dark mood was radio commentator Fulton Lewis, Jr., who delivered his ultrarightist diatribes to a national audience. A master of falsehoods and half-truths, he said, "The actual authorship of the Brannan Plan lies in the National Farmers Union, in an extreme left-wing and dubious little side runt of the family of National Farm Organizations controlled for all practical purposes, if not by actual direction, by the CIO."[8]

Charles F. Brannan, US Secretary of Agriculture, 1948-1953.

Another purveyor of falsehoods, who attempted to defame the Farmers Union, was Herbert A.

Philbrick, author of *I Led Three Lives*. He had attempted to smear the Union by associating it with communism. Finally, on October 15, 1957, he sent a letter of retraction to President Patton, stating that the FU was not a communist-front organization or communist contaminated and that Patton was not a Red sympathizer.[9]

Malicious attacks upon the Union also took place within South Dakota. An Aberdeen pastor said the FU in North Dakota and South Dakota was involved in communism.[10] A professor of economics at Augustana College, Robert C. Teuting, when speaking to a Davison County Republican rally was quoted as saying: "The Farmers Union is the next thing to a Communist organization, I think." Dr. Lawrence Stavig, president of Augustana, quickly disavowed any connection between the college and Teuting, who spoke as a private citizen. Teuting quickly issued a retraction, but his tenure at the college was very brief.[11] Homer Ayres, the Union's political pundit from Sheep Creek Local, in a letter to Emil Loriks on November 9, 1955, said: "It looks to me as if you about extracted the little tooter from tutor McTeut's tooter. I doubt if he will be tooting as loud for awhile as he has in the past."[12]

Despite the adverse votes in Congress and political witch hunters, the SDFU could take pride in its significant accomplishments during the 1950s, but it could not rest upon its laurels. The momentum gained immediately after World War II could only be maintained through action and innovation.

OPSAHL'S ADMINISTRATIVE TEAM

Late in the summer of 1947, Oscar Fosheim announced that he would not run for reelection. Delegates to the annual convention at Pierre chose Paul W. Opsahl of Aberdeen as his successor in a contested election with Ed Backlund of Davison County by a vote of 153 to 126. In the preceding campaign, Backlund had been identified by the opposition as the candidate of the FU Central Exchange.[13] Roy Brazell of Volga and Roy S. Glover of Frederick were elected vice president and secretary-treasurer respectively. The by-laws were changed in October of the following year, providing for the appointment of the secretary-treasurer by the president. In addition, the power to set the salaries of state officials was vested in the Board of Directors.[14] The district chairmen for 1948 were Sam Ulrickson of Canton, Henry Vinz of Spencer, Loyd B. Larson of Brookings, Harry E. Caslin of Kennebec and John R. Riedy of Thunderhawk.[15]

Paul Opsahl,
President of the SDFU, 1947-1958.

The administrative team underwent two significant changes within two years. Harold Golseth of Ervin triumphed over August Dahme in the vice presidential election in 1949, and Emil Loriks replaced Roy Glover as secretary-treasurer in mid-1950. With the resignation of Jack Bailey as director of publicity, Loriks also assumed that post. Glover was to become legislative counsel for the SDFU. Norman Olsen of Canton represented District I after 1951, and Fred Ferwerda of Kadoka replaced Riedy in District V.

Loriks gave up the secretary-treasurer position on July 31, 1952, but continued to work in the

Mrs. Paul Erickson and Marcella DeNure greeted visitors at the 1951 Farmers Union State Fair booth. The Farmers Union logo and the Union's major activities are emphasized.

Opsahl administration as administrative assistant, director of publicity and editor of the *Union Farmer*. Mrs. Gene (Anderson) Olson, who had been office manager of the SDFU, was Lorik's replacement as secretary-treasurer.[16] Within a year, Loriks was giving his full attention to administrative matters, and Theodore Nickisch was employed to edit the SDFU paper and handle publicity.[17] Loriks left the state FU staff in October 1956 to give greater attention to his responsibilities with the FUGTA. He had been reelected to a three-year term as a GTA board member, and he was vice president of the regional cooperative.[18]

In 1950, the Board of Directors had changed the size of the Union's publication, moving from the traditional newspaper format to a popular tabloid form of conveying the organization's message to its readers. Jack C. Seigle became the new editor of the *Union Farmer* and director of public relations on January 1, 1957. Seigle, who had a Master of Arts in journalism from Mankato State University, succeeded Robert Nelson, who resigned to accept a position on the staff of newly elected Congressman George McGovern. In the last year of Opsahl's administration, Roland W. Olson, a native of Buxton, North Dakota, and a graduate of North Dakota State University, became the new editor of the *Union Farmer* and director of public relations. He succeeded Seigle, who had resigned on July 1, 1958, to return to graduate school.[19] In addition to editing the paper, the public relations person prepared press releases, spoke on behalf of the Union in various parts of the state and arranged radio programs over stations WNAX at Yankton, KSOO at Sioux Falls, KIJV at Huron and KORN at Mitchell.

President Opsahl was born March 2, 1896, at Annandale, Minnesota. Following service in the U. S. Army during World War I, he moved to South Dakota in 1918 and began farming near Brentford, in Spink County. He became a charter member of Tetonka Local during the depth of the Great Depression. Opsahl served in the South Dakota House of Representatives from 1932 to 1938. He was one of the sponsors of the Moratorium Law and a vigorous supporter of the ore tax. In the early 1940s, he was appointed county director of the Farm Security Administration for Brown County. He also had been a fieldman for the FUGTA, a fact not lost on his opponents, many of them belonging to the anti-regional cooperative old guard, in the balloting for president in 1947.[20] After serving as president of the SDFU for eleven years, Paul Opsahl announced in July 1958 that he would not seek reelection. During the National Farmers Union convention in 1959, he received the 1959 Award for Meritorious Service. The citation read:

> Paul W. Opsahl, who voluntarily ended his presidency of the South Dakota Farmers Union in 1958, had served in the position for eleven years. This is the longest time ever served by a president of this state union, and it shows the popularity of the tall, silver-haired orator who has been reelected each year since 1947.[21]

A RECORD MEMBERSHIP

There were 11,255 members of the South Dakota Farmers Union when Opsahl became president. The delegates committed themselves to a membership of 20,000 by the time they assembled again. Despite blanketing counties with recruiters during the pre-planting period in late winter, the Union added only 2,500 new members during 1948. With renewed effort, the Union organizers assisted by fieldmen from the Bank for Cooperatives and the FUGTA, topped the 21,000 mark in membership the next year.[22]

Solicitors used a variety of techniques to recruit members during the fifties. There were the Junior Incentive Campaign and Bus Trip awards.[23] In 1958, locals that surpassed their prior year's membership total as of December 15 became eligible to send two delegates for a three day trip to visit regional coops in St. Paul. Lodging and transportation were paid by the SDFU. Almost a decade after a raise in the membership fee to $5.00, the state Union, in 1958, introduced a five year plan whereby a farm family could join the organization for five years for twenty dollars.[24] A solicitor who signed up fifteen or more five year members received free transportation and lodging to the National Farmers Union convention. A program was introduced in 1959 whereby solicitors received some compensation for their membership work. The SDFU offered to pay twenty cents to a local for every signed-up member in 1960, providing that the local and the county Union would match the amount from the state. Thus, the local solicitor received sixty cents for every member he recruited.[25]

The 1949 total was not surpassed in ensuing years, but membership remained above the 17,000 mark during the 1950s. For five consecutive years, from 1955 to 1959, the total was above 19,000. In 1954, three more counties in South Dakota received charters — Bennett, Butte and Harding. The Union had membership in every county in the state except Washabaugh County which had a county Union before the end of Opsahl's tenure.[26]

With an increased membership, the SDFU considered redistricting the state into more than five Farmers Union districts. A motion by the Board of Directors to create six districts was rejected at the convention at Mitchell in 1949.[27] Five years passed before the Board reintroduced the proposal, this time asking for seven districts. Before the annual statewide

Farmers Union pioneer Harry Caslin, at left, and Manager Jack McBreyer view the Farmers Union Cooperative Elevator, Kennebec.

meeting at Huron in October 1954, the Board had its redistricting proposal approved by all of the existing district organizations at their annual meetings.[28] The proposal, approved by the convention delegates, left Districts I, II and IV intact with the same numerical classification. District III was split, following a north-south line along the western boundaries of Marshall, Day, Clark and Kingsbury counties. The district also included Roberts, Grant, Codington, Deuel, Hamlin and Brookings counties.

District V was divided along an east-west line. Butte, Custer, Fall River, Haakon, Jackson, Jones, Lawrence, Meade, Pennington and Stanley counties made up District V. Included in District VI were Corson, Dewey, Harding, Perkins and Ziebach counties. District VII was composed of counties formerly in District III. Campbell, McPherson, Walworth, Edmunds, Potter, Faulk, Sully, Hyde, Hughes, Hand, Brown, Spink and Beadle counties comprised the new district.[29]

Districts VI and VII were permitted to organize and elect officers immediately, but the bylaws prohibited them from electing directors until the following year.[30] As of October 30, 1955, the reconstituted Board consisted of:

District I ...... Norman Olsen, Canton
District II ...... Henry A. Vinz, Spencer
District III ..... Loyd Larson, Brookings
District IV ..... Harry Caslin, Kennebec
District V ...... LeRoy Hardy, Sturgis
District VI ..... Frank Butler, Shade Hill
District VII .... Ben Radcliffe, Hitchcock[31]

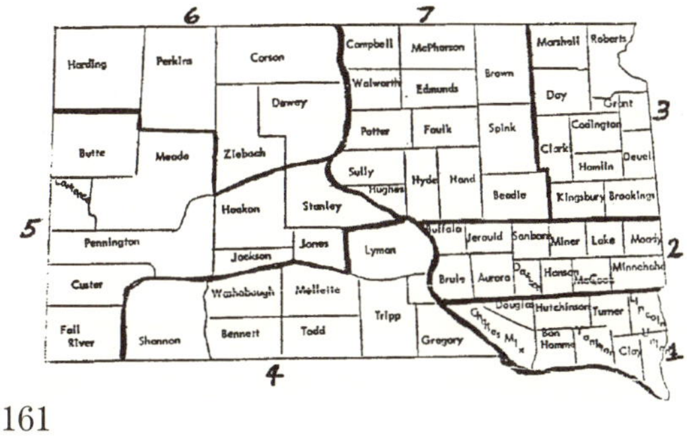

A PERMANENT HOME FOR THE SDFU

Discussions concerning a permanent home for the State Union began during World War II. Because of high property values and a lack of suitable building sites, the Board of Directors decided that it would be best to wait until after the war to begin construction.[32] No significant action was taken for almost three years after the conclusion of the war, and the Union continued to operate out of six rented rooms on the fourth floor of the Western Building at Mitchell. At a meeting of the Board, on June 2, 1948, the state representatives for the GTA and Central Exchange, Archie Erickson and Joe Fortner, proposed that a permanent building for the SDFU be constructed on the State Fairgrounds at Huron. President Opsahl appointed Board members Harry E. Caslin and Loyd B. Larson to study the possibilities for such a structure.[33] After consultation with the State Fair Board, Caslin reported that a permanent home on the fairgrounds required action by the State Legislature.[34]

Delegates to the 1948 state convention instructed the Board of Directors to secure a state office site at a reasonable cost. The government bonds being held as an investment by the State Union were to be transferred to a building fund account. The delegates did not stipulate a specific city for the state office, but indicated it should be in a central location.[35] Following offers by the mayors of Mitchell and Huron of suitable locations for a building, the Board decided the headquarters for the SDFU should be at Huron. Representatives of Henry Carlson, a Sioux Falls contractor, met with the Board on September 1, 1949, to present cost estimates, details of floor plans and construction data.[36] Plans of the building committee and the Board of Directors were approved at the 1949 annual convention, and the Board was authorized to proceed with construction of a $100,000 headquarters building at Huron.[37]

The Board requested bids for construction of the new building in mid-July 1950.[38] When the bids were opened in late November, the Board was shocked by the increased cost of building material. Despite having to raise an additional $40,000, a contract was awarded to the Gust Jacobson Construction Company to break ground in early spring of 1951.[39] The Board decided to borrow $75,000 at 3.5 percent interest from the GTA to complete the structure. Thirty cents of each member's annual dues would be used as collateral. The loan was to be repaid within twenty years following April 2, 1951.[40]

LeRoy Hardy, Sturgis. A member of the Board of Directors from District V, 1955-1966. Hardy also served as a SDFU lobbyist at Pierre and as a leader of Bus-Ins to Washington, D.C.

One other unforeseen obstacle remained in the path of the construction of the new building. On April 21, 1951, the SDFU received a notice of denial for a permit for construction from the National Production Authority in Minneapolis, Minnesota. With the outbreak of the Korean War in 1950, a war mobilization emergency had been established. Paul Opsahl and Emil Loriks traveled to Minneapolis, Loriks later going to Washington, D.C., to request a reconsideration of the permit. Their request was approved, and on April 30, 1951, sod was broken for the new building between 14th and 15th Streets in Huron.[41]

Laying the cornerstone for the Farmers Union State Office, Huron, September 6, 1951.
From left: President Paul Opsahl, former presidents Emil Loriks, Oscar Fosheim and John Skage.

Several hundred people gathered for the cornerstone laying ceremony on September 6, 1951. The featured speaker was Governer Sigurd Anderson. John Skage, on behalf of the ex-presidents of the SDFU, spoke briefly to the audience and sealed a copy of the Farmers Union's history in the cornerstone. The final cost of the new structure was approximately $135,000. Site and building costs totaled $125,000 and about $10,000 was paid for new equipment and furnishings. Staff and SDFU officers, after transferring records and other materials from Mitchell, moved into their new headquarters on February 1, 1952.[42] The Board of Directors held its first meeting in the new office building on February 6. Secretary-Treasurer Loriks reported that $90,000 had been paid to the contractor.[43] The formal dedication of the SDFU's permanent home took place on May 10, 1952, during the golden jubilee year of the National Farmers Union.

The Farmers Union paid off the building debt very quickly. By October 1954, it had been reduced to $12,000 and thought was given to enlarging the office building.[44] On June 24, 1955, the Board of Directors authorized the construction of a second floor addition. They accepted the bid of the Jacobson Construction Company to build the addition for $34,411, with occupancy planned for December 1. The Farmers Union Grain Terminal Association agreed to finance this portion of the building as it had the original building. A promissory note was to be held by the GTA in the amount of $30,000 at 3.5 percent interest for a twenty-year period. The Board proposed annual payments of $3,000 to amortize the loan.[45] The loan was to be repaid from rental income only. The rent money coming from the Farmers Union Insurance agency which occupied the second story upon its completion.[46]

South Dakota Farmers Union state headquarters building, Huron.

Shortly thereafter, the Union's state officers appeared in circuit court to secure tax relief for at least a part of the new office building. The court agreed, in December 1956, that the portion of the building that housed the educational organization would be tax exempt.[47]

Huron also became the home of the annual convention during most of the fifties. Beginning in October 1953, a spacious new arena in the "Fair City" became the headquarters for the SDFU convention and the site of all sessions of the general assembly.[48]

At the time Union members began to enjoy their new facilities, South Dakota's farmers were experiencing a significant change in their lifestyle. Revolutionary changes came to farms using electrical power.

MISSOURI BASIN DEVELOPMENT AND ELECTRIFICATION

Fort Randall, the first of the Missouri River dams in South Dakota, was dedicated in August 1956. High voltage lines, built and operated by the Bureau of Reclamation, were to carry electric power across South Dakota and interconnect with public power districts in Nebraska. The first current from Fort Randall Dam was released upon an electric signal from President Dwight D. Eisenhower at the White House on March 15, 1954.[49] The first power generated at Fort Randall went to REA cooperatives.

Construction of Oahe Dam, the key structure in the Missouri River Basin program, began in September 1948. The Gavins Point Dam was built between 1952 and 1956. Ground was broken for the construction of Big Bend Dam in May 1960. The last of the mainstem dams on the Missouri River in South Dakota, it was dedicated on September 15, 1966.

The Oahe project appeared in trouble in the spring of 1953, when Republican budget slashers in Congress eliminated any further funds for the dam's construction. In response, the South Dakota Farmers Union acted on two fronts. The Union hosted a mass rally at Huron, attended by 1,500 people. A huge banner on the stage read "Don't let them destroy your future." President Opsahl gave the audience another slogan — "Keep the farm lights burning," and the way to do that was to "Turn the heat on Washington." Present at the meeting were A. C. Hauffe, president of the South Dakota Rural Electrification Administration, Governor Sigurd Anderson, Opsahl and Kenneth Holum, Sr., the South Dakota member of the eleven man Missouri Basin Survey Commission.[50] Secondly, a host of South Dakotans, including Paul Opsahl, appeared before the House and Senate Civil Functions Subcommittee to speak for development of the Missouri Basin and the need for funds for Oahe. The lobbying had its reward. A large portion of the funds for the fiscal year was restored, $8,250,000 of the original allotment of $14,800,000.[51]

By the fall of 1953, $18 million had been spent, all in Phases I and II of earthmoving at the Oahe site. Competitive bids were opened by the Army Corps of Engineers, in October 1953, for outlet works and six tunnels on the dam. Acceptance of the low bid of $9,500,000 marked the first large stage of work on Oahe.[52] The final cost of the Oahe project was $350 million. President John F. Kennedy was the featured speaker at the Power-on-the-Line ceremonies at Oahe Dam on August 17, 1962. Kennedy pointed out that in 1932 only 10 percent of the nation's farms had electric power. At the time he made his visit to Pierre, over 90 percent of the farmsteads enjoyed electrical service.[53] The catalyst had been the creation of the Rural Electrification (REA) in 1935.

The Rural Electrification Administration was established by Executive Order 7037 on May 11, 1935, under powers granted by the Emergency Relief Appropriation Act of 1935.[54] One hundred million dollars was appropriated to formulate and to administer a program of generating and distributing electricity in isolated rural areas, areas excluded from service by private utilities. The REA became a permanent agency when Congress passed the Rural Electrification Act of 1936, signed into law by President Roosevelt on May 21, 1936. On August 1, 1935, President Roosevelt had issued Regulation Four. It established the REA as a lending agency and gave it the authority to lend the entire cost of constructing light and power lines in rural areas, on liberal terms of 3 percent interest with amortization over a twenty-year period. Although the legislation gave priority to publicly-owned plants distributing electricity, it was assumed that existing private power companies would use the federal loan program.

When existing power companies did not respond, farm organizations that championed cooperative principles came to the forefront as borrowers under the REA program. In South Dakota, the first REA co-op meeting was held in the Manning-O'Connor store in Burbank at which seventeen farmers from Clay and Union counties organized as the Fairview Rural Electric System. Their first application for a loan was rejected because the co-op was too small. It was difficult finding sufficient farmers who were willing to risk the $5.00 signup fee to qualify for an REA loan.[55] Rangvald "Sweeny" Sveningson, whose farm home near Lane received electricity in 1994, said, "After the depression, farmers didn't want to part with $5 for a membership fee. They thought it would take too long to

Kenneth Holum, Sr., a member of the Missouri Basin Survey Commission.

The impact of rural electrification (1950s). Yardlights became a part of the rural landscape.

get electricity or the cooperatives wouldn't amount to anything."[56] The farmers who had gathered at Burbank did not give up. In the months that followed, they went door to door, increasing their membership and the area to be serviced by the cooperative. The first REA-financed line in South Dakota, the Union Electric Cooperative, went into operation on May 21, 1937.[57] In the following year, 1938, Hutchinson Electric and Union County had functioning electric cooperatives. By the end of 1939, there were five electric cooperatives in the state, bringing power to only 4 percent or approximately 3,000 of South Dakota's farms.

Darkness persisted over much of the South Dakota countryside until after World War II. Farm women continued to be slaves to the washboard and the coal or wood-burning range. Cows continued to be milked by hand by the dim light of lanterns. Kerosene lamps, with their sooty chimneys, provided the only light in many farm homes. Following the end of hostilities in 1945, members of the South Dakota Farmers Union, county agents and other interested rural people began to go up and down country roads seeking the signatures of new members and asking for the hard-to-come-by $5.00 signup fees.[58] Electric cooperatives were organized and loan applications prepared for submission to the Rural Electrification Administration.

Rural electrification, however, got off to a slow start in South Dakota. It was difficult to build lines because there was no state law giving cooperatives the authority to condemn land. Beginning in 1939, the South Dakota Farmers Union and other like-minded organizations lobbied the State Legislature to establish uniform procedures for electric co-ops and to give them the power of condemnation. The proposed legislation was rejected in four successive legislative sessions, 1939-1945, because of strong opposition from private power companies. After the cooperatives threatened to establish public power districts, the opposition ended. On February 26, 1947, Governor George T. Mickelson signed Senate Bill No. 83, the South Dakota Electric Cooperative Act; a measure heartily supported by the Farmers Union. The act, having established uniform procedures for electric cooperatives, cleared the way for the construction of transmission lines. Four years later, the last electric cooperative to be formed in the state, West Central Electric of Murdo, was energized.[59]

Until the Pick-Sloan facilities began producing electricity, the amount of power available for cooperatives was limited. They decided to build their own transmission systems from the Missouri River dams. As of 1949, East River Electric would provide power for the eastern half of the state. Rushmore Electric did the same for the western half of South Dakota beginning in 1952.[60]

Ninety-six percent of all South Dakota farms enjoyed electrical service in 1960. Nine-tenths of these electrified farms were serviced by rural electric cooperatives. About 45 percent of the power used by co-ops in eastern South Dakota during the 1980s was generated by the mainstem dams on the Missouri River. The remaining 55 percent of electrical power was generated from coal. In 1985, thirty-five rural electric cooperative systems represented more than 80,000 metered members, served by nearly 62,000 miles of line.[61]

THE UNION AND THE STATE LEGISLATURE

From the late 1940s until 1961, the South Dakota Farmers Union emphasized three issues in the biennial meetings of the Legislature. The issues were public power districts, taxes and the defense of the family farm. When the mainstem dams on the Missouri River became a reality, concern was expressed over the allocation of the electrical power generated by these dams. It appeared likely that demand might exceed supply. Federal law stipulated that public bodies such as municipalities, REA cooperatives and public power districts had priority over private companies in the use of public power. Many South Dakotans, including members of the FU, feared that much of the electric current might go to public power districts in neighboring states. Farm and political leaders believed South Dakota was entitled to a major share of the power because it was being generated at dams located within the state's borders. The Union's proclaimed goal in 1948 was the continuation of the federal law's priority on electric power to public power organizations and the enactment of a state power districts' law.[62]

During the 1949 session of the State Legislature, a Consumers Power District Bill passed the House of Representatives by a vote of 48-27, but it was tabled in the Senate, because of a parliamentary maneuver, by the slim margin of 18 to 17.[63] Merrill Q. Sharpe, who continued to have a vital interest in the Missouri Basin after his term as governor, responded to the Senate's action with a vigorous plea for the enactment of public power districts' legislation, pointing out that South Dakota could have saved more than $6 million in 1948 if it had had public power rates comparable to TVA and those obtained in the Pacific Northwest.[64]

Governor George T. Mickelson summoned legislators to a special session in February 1950. They created the authority for creation of public power districts, ensuring that South Dakota might share equally with other states the benefits of the development of the Missouri River. Two months later, the SDFU decided that there must be closer coordination between the Union and REAs toward the goal of enacting in South Dakota a workable public power districts' bill. Farmers Union leaders created the Committee for Organization

A unit in the parade at the annual SDFU convention, Mitchell (1952).

Friendly Toward a Power District Law. The great majority of the FU's legislative leaders were members of REA cooperatives.[65] South Dakota, however, had no public power districts as of 1961.

During the public power debates, the Farmers Union bolstered its lobbying efforts in the state capital. The Union, traditionally, had two legislative consultants in Pierre during the sessions of the Legislature. Beginning with the 1955 session, the SDFU staged a social hour. It was used to promote FU membership and policy, and to convince legislators that the Union was not the "rabble-rousing" organization that some people pictured it to be.[66] Five years earlier, in early April 1950, the SDFU had purchased a thirty-seven passenger bus. Beginning with the 1951 legislative session, the Union had its own bus to carry members to Pierre to meet with legislators from their districts and to lobby for the passage of pro-family farm legislation.[67]

A veteran legislator became the South Dakota Farmers Union's man in Pierre in 1951. Roy S. Glover, a native of Brown County, was named the Union's Legislative Counsel, the full time representative for the organization. He became the archetype for future FU lobbyists. Glover had served one term in the State House of Representatives in the late 1920s as a Republican. He returned to the House as a Democrat beginning with the 1935 session and was one of Brown County's representatives for the next eight years. During the 1943 session, he authored House Bill No. 169, which created a division of cooperatives in the Department of Agriculture. He believed that the cooperative movement provided the best mechanism for the small family farmer to compete with expanding corporations.[68] In 1947, he lobbied for the first time as the representative of the Farmers Union.

For twelve years, Glover represented farmers. When the Legislature was in session, he worked out of Room 240 of the St. Charles Hotel. The normal working day for the FU lobbyist began at 8:30 A.M. when he made his way to the State Capitol Building. His assignment was two-fold. First, he studied new bills to determine whether they were favorable or unfavorable to farmers, appeared before committees considering farm-related measures and kept track of upcoming legislation. Second, Glover spent a large amount of time visiting with legislators, soliciting their views and rsponding to questions about how farmers would react to proposed legislation.[69] It was often late in the evening before he returned to his quarters at the hotel.

The lobbyist also had to communicate with members of the Farmers Union. Glover informed them of what was occurring at the Capitol through articles in the *Union Farmer*. During the even-number years, when the Legislature was not in session, he traveled the state to address Union meetings at the local and county level. Glover also worked with membership drives, believing that a larger membership was needed to carry forward the Farmers Union's goals for parity and to maintain the family farm.[70] The 1963 legislative session was Glover's last as counsel. He died of a heart attack in December 1963.

The Union had its share of legislative triumphs and setbacks during the 1950s. In every instance, its representatives were in Pierre to protect the interests of the farmer and to block detrimental legislation. The SDFU disagreed with Governor George T. Mickelson on the sources of revenue to pay a bonus to veterans of World War II. The Governor wanted an increase in the sales tax. The Union's Board of Directors recommended that the first $15 million of the bonus be paid from the state's general fund. The Board believed the remainder of the money should come from a graduated net income tax and a 2 percent increase in the ore tax.[71] Mickelson had his way as the Legislature appropriated $7,800,000 from the general fund. Supplemental revenues were provided through an increase in the sales tax from 2 to 3 percent, an increased tax on alcoholic beverages and cigarettes and a special 2-mill property tax levy.[72]

The Union fought for the rural consumer for over three years when Northwest Bell Telephone Company wanted to increase telephone rates in 1948. The Public Utilities Commission denied the increase, but Bell Telephone, despite the strenuous objections of the Farmers Union, carried the case to the South Dakota Supreme Court. When the justices, by a four to one decision, favored the rate increase, the SDFU, through its attorney Joe Robbie, petitioned for a rehearing of the case. After the State Supreme Court denied a rehearing, Robbie, on behalf of the Union, appealed the case to the U. S. Supreme Court in the fall of 1950. The Public Utilities Commission made no attempt to share in such an appeal. In late winter, 1951, the nation's highest court refused to hear the appeal.[73]

Roy Glover, SDFU Legislative Counsel, 1951-1963.

One other battle that was fought and lost in the statehouse involved a reduction in the ore tax. A bill was introduced in the Senate, during the 1957 session, to reduce the state ore tax by 75 percent. Homestake Mining Company claimed to be in financial difficulties. The only opposition to the proposed reduction came from the Farmers Union.[74] The Senate rejected the proposal, but a bill emanating from the House of Representatives, calling for a 25 percent reduction in the ore tax was passed by the House and the Senate in the last week of the session. Passage came despite the circulation of Homestake's annual report which showed the company made a profit in 1956.[75] The FU succeeded in defeating a bill during the 1959 session that would have provided for an 8.5 percent net income severance tax on ore instead of the prevailing 2.5 percent gross tonnage tax.[76]

The Pietz-Bierwagen bill, referred to as the family farm bill, received avid support from the Farmers Union. The legislation, calling for a graduated land tax in South Dakota, was introduced during the 1957 biennial session. The bill, designed to discourage corporative type farms, was referred to the 1959 session, where it was killed.[77]

One of the Union's most significant legislative triumphs was protecting the integrity of the cooperatives. The well-funded National Tax Equality Association (NTEA) launched an assault on cooperatives in the early 1950s.[78] The 1957 Legislature rejected three bills that threatened the very existence of cooperatives and their educational programs. Led by Glover and Opsahl, the SDFU turned back the NTEA-generated legislation. One of the bills made it a misdemeanor for cooperatives to use the dues check-off system for collecting membership fees. Another of the rejected bills would have required co-ops to distribute all earnings in cash each year, eliminating opportunities for cooperatives to hold some of their earnings and apply them toward expansion.[79]

South Dakota's railroads fell victim to the technological revolution following World War II. By the end of the 1950s, their plight drew the attention of the Farmers Union and the politicians in Pierre. Between 1928 and 1958, nearly 300 miles of railroad were discontinued in the state. In 1961, only three passenger trains were operating regularly. When the Chicago and Northwestern Railroad sought permission from the Public Utilities Commission, in January 1958, to withdraw agency service and depots from sixty-nine towns in the state, President Paul Opsahl strongly objected. He believed the railroad's action would do incalculable harm to South Dakota's economy and that the trucking industry would gain a virtual monopoly over freight transportation in the state.[80]

Available data supported his assumption. The proportion of livestock carried to Sioux Falls by truck increased from 12 percent in 1930 to 98 percent in 1950. In 1960, all shipments of hogs to the Sioux Falls livestock yards were made by truck.[81] Salvaging railroad service in South Dakota became a Union cause during the next three decades. The Legislature, urged on by Governors Harvey Wollman and William Janklow, established a South Dakota rail authority in 1980 to ensure essential rail service.

Other significant changes within the SDFU occurred during the 1950s. There were new programs in education, in labor relations, in lobbying, and in international relations. At the same time, the Farmers Union fought a valiant external battle to maintain price supports against the free market advocates of the Eisenhower administration.

Chapter 12

POLITICS
1948 - 1961

"Wheat prices in a free market would go down faster than a combine cuts wheat."
—*Union Farmer*, 4 July 1955.

Approximately 100,000 farmers gathered at Kasson, Minnesota, on September 6, 1952, to hear speeches by Dwight D. Eisenhower and Adlai Stevenson, the two men vying to be president in the 1952 election campaign. Eisenhower's speech, sometimes called the "golden promise," was filled with ambiguities. A widely held impression was that the key phrase in his speech was:

> I firmly believe that agriculture is entitled to a fair, full share of the national income and it must be a policy of the Government to help agriculture achieve this goal.... And a fair share is not merely 90% of parity — it is full parity![1]

A careful reading of the remainder of his speech clearly indicated that he did not promise full parity. After all, the two-year guarantee of supports for basics at 90 percent of parity was passed in June 1952 over the opposition of a majority of the Republican party in Congress. In the same speech, he savagely attacked present programs "which make the farmer a political captive." The goal of his administration, Eisenhower said, would be programs which "do not regiment you, do not put the Federal Government in charge of farms." He disparaged the principle of price supports by stating that "Farmers would rather earn their fair share than to have it as a government handout."[2] With Eisenhower's elevation to the presidency in November 1952, the Farmers Union began an eight-year struggle to save the remnants of the farm program. Pessimism dominated the farmers' mood during the fifties. In the decade's conservative climate, they experienced another farm recession.

President Eisenhower appointed Ezra Taft Benson as Secretary of Agriculture. Benson, an agent for farm cooperatives and a conservative who had supported Robert Taft for the Republican nomination, was little known on the Prairie Plains. He was not the choice of Union members. They had called for the appointment of Clifford Hope, a Congressman from Kansas, as the new head of the Agriculture Department.[3] The Republican administration believed lower, flexible price supports would discourage production. By 1954, there would be enough wheat in storage bins to supply all of the needs of the market for a full year. The surplus depressed prices, and the government had an ineffective program for selling the

NFU President James G. Patton addressing a local meeting somewhere in South Dakota in late 1940s.

surplus. The President said he would be willing to give it away through school lunches, disaster relief and emergency assistance to foreign countries.

The Farmers Union rallied to save a farm policy that had been in place for twenty years. On June 2, 1953, M. W. Thatcher spoke to a huge rally of more than 3,000 people at the Huron Arena. Challenging advocates of free enterprise for agriculture to debate the issue, Thatcher said farmers must have parity and price support assurance.[4] In July, Secretary Benson announced marketing quotas on the 1953 wheat crop. The current farm law stated that farmers were required to vote on quotas when the supply of wheat was more than 20 percent above normal. Benson said the 1953 crop would provide a supply of 48 percent above normal. Wheat farmers voted to approve or reject quotas by secret ballot on August 14. The Farmers Union mounted a major campaign to get farmers to vote yes as all wheat marketed under the quota system would be supported at 90 percent of parity. If farmers voted no, it would mean encouraging flexible price supports with self-determined controls. A two-thirds vote was required for marketing quotas.[5]

South Dakota's farmers voted 98 percent in favor of wheat quotas. On the national level, 87 percent of the wheat growers voted yes.[6] Farmers would get $2.21 per bushel on quota wheat. Benson reduced the planting allotment for the next year's wheat crop from 77 million acres to 62 million acres, a 20 percent reduction in the acres on which wheat could be sown.[7] Farmers would continue to vote in favor of marketing quotas until 1963.

NFU President Patton warned that prices faced a "pile-driver blow" with the expiration of the Agricultural Act of 1949 on January 1, 1954. Fearing the worst for family farmers, Patton said "powerful combinations of political forces" sought to destroy the last twenty years of agricultural gains. Among these political forces, he included the American Farm Bureau. The Bureau, he said, supported "... new, low parity and sliding price supports...." He accused the Farm Bureau of "spending its time meeting with big business on LaSalle Street in Chicago and on Wall Street in New York City, doing their farming that way.[8]

Nearly 200 young people participating in the youth program at the 1952 State Convention at the Corn Palace in Mitchell.

President Eisenhower announced his flexible price support program for agriculture in January 1954. He proposed sliding scale price supports from 90 to 75 percent of parity beginning in 1955. Throughout much of the summer, a major battle took place in Congress over the parity issue. The Farmers Union campaigned for firm price supports of 90 percent of parity on wheat, corn and other basic crops for the next two years. The Union also favored a food disposal program that would get surplus food into the hands of needy people in the United States and into those areas overseas that needed help.[9] Governor Sigurd Anderson took the side of the SDFU; he favored full parity. He said farmers were entitled to the same subsidies and supports as the rest of the economy. "If you are going to have a tariff for industry," Anderson stated, "then farmers should have it too."[10] Why, he asked, should farmers be absolutely free competitors in a society that has props for most other economic segments.[11] In mid-August 1954, the House and Senate agreed on a 82.5 to 90 percent level of support for basic crops. The President signed the legislation on August 28, 1954. As of December 15, 1954, the parity index stood at 86 percent, a drop of fifteen points in little more than two years.[12]

The Farmers Union worked diligently to get out the vote for the wheat referendum on June 25, 1955. More than forty wheat meetings were held in the state by the FU, GTA and Central Exchange officials. A yes vote would indicate disapproval of the Benson-Eisenhower flexible price support program. A no vote would have meant automatic supports on wheat for 1956 of 50 percent of parity or $1.19 per bushel rather than $1.81 per bushel.[13] With a no vote, farmers could have raised unlimited acreage, but they had to gamble on a free market price. Price supports were favored by 93.4 percent of South Dakota's farmers. Seventy-eight percent of the nation's wheat farmers voted yes.[14]

By this time, the difference between farm prices and manufactured products purchased by farmers had reached their widest spread since the end of World War II. In 1951, the indexes of the two price levels stood just a fraction apart. Farm prices plummeted 23 percent in the ensuing four years while the prices on manufactured products increased a point or two.[15] Farm buying power hit a sixteen-year low on August 15, 1956, as the parity ratio stood at 82 percent.[16] Between 1947 and 1957, retail food prices rose 16 percent, while prices paid to farmers fell by 14 percent.[17] Mrs. Erick Johnson of Trail, Minnesota, composed "A Farmer's Lament." Verse seven of her pastoral poetry expressed the Union's sentiments regarding Secretary Benson. Mrs. Johnson wrote:

> But Ezra sits in his office so bold; and "90 percent" he does turn cold. Flexible price supports he wants most, And of his wonderful policies he does boast.[18]

Eisenhower reintroduced an old New Deal program in 1956, when he advocated that the Soil Bank be reinstated in an effort to reduce the surplus. There had been a sudden drop in export demand following the Korean War. The Soil Bank Act, a massive soil conservation program, enabled farmers to lease their land to the federal government on three to ten year contracts, taking the land out of production. They would otherwise plant wheat or corn, and the government would pay them in either cash or kind the value of their probable yield. Benson mustered no enthusiasm for the proposal, as farmers would be paid for not producing. The President had wanted the Soil Bank tied to flexible price supports. The Democrats balked at this because of the depressed economic conditions in the countryside. Congress passed a farm bill that tied the Soil Bank to 90 percent of parity for all basic crops. Eisenhower vetoed the bill on April 16, 1956, arguing that rigid supports had created the surplus problem.[19] The President would get his Soil Bank, 90 percent of parity was rejected, and the surplus of agricultural products continued to mount. Nationwide, 28.6 million acres were removed from production because of the Soil Bank legislation. Many of the farmers who entered the program never returned to farming. Technology, however, grew so rapidly that the program did not lead to a decline in production.

The Eisenhower administration also tried to reduce surplus commodities through a food stamp program at home and Public Law 480 for countries in need of food but unable to pay for it in hard currency. One of the positive benefits of Public Law 480 was that many of these countries became customers for U.S. agricultural products as their economies developed.

M. W. Thatcher blamed Benson for the mounting surplus. In January 1953, all agricultural products in government storage totaled about $1 billion. As of July 31, 1956, produce in storage amounted to about $6 billion. The general manager of the GTA said the Administration had forced farmers to produce more to stay in business. According to Thatcher, Benson should have cut production when he took over.[20] The Secretary of Agriculture did set acreage allotment at the lowest point allowable by law, 55 million acres, in 1956.[21]

Five irate South Dakota farmers vented their frustration with the Administration's farm policy on October 10, 1957. They threw eggs, "hen grenades," at Secretary Benson before he addressed about 1,000 people at the state corn picking contest south of Sioux Falls. They wanted Benson fired plus firm price supports. Benson ducked the barrage before he began to speak. Each of the egg throwers was fined $50.00.[22] The following day, Paul Opsahl issued a news release expressing regret over the incident. He said the SDFU works for farm parity through "democratic means" and not violence. He went on to say, "Throwing eggs at a Secretary of Agriculture is not the way South Dakota farm families meet the problems of agriculture." "Regardless of how we may feel about the way Secretary Benson has done his job," Opsahl added, "there is no excuse for such violence as occurred yesterday."[23] Privately,

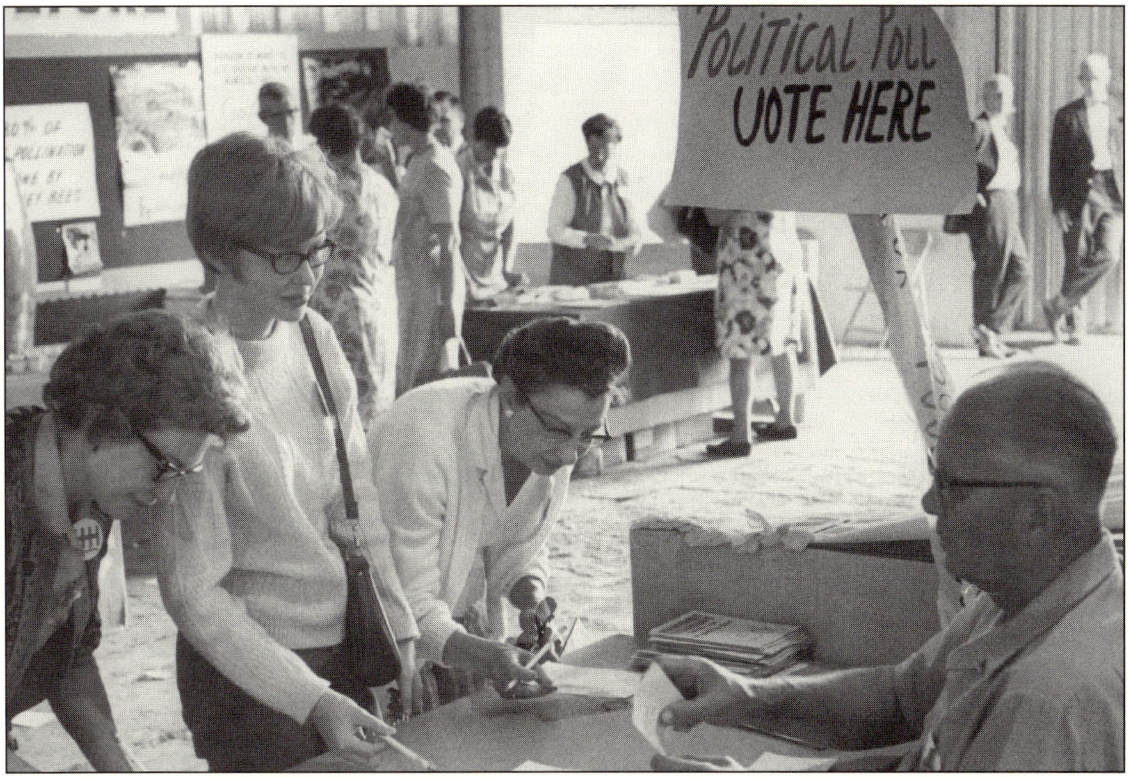

Polling in the SDFU booth at the State Fair in Huron.

Union leaders expressed amusement over the incident. Loriks wrote to Art Stern of Emery, "Were you down to Benson's egg party? Too bad they wasted fresh eggs, some one remarked. This will be remembered much longer than Benson's speech!"[24] In correspondence with Opsahl, Loriks said that the farmers had dramatized the importance of a farm commodity — eggs.[25]

When the 1958 farm bill came up for discussion, farm-state Republicans joined with Democrats to offer a bill that maintained high federal price supports for one year. Some Prairie Plains Republicans urged Eisenhower to support the proposed legislation and to fire Benson. In late March, the bill passed both Houses of Congress. Eisenhower reluctantly vetoed it, and he refused to dismiss the controversial Secretary of Agriculture. The Agricultural Act that finally became law on August 28, 1958, modified price supports on basic crops for 1959 and 1960. It gave the farmers a choice between modified price supports and an increase of crop allotments. The price support for some crops was reduced to 65 percent of parity.

Approximately a month before the 1958 farm bill became law, delegates and economists from fifteen nations met at South Dakota State College to develop a plan for the better distribution of surplus wheat throughout the world. SDFU President Paul Opsahl told the delegates that there was no surplus when there were people in the world who did not have enough to eat. If the produce of American farmers was distributed effectively, political tensions in many parts of the globe would be relieved. In general, the conferees approved of the policies carried out under Public Law 480, which spelled-out methods for the disposal of surplus wheat.[26] Federal methods used to boost the export of wheat included donations overseas, sales for dollars at modified price schedules, taking foreign currencies in payments and bartering for defense materials. Even though the Farmers Union would have preferred a five year extension of P. L. 480, the organization's leaders praised the eighteen month extension authorized in 1958.[27]

The irony of the Eisenhower-Benson farm legislation was that none of the measures worked. They had hoped to get the government out of farming and to reduce the surplus. The government, in 1961, was more closely involved in agriculture than it had been in 1950, and the amount of money spent to counter a growing surplus also increased. At the same time, the farmers' purchasing power fell, and the number of farm families dropped precipitously. Farm numbers dropped 1.7 million (31.1 percent during the 1950s) and the farm population declined by about 8 million people. There were 66,452 farms in South Dakota in 1950. Average acreage per farm was 674 acres. Nine years later, the number of farms in the state had decreased to 55,727 while the average farm size had increased to 805 acres. Secretary Benson's personality, adverse policies and seeming disregard for the family farmer, forced many Union members, who professed nonpartisanship, into more of a political stance. Some of this feeling was found in "The G.O.P. Psalm," which circulated in Union ranks in 1956. It read:

> Eisenhower is my shepherd. I am in want.
> He leadeth me through still factories.
> He restoreth my doubt in the Republican Party.
> He guideth me in the path of unemployment
> For his party's sake.[28]

John F. Kennedy's triumph over Richard Nixon in the 1960 presidential election heartened many Union members. Kennedy, during the pre-election campaign, in his pronouncements on agriculture, had used Farmers Union sources.[29] His appointments of Orville Freeman as Secretary of Agriculture and George McGovern as director of the Food for Peace program were viewed with optimism. Freeman was a longtime foe of Benson's policies.[30]

BUS TRIPS AND LOBBYING

From the outset, the Farmers Union lobbied legislative bodies, believing that lobbying was the only way to get what the membership wanted. President James G. Patton said, "We live in a legislated economy" and that "more farm prices are made in Washington than are made in all the market places in the United States."[31] Under President Charles S. Barrett. a powerful lobbyist, the Union was very successful in its lobbying efforts through 1915. During the 1950s and early 1960s, the Farmers Union's influence in the nation's capital was as great as in Barrett's time.[32]

Paul Opsahl played a key role in strengthening the lobbying efforts of the SDFU in Washington, D. C. Approximately two and one-half years after his election, the Board of Directors, on March 20, 1950, authorized the purchase of a thirty-seven passenger bus.[33] With the acquisition of the bus, the SDFU hoped to cut the costs of transporting lobbyists to Washington by 50 percent. The NFU had begun a grassroots lobbying program in 1947, using car caravans. In 1949, the SDFU sent its first contingent to Washington, spending more than

The SDFU purchased a 37-passenger bus in 1950. Pictured are State Secretary Emil Loriks, President Paul Opsahl and Education Director, Hazel Brekke.

$13,000 on the caravans that made the journey in January and February.[34] In 1951, a smaller bus was added to the Union's transportation service. The smaller vehicle provided continuity when road limitations, particularly in the spring of the year, prevented the large bus from operating.[35]

By 1959, more than 1,000 of the SDFU's membership had participated in the lobbying effort in Washington. Many of the lobbyists also went to New York City and toured the United Nations' facilities. Customarily, the participants drove straight through from South Dakota to the nation's capital. Headquarters for several years was the Dodge Hotel. On the first day, the Dakotans visited historic sites, including Arlington National Cemetery, Mt. Vernon and the Washington National Monument. Back at the hotel that evening, the group received a briefing on the important farm bills before Congress. Usually a member of the SDFU Board of Directors, Harry Caslin, LeRoy Hardy, Norman Olsen, et. al., served as the liaison person. The next day the group divided into teams of two, and each team was assigned to visit particular Congressional offices. On one trip, Henry Bell of Flandreau teamed up with Harry Caslin, and they made fourteen calls.[36] According to Norman Olsen, a team often met with the clerks or aides to Congressmen, who relayed the gist of the visit to the Senator or Representative. Before returning to South Dakota, the grassroots lobbyists were usually briefed by one or two people from the Department of Agriculture.[37]

The Washington trips were usually three or four day affairs. The SDFU conducted several bus tours in the two month period. Union officials believed that the greater the number of farmers who showed up in Washington, the greater the publicity and the greater the impact on members of Congress. Throughout the fifties, the lobbyists tried to convince Congressmen that the price support program should not be scuttled by the use of flexible price supports. Mrs. I. Ray Howard of Huron said her feet grew to a size nine because of all that "pounding the pavement" for parity.[38]

Bus-In to Washington, D.C., February, 1953.

The long and hurried trips were not devoid of humor. One group on their way to Washington formed South Dakota Traveling Local No. 1. Another group conducted light-hearted quizzes concerning the Farmers Union, awarding candy bars to the winners. The 1955 lobbyists laid down several rules, including "No snoring ever, nor talking in your sleep." Drivers were instructed "not to wink at ladies, otherwise, they would be responsible for a cup of coffee at the next stop."[39]

Overall, the Washington caravans had a desirable effect. The Union took part in an effective delaying action of the Eisenhower administration's effort to quickly eliminate price sup-

Bus-In participants. SDFU Secretary Emil Loriks, in front row, 4th from left, was the trip leader.

ports. Ardent supporters of the St. Lawrence Seaway, the SDFU began to pressure its Congressmen to support the project in 1947. The Seaway would lower freight rates, improve grain markets, make more box cars available and contribute to national defense.[40] In 1954, Congress passed legislation to begin the St. Lawrence Seaway project, resulting in the construction of a canal between Montreal and Lake Erie. At its completion, farm produce would be conveyed by water transport from Duluth, Minnesota, and Superior, Wisconsin, to the Atlantic Ocean.

The SDFU did approve of one of Benson's actions. In 1954, he issued an order divorcing all government agricultural agencies from all farm organizations. The order had a major effect in Illinois and Iowa where extension agents continued to organize for the Farm Bureau. As noted in chapter six, the South Dakota Legislature, in 1937, cut the bonds between the Extension Service and the Farm Bureau.[41]

In its relations with Congress and the President, the Union strove mightily to fulfill its basic aims — 1) maintaining the family farm, and 2) insuring the security of family farmers in their income, tenure and happiness.[42] NFU President Patton believed these objectives could only be accomplished in a peaceful world, and the United Nations was the key to a peaceful world.[43]

THE FARMERS UNION AND THE UNITED NATIONS

In contrast to a quarter-century earlier, international affairs became an important continuing preoccupation of the Farmers Union after World War II. Involvement rather than isolation became the best solution for securing peace. The proper foreign policy for the United States was to become involved through the United Nations. Paul Opsahl said, "We are all looking to the day when the United Nations will be the dominant power in the world for permanent peace."[44]

The Union continued to oppose any military action or alliance in which the United States acted unilaterally. Thus, it opposed the North Atlantic Treaty Organization, the Greek-Turkish aid program and universal military training. The outbreak of the Korean War in June 1950 was another matter. In supporting American involvement, NFU President Patton said that "the central act is that the flag of the United Nations is flying in a war against aggression." He had some reservations about the government of the Korean Republic, but "the fact remains that South Korea was invaded and that the United Nations clearly has stamped that action as an aggressive action...."[45] Soon after the outbreak of the Korean War, Opsahl stated that the Farmers Union opposed communism, fascism or totalitarianism in any form.[46]

American military involvement on the Asian mainland caused the Truman administration to seek means to strengthen the country's armed forces in the shortest time possible. A Universal Military Training bill was introduced in Congress in 1951. It would have subjected all able-bodied youths to six months military training as soon as they reached age eighteen. Following their training, they would be placed in reserve units and subject to recall until they reached age twenty-six.[47] The SDFU took a strong stand against Universal Military Training. The Union believed it would have meant a permanent military bureaucracy, something inimical to democracy.[48]

The American Association for the United Nations (AAUN) was established to inform the American people of the beneficial nature of the United Nations. The Farmers Union became a staunch supporter of the AAUN. South Dakota, in 1954, became the thirty-seventh state to form a United Nations Association. Paul Opsahl and August Dahme were South Dakota AAUN board members during the 1950s. Eleanor Roosevelt, a major supporter of the UN since its inception, came to Huron, on April 21, 1954, to promote the AAUN. Speaking to an estimated crowd of 3,500, at the Huron Arena, she gave her audience an elementary lesson in human relations and the meaning of the United Nations. Mrs. Roosevelt was accompanied by Clark Eichelberger, the executive director of the AAUN. He conducted a series of workshops following the former First Lady's speech.[49] On the morning of April 22, Mrs. Roosevelt and Dr. Eichelberger visited the state offices of the Farmers Union. After being made honorary members of Headquarters Local 1047 of the SDFU, they had coffee and toured the building.[50] In the ensuing years, an AAUN luncheon, with a guest speaker, was a part of the annual state convention. Information about the United Nations became an integral part of the Union's youth education program.

In addition to its support of the UN, the Farmers Union repeatedly reaffirmed its support of long-term foreign economic aid, a stronger Food and Agricultural Organization and a world food bank, with distribution of crop surpluses to underdeveloped areas. The Union's new internationalism was also displayed through its membership in Farmers and World Affairs, Incorporated, and the International Federation of Agricultural Producers, a world organization of farm groups.[51] From extreme isolationism in the 1930s, the Farmers Union, by 1961, had swung far in the opposite direction. The 1961 NFU program stated that the organization "will continue to strive for the earliest possible attainment of a democratic world brotherhood of nations...."[52]

While the Union pursued peace and brotherhood on the international scene, all was not tranquil on the state scene in 1958. A hotly contested election for the presidency of the SDFU followed President Paul Opsahl's announcement that he was stepping down after eleven years at the state Union's helm.

POLITICS — UNION STYLE IN 1958

Paul Opsahl had his critics within the SDFU by 1957-1958. There were complaints about favoritism, improper record keeping at the state office and Union expenditures.[53] Former President John Skage was one of the leading dissidents. He and Opsahl had disagreed on many issues since Opsahl became president in 1947. Skage had his eye on regaining the leadership of the SDFU, and he began to work toward that goal in 1957. In his correspondence with other Union members, he accused Opsahl of breaking promises. In the summer of 1958, Skage maintained a steady stream of correspondence with Emil Loriks, who became the SDFU's elder statesman, on what he believed to be Opsahl's perfidy.

Skage said Opsahl had promised him that he would announce in 1957 that he would not be a candidate at the 1958 convention, and that he would take no part in the selection of his successor. He did neither. A sympathetic Loriks wrote to Skage on July 30, 1958, informing him of Opsahl's plans. Loriks wrote, "Apparently P. W. [Opsahl] has decided to step out as President of the Farmers Union at the next convention. However, it appears he is going to use the *Union Farmer* to promote the candidates of his choice for President and Vice President, i.e. Golseth and Radcliffe respectively. This may be in the next issue. While I reminded him of his last year's pledge to you, he obviously has had another lapse of memory."[54]

Opsahl's announcement was in the late July issue of the *Union Farmer*. He endorsed Harold Golseth, who had been his vice president since 1949 to succeed him. Golseth indicated he would run, and Ben Radcliffe, the Board representative from District VII, said he would be a candidate for vice president.[55] August Dahme, who had lost to Golseth in 1949, wrote to Loriks a week later and asked, "When are the fellows [the opponents of Golseth and Radcliffe] going to make their move?"[56]

John Skage formally announced his candidacy in early September. His running mate was Norman E. Olsen of Canton. Skage, at that time, continued to manage the highly successful Deuel County Farmers Union Oil Company. Its volume of business was nearing a million dollars, and it checked-off more than 700 membership dues to the Farmers Union annually.[57] Olsen had been District I's representative on the Board of Directors for seven years. He was on the Farmers Union Livestock Commission boards in Sioux City and Sioux Falls, having served on the latter board for seven years. He also had been president of Dayton Local for five years.[58]

Accusations and rumors abounded in the weeks before the annual convention. Skage's partisans attacked Golseth's voting record as a state senator. They claimed he was anti-labor, going against the Union's emphasis on farmer-worker cooperation.[59] Golseth's supporters said that Norman Olsen did not patronize Farmers Union retail outlets. The charge that he was allied with the CCA was quickly rebutted by Olsen's supporters in Lincoln County.[60]

President Paul Opsahl meets with the Board of Directors in 1951. L to R: Harry Caslin, District IV; Fred Ferwerda, District V, Emil Loriks, State Secretary; President Opsahl; Harold Golseth, Vice President; Sam Ulrickson, District I, Henry Vinz, District II; Norman Olsen, incoming District I director. District III Director Loyd Larson is not pictured.

The most contentious issue in the election was the role of the Farmers Union Central Exchange's fieldmen. Arthur Sterns of Emery accused the fieldmen of associating Olsen with the CCA. In a letter to Loriks, Sterns wrote, "You can tell by the odor of that sort of propaganda where it comes from." He went on to say that field representatives of a certain regional were busy disseminating this stuff.[61] Skage protested the fieldmen's actions to Thomas Steichen, the general manager of the Central Exchange.[62]

On the day when balloting took place at the Huron Arena, Skage charged that the Exchange's fieldmen were at each door of the convention hall, buttonholing delegates as they entered, urging them to vote for Golseth. Skage said the fieldmen opposed the checking-off of a cooperative member's dues.[63] The team of Golseth and Radcliffe emerged victorious from the election. There were several new faces among the Board of Directors. In District I, Arthur Stern replaced Norman Olsen. R. V. Fitzgerald of Plankinton replaced longtime Board member Henry Vinz of Spencer. Roy Nelson of Bowdle took Ben Radcliffe's place in representing District VII. The returning directors were Loyd Larson, Harry Caslin, Frank Butler and LeRoy Hardy.[64]

Immediately after the election, Loriks wrote to Skage, expressing his displeasure with the role played by the Central Exchange in the election. The first three paragraphs of the letter he wrote on October 10, 1958, expressed his concerns. They read as follows:

> Dear John:
>
> Just re-read the Convention statements in behalf of you & Norman, and I trust many of the members and delegates got them and preserved them

> for future reference! Possibly it takes awhile for the truth to sink in, and that all they now have is a Company Union!
>
> GL [Moseson] is the symbol of that - and so long as the FUCE continues him in his present exalted position, that conviction stands! Steichen promised non- interference a year ago, and Steichen reaffirmed this position now. Continuance of GL brands all these assurances as false!
>
> I believe there are ways of impressing these facts on the FUCE in a manner they will understand without injury to the F.U. Failure to arrive at a policy of non-interference, can easily mean damage to both. The sooner this is understood the better.[65]

The controversy surrounding the 1958 election had its roots in the past. Employees of the Central Exchange had played a role in Fosheim's narrow victory over Skage in 1944. Dahme had lost the vice presidential election to Golseth in 1949. The issues that divided South Dakota's Union members in 1958 centered more on personalities than on major philosophical differences. Thus, the Union did not experience the lasting animosity that accompanied the struggle between the radicals and progressives during the 1930s. Following the 1958 election, Union members quickly returned to advancing the cause of the family farmer in a time when more and more people were leaving the land.

Chapter 13

PROMOTING GROWTH AND UNDERSTANDING 1948 - 1961

Farmers Union education is designed to give its members the knowledge they must have to make economic and political democracy really work.
— James G. Patton, 1952.

In his final presidential report, Paul Opsahl summarized the significant actions of his administration. Some of these accomplishments were recounted in Chapter Twelve. Additional achievements of his administration and that of his successor, Harold Golseth follow. These gains included closer relations with urban labor, an annual livestock marketing day, strengthening the insurance program, and enlarging the scope of the state workshops.

THE FARMER AND THE URBAN LABORER

The first farmer-worker conference took place at Huron on March 26-27, 1954. The theme of the session was "Agriculture, Labor and the Future — Food, People and Resources." Realizing that the farmers and urban laborers had similar economic and social problems, the goal of the meeting was to promote unity and understanding between the two groups. Albert Maag of Huron, state president of the American Federation of Labor, played a prominent role in the conference.

An effort was made, in each ensuing year, to bring in at least one speaker from a major metropolitan area to represent labor. Mark Starr, Education Director of the International Ladies Garment Workers of America, from New York City, spoke to the 200 conferees in 1954.[1] One of the most prominent people to address a Farmer-Worker Conference was Dr Leon Keyserling, the former chairman of President Truman's Council of Economic Advisers. He spoke at the fourth annual conference in 1957.[2]

The SDFU worked with the state's labor organizations in attempting to repeal the right-to-work law. The policy program for 1959 called upon the Legislature to protect the right of laborers to join unions of their choice and to bargain through their unions.[3] Although they had the ear of a sympathetic governor, Democrat Ralph Herseth in 1959 and 1960, this effort came to naught. One of the results of the annual Farmer-Worker conferences was the creation of a regional Farmers-Workers Association on July 21, 1961. The Association included FU members and urban workers from Clay, Lake, Lincoln, Minnehaha, McCook, Moody, Turner, Union and Yankton counties.[4]

Farmers Union Market Day at the Sioux Falls Stockyards (1951).

THE COOPERATIVE TERMINAL MARKET DAY

In June 1948, the Board of Directors approved a Cooperative Terminal Market Day for the Sioux Falls livestock yards for October 13, 1948. Designed to accommodate Union youths who raised baby calves for sale, the Co-op Marketing Day became a valuable educational program for displaying and explaining terminal livestock marketing.[5] All participants were invited to consign one or more head of livestock to the Farmers Union Livestock Commission in what became an annual event. The program gave sellers an opportunity to follow the steps in marketing — 1) handling livestock on the terminal market, 2) selling them to the highest bidder, 3) weighing the livestock under government supervision and 4) completing the sale. During the course of the day, Juniors and Junior Reserves listened to a guest speaker and participated in demonstrations on such topics as the grading of beef and pork on the hoof plus learning how to select and cut different types of meat. The day usually climaxed with a free picnic lunch.[6] Between 150 and 200 young people registered for this event each year during the 1950s.

The annual Market Day received a boost during the second year of Harold Golseth's administration. It was announced, in January 1960, that Juniors and Junior Reserves could secure loans for the purchase and feeding of beef cattle. Interested youngsters obtained loans from the FU Agricultural Corporation, a subsidiary of the Farmers Union Livestock Commission at Sioux Falls. Conditions for securing the loan included: 1) A parent or legal guardian must co-sign the note that required 6 percent interest. 2) A youth might secure one to five head of beef cattle with the assistance of the FU Livestock Commission or an authorized person from the young person's locality. 3) The livestock must have proper care. 4) Livestock secured through loans had to be sold through the FU Livestock Commission at the annual Market Day in August. 5) The person receiving a loan had to keep accurate cost records on the livestock.[7] The loan program gave Union youths practical experience in feeding and marketing cattle as well as an incentive for personal financial gain.

INSURANCE AND MEMBERSHIP

Following the licensing of the National Farmers Union insurance service in South Dakota in July 1948, Paul G. Erickson had become the manager of the newly founded South Dakota Farmers Union Service Association, and he guided the company through its first thirteen years of existence. A Farmers Union member for thirty-five years, he had begun his work on the state level in 1936, when he was hired as education director. Erickson sold his farm in Tripp County in 1937, and at that time, added to his duties, being appointed secretary-treasurer for the SDFU.[8]

In addition to providing insurance coverage, the state agency helped to defray costs of the educational program by stabilizing membership through its check-off requirement for policy holders. Beginning on April 1, 1952, membership in the Farmers Union was required of all policy holders. The requirement, approved at the annual meeting of NFU Insurances' stockholders, brought dues for 1,600 members to the state office, a significant increase in membership.[9]

The strengthening of the insurance program played a key role in building a stronger Farmers Union. Each program of the Union reinforced the other programs. For example, if there had been no insurance program, there might have been no fertilizer program. Agents explained to farmers that to invest in FU insurance was to invest in all of the objectives of the Farmers Union.

Auto and truck insurance led in volume during the fifties. After the South Dakota Legislature enacted a Financial Responsibility Law in 1953, there was an increased demand for auto insurance. Midway through the decade, the SDFU Service Association sold auto, life, crop-hail, hospital, school accident, farm liability and property insurance. There were 203 part-time insurance agents throughout the state, excluding the office staff. In December 1954, a new claims office was opened at the state headquarters.[10]

President Opsahl reported that on June 30, 1957, paid-up life insurance for Farmers Union members in South Dakota totaled $5,950,136. The life insurance was carried by about 4,500 policy holders. Casualty insurance was held by approximately 13,000 policy holders with premiums totaling $700,000 on June 30.[11] Sales drives for life insurance had a special association with Farmers Union history and leadership. One such drive was staged on August 2, "Founders Day," observing the founding of the Farmers Union on August 2, 1902, at Point, Texas. Beginning in 1957, a two-week campaign was conducted in early November, celebrating NFU President James G. Patton's 55th birthday. In that period of time, and in future observances of Patton's birthday, the insurance agents were known as "Patton's Platoon." Patton had begun his Farmers Union career as an insurance agent.[12] In 1961, South Dakota's agents wrote $487,500 of life insurance during the annual two week drive. Writing 30 percent of the total national figure, South Dakota's Union finished ahead of the other state Unions.[13]

A Winter Workshop (1948). Participants included from the left: President Paul Opsahl, Omaha Bank for Cooperatives Vice President, Wilbur Thompson, Friends Service Committee Legislative Representative Chester Graham, SDFU Education Director Hazel Breeke and NFU Fieldman Stanley Vogt.

Fire insurance transactions were moved from the national office at Denver to the SDFU headquarters at Huron in June 1959. South Dakota became one of three states where Farmers Union insurance agents did their own underwriting. With the consolidation of all FU insurance activities at the Huron office, the Service Association provided more efficient and faster service to policy holders.[14] On September 1, 1961, Paul Erickson retired as manager of the FU Service Association. As the only manager of the Association since its inception, he oversaw the movement of the firm's records from Mitchell to the new state office building at Huron in 1952. Between 1951 and 1961, the Service Association had experienced a ten-fold increase in its business volume, making South Dakota the fourth ranking insurance state for the National Farmers Union. The sixty-four year old Erickson was replaced by Robert Kudrna, the district manager for Farmers Union insurances in southwestern North Dakota.[15]

THE MIDWINTER EDUCATION CONFERENCE

The state workshops, that began in 1948, became the outstanding event between the annual state conventions. The event soon became known as the Midwinter Education

Conference. At first, a three day session was held at Huron. In 1951, the theme was "Unity in Democracy."[16] What began as one conference, soon became a series of conferences, held at different locations throughout the state in January and February. Themes often related to the current status of agriculture. Responding to the assault on parity and the continued decline of the farmer's purchasing power, the theme for 1958 was "The Fix We're In and What Do We Do Next."[17] In 1956, there were thirty-two area conferences throughout the state. Attendance at the conferences averaged between seventy-five and one hundred people. At every session, special teams composed of state office staff acquainted the membership with the total Farmers Union program.[18]

GRASSROOTS INVOLVEMENT

Three other means were used to involve members at the grassroot level. Beginning in the winter of 1955, county FU presidents were called to Huron for a two day conference, at which staff members briefed them on various aspects of the Farmers Union program. Emphasis was placed on how state personnel could be used at the local level and how programs could be made more effective.[19] Secondly, through an amendment to the SDFU's bylaws, delegates to the 1957 state convention designated county presidents to serve as an advisory committee to the state Board of Directors. In this new role, the presidents would participate in formulating the policy program for each year.[20] Finally, the "action official" program was inaugurated in 1954. The program recognized those Union members in South Dakota who actively carried out their official duties during the year. For stimulating interest and attendance at local meetings and for contributing to increased membership and an expanded educational program, action officials received awards at the state convention. Beginning with thirteen award winners in 1954, the number of honorees tripled within two years.[21]

GOLSETH'S SHORT ADMINISTRATION

As vice president of the South Dakota Farmers Union since 1949, Harold Golseth shared in the achievements of the organization in the 1950s. In becoming the ninth president of the state Union in 1958, he had the opportunity to lead the farm body into the next decade. Unfortunately, he became the only SDFU president to die in office.

Harold Marvin Golseth was born on November 27, 1910, at Hayti, South Dakota. When he was two, his father, a store owner, moved the family to a farm seven miles northwest of Erwin, South Dakota. After graduating from Erwin High School in 1927, he enrolled at Gustavus Adolphus College at St. Peter, Minnesota. He graduated with a major in liberal arts. He returned to the family farm in 1942 and resided there until his election as president of the SDFU. Golseth served as a director and as chairman of the Farmers Union Oil Company at Lake Preston. He represented Kingsbury and Hamlin counties as a state senator during the 1947-1949 and 1953-1955 legislative sessions.[22]

His style of leadership was quite different from that of his predecessors. He was a quiet and modest person. Knowledgable and perceptive, Golseth engaged in thoughtful deliberation before taking action on administrative and policy matters. He could make difficult decisions when necessary.[23]

Harold Golseth,
President of SDFU, 1958-1961.

Ray Fitzgerald,
State Secretary of Agriculture.

There were few changes in the administrative team during his short tenure. In mid-1959, the membership in District II elected Ewald Osterberg of Salem to replace Ray Fitzgerald, Jr., of Plankinton on the state Board of Directors.[24] Governor-elect Ralph Herseth had named the thirty-five year old Fitzgerald as State Secretary of Agriculture in December 1958. A native of Aurora County, Fitzgerald graduated from Xavier University, in Cincinnati, Ohio. He served for three years in the army, engaged in the implement business and farmed prior to his appointment by Herseth.[25] Herman Pietz of Parkston represented District I after the 1960 convention. Mrs. Gene Olson, the SDFU secretary-treasurer for six and one-half years resigned as of July 1, 1959. She was replaced by Mrs. Joy L. Johnson, a lifelong resident of Huron.[26]

The South Dakota Farmers Union held its 44th annual convention at the Huron Arena in mid-November 1959. Attendance was reduced as the meeting opened with a snowstorm. It was eighteen degrees below zero when the delegates met on Saturday morning. After enduring such conditions, the decision was readily made to return to an early October meeting in 1960.[27] The county councillors recommended a significant change in the by-laws at the 1960 convention. Starting in 1961, the president, vice-president and the Board of Directors of the SDFU would be elected for a two-year term rather than one.[28]

The State Union had mixed returns on the legislation it supported during the 1959 legislative session. As indicated earlier, the Farmers Union was able to stop any change in the conditions for the collection of the ore tax. Three other recommended pieces of legislation were approved by the Legislature. They included: 1) the creation of a twenty-one member citizen's tax study committee, 2) the exemption of LP gas from sales tax when used for non-highway purposes and 3) a water conservancy district bill. A proposal to reorganize the State Department of Agriculture was rejected.[29] Wanting to claim a portion of the Missouri River water impounded in South Dakota, the SDFU, at its 1960 convention, went on record in support of water conservancy subdistricts.[30]

Ewald Osterberg, Salem, represented District II on the SDFU Board of Directors from 1959-1974.

Teresa Schlenker, the singing cowgirl. She was heard on Saturday afternoons over KWAT radio, Watertown.

For the first time in its history, the state Union sponsored a television program, advertising the benefits of cooperation, fair farm programs and the interdependence of farmers and city people. The first show aired over KSOO-TV, Sioux Falls, on July 18, 1960. At a later date, the same program was transmitted over KORN-TV, Mitchell. Co-sponsors with the SDFU of the evening news were the FU Livestock Commission of Sioux Falls and the Farmers Union Grain Terminal Association.[31] A popular SDFU radio program between 1949 and 1960 was broadcast over KWAT-Watertown on Saturday afternoons. The program, hosted by Mabel Stormo, featured Teresa Schlenker, the singing cowgirl.[32]

Golseth, who had had a mild heart attack in the early 1950s, suffered a coronary occlusion while attending the national convention at Springfield, Illinois, in March 1959. After several weeks of convalescence, he fully resumed his presidential duties, and just prior to the next national convention, at Denver, he was elected to the nine man executive committee of the board of directors of the NFU.[33] Golseth appeared in good health, when he died of a heart attack while having dinner with friends at a farm home near Huron on January 22, 1961.

One of his administrative tasks had been to find competent people to head the State Union's education program. Between late 1956 and October 1959, several people served briefly as education director or Junior leader, creating a brief period of instability in one of the Union's most important programs.

SDFU'S EDUCATION PROGRAM BECOMES INTERNATIONAL

In the spring of 1948, 1,732 youth were enrolled in the Junior program of the South Dakota Farmers Union. The educational program continued to expand at the Junior and adult levels. Five people were employed for purposes of educational organization and cooperative work. John Sieh, the Junior leader, was reassigned as one of the fieldmen. In his new position, he continued to work with young people as he had charge of the FU youth camps.

Hazel Brekke, the Director of Education, assumed Sieh's responsibilities as Junior leader.[34] She relinquished the latter position to Mrs. Henry (Leona) Ravenberg of Canova on January 1, 1950.[35] As the Junior leader, Mrs. Ravenberg became the field worker for the Education Department, speaking to locals, providing them with information and helping them to organize activities. She became the Director of Education on January 1, 1951; the preceding December, Hazel Brekke had resigned because of illness in her family.[36]

Youth activities were only one part of the education director's responsibilities. The director also facilitated the distribution of publications and arranged for and spoke at many educational meetings that were held annually throughout the state. These meetings included the mid-winter area conferences in January and February, the annual Farmer-Worker Conference in early spring and county rallies. During the course of the year, the director would attend two or three national conferences. The education director received assistance in the planning of educational work from an educational council, consisting of the educational directors of the seven districts (five districts prior to 1954) and two junior members selected from the torchbearers of the year. The educational council met at least twice a year to plan youth activities and camps and to outline needed adult work in the respective districts. A film library was begun in 1956 under the direction of Mrs. I. Ray Howard. Within a few months, approximately fifty films were available for use by the local, county and district Farmers Union organizations.[37]

The bulk of the director's time was spent in conducting an expanding youth program. In 1955, twenty-five years after NFU President C. E. Huff appointed a committee to coordinate the Junior program, 12,215 membership cards were issued to the SDFU's youth. Cards were awarded to Juniors and Reserves in the following classifications:

4042 cards to voting Juniors (ages 16 - 21 years)
1,655 cards to non-voting Juniors (ages 14 - 15 years)
6,518 cards to reserve members (ages 6 - 14 years).[38]

Cooperatives continued to make significant contributions to the educational fund. Funds for 1956 totaled $78,581.52, an increase of $8,000 over the previous year. The GTA and the Central Exchange contributed the majority of this amount, $68,214.71, with local co-ops contributing the remaining amount of $10,366.81.[39]

Despite the impressive numbers, Mrs. Ravenberg informed delegates to the 1955 convention of certain weaknesses in the education program. Too many locals were inactive or holding poorly attended meetings. There was growing competition from other activities. Television was one factor contributing to poor attendance at local sessions. Greater attention had to be given to the youthful members of the organization. She concluded her remarks by saying that "The local with a lively youth group is an active local."[40]

The Junior and Reserve programs in the 1950s included a study topic and a workbook. Completion of the book assigned for the year, attendance at four or more FU meetings, and participation in two special activities constituted the requirements for achievement awards. The most significant of these awards was a $1,000 insurance policy provided by the National Farmers Union. Other awards included camp scholarships and special pins.[41]

SDFU youth contributed to the financial well-being of the organization. In the early 1950s, earnings from the camp co-op store were given to help pay for the new state office building. Campers donated money to place a flagpole on the state office grounds and to help

Family campers at Nemo, July 1-5, 1951.

pay for the new Farmers Union busses.[42] A used piano was purchased by the Juniors and Reserves for the auditorium in the state headquarters.[43]

A cooperative grain project for FU Juniors was introduced in 1949. Originating in the Nunda area, its purpose was to teach young people the value of buying and selling through a cooperative market. Youths who took part in the project planted and cared for a specified acreage of grain. Participants were required to keep records of all purchases, work, seeds and supplies, sales and income. The grain produced had to be sold through the sponsoring co-op elevator. The Juniors who took part in the program received a three day tour of the Union's regional cooperatives at St. Paul, Minnesota. The bus tours were sponsored by South Dakota's cooperatives.[44]

Shirley Dickinson of Greeley, Colorado, succeeded Leona Ravenberg as Director of Education on December 17, 1957. Within ten months, she had resigned to work on a master's degree in social work at Denver University. Before relinquishing the directorship on September 6, 1958, she had established a South Dakota Junior Advisory Council. The council was formed to provide talented young people with an opportunity to develop leadership ability.[45] Dickinson was replaced by Mrs. Berdyne Halse of Clear Lake, South Dakota. A former teacher in northeastern South Dakota, she and her husband, Lloyd, had farmed and operated a dairy and locker plant at Clear Lake. A world traveler, she appeared to be ideally suited to take over an education program that had an international emphasis.[46]

Halse's tenure, too, was short. She resigned on October 1, 1959, to teach adult classes in English at Stavanger, Norway.[47] There had also been a change in the SDFU Junior Leader

Arlene Schley,
Director of Education, 1959-1967.

position. Marvin G. Winter, head of the Junior program since 1955, resigned on December 1, 1958, to join the Farmers Union insurance staff. In insurance, he worked primarily with the Junior award life insurance program.[48] His replacement was Arlene Schley of Stratford, South Dakota. She became the state Junior Leader on June 7, 1959. Four years earlier, she and Richard Bong, of Canton, had been selected by the Education Council to be Junior members of the Council for 1956.[49] Schley graduated from Northern State Teachers College at Aberdeen with majors in music and education. A former torchbearer and assistant at district and state camps, she had taught elementary school music at Sioux Falls before becoming Junior Leader. When Halse resigned on October 1, 1959, Arlene Schley was named SDFU Director of Education. Stability and a renewed sense of purpose returned to the SDFU educational program with her accession to the office.

There had also been personnel changes in the education program at the national level. Mrs. Flossie Harris of Denver replaced Gladys Talbott Edwards as NFU Director of Education in the fall of 1953. Edwards, the Director of Education for twenty-three years, was named associate editor of the *National Farmers Union*, the organization's monthly publication. Harris had been associated with the Farmers Union since 1947 as an educational field worker.[50] A year later, Dr. John M. Eklund was appointed the national Director of Education. Possessing a doctorate in education from Columbia University, he had been very active in international affairs. He brought this interest to the world affairs program of the Farmers Union.[51]

In addition to the FU being very active in United Nations' activities, the National Farmers Union, working with the International Cooperation Administration, sponsored an exchange program between the United States and Great Britain, beginning in 1956. Each trainee worked for about nine months on the host farm with provision for a month of travel. The trainees were compensated at the current wage rate. The young farmers paid their own travel expenses.[52] The program was expanded to include trainees from nations in Western Europe. Seven of them came to South Dakota in February 1957 to begin a nine month residence on the farms of Union members, who offered to serve as hosts.[53] During the remainder of the decade, an average of six to eight trainees were placed annually in South Dakota. Farm organization leaders from third world nations (i.e. India) spent brief periods of time in South Dakota as part of a tour of the United States.[54] The general purpose of the tours was to observe farm and community life, organization and activity. In South Dakota, they learned how the Farmers Union worked on the state and community level, and they studied the development, operation and management of the FU's cooperatives.[55] In the 1960s, Arlene Schley began a successful program to bring young farm leaders from Latin American countries to spend several months on South Dakota farms.

The emphasis on international understanding was also found in the study texts of the Juniors and Junior Reserves. The Junior texts for 1960 and 1961, *Tools For Peace* and

"Its Program Time." A talent program staged by the Tripp and Gregory County Camp, held at the Farmers Union elevator at Dallas (1956). L to R: Duane and Marlene Schweigert, Freddie and Carol Ann Thoene from Melrose Local.

Working For Peace, were authored by John Eklund. They were studies of how the United Nations, the specialized agencies of the UN in education, food and health and the Farmers Union worked toward building a positive peace.[56] The Junior Reserves read *A New World On Your Doorstep*, which stressed the importance of all people in the community.[57] The United Nations was often a subject for study at district and state camps. Youths studied the means by which neighbors could help one another through the Farmers Union local, cooperatives and UNICEF, the UN organization for children. The young Union members did more than study ways to help their neighbors. For several years, the Junior Reserves made stuffed toy animals, dressed and stuffed dolls and created doll furniture to be given to needy children and orphanages at home and abroad. Shortly before Christmas in 1959, more than 130 stuffed toy animals and dolls and about thirty TV stools made by the FU Juniors and Junior Reserves were given to children at the State Hospital at Redfield.[58]

"One World" was the theme of the 1948 All States Camp. The All Staters met for the first time at Red Rocks Camp, outside of Morrison, Colorado. It was the first year that the former Civilian Conservation Corps (CCC) facility had been used in seven years. Located in a valley, the camp was surrounded by red rocks and mountains, with Bear Creek flowing just below the barracks.

Eleven South Dakota Juniors registered for the camp. Making the journey to Colorado were Clifford Bailey, June Bult, Eugene Gisselbeck, Mary Lou Kistler, Marian Kleinsasser, Lyle Kruger, Florence Lynch, Jimmie Morris, Duane Schmidt, Dorothy Stellwag and Marvin

South Dakota's campers at Red Rock, Colorado (1948) in front of van. Third from left is John Sieh, Youth Director. The three women in the picture from left to right are: June Bult, Wessington Springs; Marian Kleinsasser, Freeman; and Florence Lynch, Woonsocket.

South Dakota campers at All-States Camp, Red Rock, Colorado (1948).

All-States campers in the Red Rock Amplitheatre, Colorado (1956).

Winter. They were accompanied by John Sieh, the director for SDFU youth camps, and Jo Erickson of Sioux Falls, the camp manager at Red Rocks.

For ten days, the bugler's reveille call resounded through the valley at 6:45 A.M. Breakfast in the mess hall was an hour later and the first assembly at 8:45 A.M. The campers attended three primary classes — Human Relationships, taught by Gladys Talbott Edwards, The Farmers Union, and The World Today. Different racial and ethnic groups from Denver were present for a special fellowship program on Sunday evening. Authentic copies of folk dance costumes from other nations were worn by several of the Juniors at camp folk dances. Through their classes and activities, the campers developed a belief that they must build a world which would offer a rich and abundant life for all citizens, beginning with the individual in his or her own community.

John Sieh served in several capacities at the '48 camp. He drove the Chevrolet bus that brought several of the South Dakota campers to Red Rocks. In the camp annual, he was described as the most eligible bachelor and the "meanest man in camp" because he was the dean of boys. His Chevrolet bus was noticed wearing Packard paint after going around a sharp hairpin curve on the trip toward the summit of nearby Mount Evans. Sieh composed the words and music for "One World," a song that summarized the goals of the ten day camp.

ONE WORLD
One world United Nations all
In peace and justice
Working for a better world I know.

One world with God on high our guide
We'll banish fear
And build upon the rocks of brotherhood.

The Rules Committee at All States Camp, Red Rock, Colorado (1948). John Sieh, SDFU Youth Director, is on the extreme left.

Bugler rouses sleeping campers, Red Rock All-States Camp (1948).

National Farmers Union Education Center, Bailey, Colorado.

We are men of toil,
Guardians of the soil,
Blood and sweat we've seen
This is what we mean —

One world, one world, one world
Just one world for you and me.[59]

A state camp was held at Placerville, approximately sixteen miles west of Rapid City, for the first time from June 14 to June 18, 1959. For a few years afterward, the two state camps continued to be held at different locations, but one of them was always at Placerville. By the late 1960s, Placerville became the permanent location for both camps.[60] The All States Camp was also relocated. After being held at Red Rocks for twelve years, the camp's water supply became contaminated, and for ten days, the Farmers Union had to buy and haul spring water for about 200 people to the camp in the middle of July.[61] It was decided in 1959 to construct a new National Farmers Union Education Center at Bailey, Colorado, fifty miles southwest of Denver.[62]

Delegates to the SDFU convention in November 1959 voted to raise $10,000 toward building administrative facilities at the Bailey location. The Board of Directors authorized that the $10,000 be released from state funds on condition that South Dakota's locals raise that amount of money and reimburse the state treasury. Within a year, nearly $8,000 of the amount had been repaid to the state Union.[63] The locals' response was so great that by

October 1961, the SDFU had contributed almost $12,000 to the FU Educational Center.[64] The first All-States Camp was held at the new facility in July 1960. Nineteen campers from South Dakota were among the 259 present to inaugurate the Bailey site.[65]

Cooperatives were one means of building a better world. Campers at Red Rocks, Placerville and other FU sites learned that for a cooperative to be successful, they must be brothers and sisters to their neighbors, working together to achieve a common goal. The keys to cooperative enterprise were sharing and working together for the common good of the group and society. In the '40s and '50s, the spirit of economic cooperation continued to pervade rural South Dakota. There were fewer new cooperatives but the co-ops that were already established flourished, building new facilities, providing new services and products and increasing refunds to their patrons.

Chapter 14

COOPERATIVES — A GROWTH INDUSTRY 1948 - 1961

It is the story of the unnecessary middleman and the profiteer, and it will certainly be the business of the laborer and farmer to eliminate him.
—Arthur C. Townley

The founding of new cooperatives in South Dakota peaked during the 1940s. By the 1950s, the Farmers Union no longer saw cooperatives as the solution to farm problems. The larger and established cooperatives, nevertheless, continued to grow and to prosper. Nationally, over 2,000 purchasing and marketing cooperatives used the Farmers Union as part of their names. These co-ops did an annual business running into the billions.[1]

In South Dakota, during the fifties, existing cooperatives, particularly the SDFU oil companies, built new facilities as their business expanded. Assets increased, dollar volume was up, and most cooperatives indicated an increase in net savings. By 1960, almost 190 cooperatives in South Dakota had celebrated their 25th anniversary. Born at the height of the drought, when low prices and high costs threatened bankruptcy, the co-ops, by providing supplies at reduced costs, had helped save many farms.[2]

The Roscoe Farmers Union Oil Company and the Corsica Cooperative Association represented success stories in the northern and southern parts of South Dakota respectively. Both co-ops had completed new buildings in 1959. The Roscoe facility, after thirteen years of operation, had a total business volume of more than $1.6 million. It had saved its patrons $152,387 in total earnings, and it had paid out $35,000 in cash refunds.[3] The Corsica co-op had a total sales volume of more than $3.7 million over its nineteen year existence. Total net savings over the years had been $455,926, with $136,000 paid in cash to patrons.[4] The Corsica Cooperative Association had checked-off more than 300 members, and the Association had remitted $12,167 to the education fund. The Roscoe FU Oil Company had about 350 members, and it had paid $4,146 in educational funds.[5] The Corsica co-op broke ground for a $750,000 feed plant in mid-summer of 1994. The plant, which became operational in September 1994, was capable of manufacturing 150 tons of feed per day. It was a boon to area farmers as it would use up to 2.5 million bushels of grain per year to produce the feed. Area livestock producers benefited from an improved feed supply.[6]

The same type of growth had been experienced by the cooperative elevator companies. The success story of the Farmers Union cooperative elevator at Kennebec, for example, had begun in 1947 when fifty farmers pledged $21,000 to buy out a private elevator. Eleven years later, the grain coop had more than 400 stockholders, who had subscribed $110,000 of their own money to enlarge the elevator's grain storage capacity to 370,000 bushels.[7]

The South Dakota Wheat Growers Association in northern South Dakota also reported growth. From 1923 to 1930, the wheat growers had operated a wheat pool. When the pool was discontinued, its assets were used to establish the Wheat Growers Association. Under the management of Charles W. Croes, the grain marketing co-op began with four elevators with a net value of $62,000. In the summer of 1955, the cooperative owned thirty-one elevators and other buildings.[8] Croes was one of the state's cooperative leaders. He succeeded Sam Ulrickson as president of the South Dakota Association of Cooperatives, founded in 1944, in late 1957.[9]

THE FARMERS UNION LIVESTOCK COMMISSION

The FU Livestock Commission at Sioux Falls celebrated its twenty-fifth anniversary in 1954. It had been the one organization that had helped keep the SDFU operative during the 1930s when the revenue from dues was inadequate to maintain the financial health of the organization.[10] John V. Carlson resigned as manager of the Livestock Commission on December 29, 1957, because of ill-health. Carlson had served the commission for twenty-years, a time of great growth for the firm. His successor was Lester D. Gravelle, who had become affiliated with the FU Livestock Commission on December 4, 1929. Following his graduation from Madison High School in 1926, Gravelle had taken courses in agriculture at South Dakota State College before beginning work at the Sioux Falls livestock yards in 1928. After joining the FU Livestock Commission, he held a variety of jobs, including assistant salesman and head hog salesman in 1935.[11]

THE UNHAPPY FATE OF EQUITY UNION CREAMERIES

One of the most venerable cooperatives, Equity Union Creameries of Aberdeen, did not fare as well as the supply and grain cooperatives affiliated with the Central Exchange and the Grain Terminal Association. The regional cooperative for dairy products began operating in 1916. Equity Union established an additional plant site at Mitchell in 1926, and later on, it also operated creameries at Jamestown and Devils Lake, North Dakota.[12] Under the management of Dwyte Wilson in the early fifties, the Farmers Union affiliate flourished. In 1952, Equity Union manufactured over a quarter of a million pounds of butter, sold thousands of gallons of ice cream and had sales of over $4 million.[13] By the end of the decade, however, a federal involuntary bankruptcy action had been brought against the corporation. A changing rural economy, smaller dairy herds and transportation and management problems caused the demise of the firm. The bankruptcy action was dismissed, and in the late winter of 1960, the liquidation of the defunct Equity Union Creameries proceeded in circuit court. This allowed the cooperative to continue its original plan to liquidate for the benefit of creditors. There were 3,000 farmers holding checks or payments due them.[14]

GREATER COMMUNICATION VIA THE COOPERATIVE

The telephone was another technological device for improving farm life. In 1920, about 44,000 of South Dakota's farmers had telephone service. With the collapse of the economy during the 1930s, only 24,821 farmers had telephones at the end of the decade. The telephone, essential to farmers in seeking help for emergency situations, checking on grain and

Equity Union Creameries.

livestock prices and inquiring as to the availability of spare parts, became readily available to rural people after World War II because of the cooperative movement.[15]

Telephone cooperatives began to spring up throughout the state during the 1950s. In the summer of 1954, there were fifteen telephone cooperatives in South Dakota, either organized or in the process of organizing. In May 1954, the first poles were planted by the Dakota Cooperative Telephone Company. Designed to serve 2,500 subscribers in Clay, Hutchinson, McCook, Turner and Yankton counties, it was the first co-op in the state to receive an REA-type telephone loan.[16]

WARDING OFF THE OPPOSITION

The South Dakota Association of Cooperatives sponsored an essay contest for high school students in 1959. The subject was "What can a cooperative do for its members?" The essay winners emphasized four important advantages of cooperatives. The advantages included: 1) cooperatives increased buying power, 2) cooperatives provided new services not previously available, 3) cooperatives increased the tax base in the local community and 4) cooperatives provided more opportunities for employment.[17]

Not everyone agreed with these youthful impressions of cooperatives. Much of the Farmers Union's energy in the 1950s was used to prevent the taxation of cooperatives and to limit the inroads into South Dakota of the Consumers Cooperative Association (CCA). The National Tax Equality Association (NTEA) had been a threat to cooperatives for decades. Since 1914, it had tried to get Congress to pass punitive tax laws on co-ops, seeking to tax co-op patronage savings of members as income to the cooperative. The NTEA had its beginnings in the Twin Cities and was headed by a private grain merchant in Minneapolis. The anti-co-op group promoted state and national legislation to restrain cooperative growth and development.[18] The NTEA found political conditions in Washington, D.C., during the 1950s, favorable for pushing its agenda.

Dwyte Wilson, NFU Headquarters (1966).

A proposal of the Eisenhower administration would have required cooperatives to issue certificates bearing at least 4 percent interest, which would have to be retired for cash within three years. An income tax would have to be paid on earnings not distributed in that form.[19] The loudest complaint of the NTEA was that regular proprietary corporations paid up to fifty-two cents of every dollar of income to the U. S. Treasury while cooperatives completely escaped paying an income tax.[20] Cooperative lobbyists reminded members of Congress that co-ops did not pay a corporate income tax because it distributed its earnings to its members. Spokesmen for the National Tax Equality Association replied that cooperatives could claim they had no earnings and still have money to expand by giving their members an IOU for their share of the profits and retaining the money for a number of years. Fortunately, cooperatives won the legislative battle as Congress dropped the Administration's proposal.[21]

Several years after the demise of McCarthyism, anti-cooperative forces continued their attempts to link the cooperative movement with foreign ideologies. One such attempt occurred in mid-February 1958. Glenn Martz, a spokesman for the political far right, held a series of ten meetings at different sites in South Dakota, speaking on "Communists in Agriculture." In sowing the seeds of hatred and distrust against cooperatives, in general, he also maligned the rural electric co-ops and the Farmers Union. Former Governor M. Q. Sharpe, after listening to Martz's presentation at Presho, called it "The most Un-American presentation I have ever heard."[22]

The struggle between the Farmers Union Central Exchange and the Consumers Cooperative Association over cooperative accounts in South Dakota intensified in February 1948 and involved the SDFU. The quarrel went beyond financial accounts when the CCA announced its intention to create an educational council within the state. It happened at the very time when the South Dakota Farmers Union had been working on a group educational plan that would have involved a number of county unions. One projected group in southeastern South Dakota would have included nine counties — Moody, Lake, McCook, Minnehaha, Turner, Lincoln, Union, Clay and Yankton counties. The North Dakota Farmers Union had successfully used area educational groupings.

For several years, the fieldmen for Central Exchange and the CCA had been competing for business in South Dakota. The CCA seemed to be gaining ground because of dissension with the South Dakota Farmers Union over the tactics of the Central Exchange's fieldmen, some sharp differences between the FU Central Exchange and the FU Grain Terminal Association, and the seeming pro-CCA stance of Walter Lassen, the secretary of the South Dakota Association of Cooperatives. It was known that Lassen did not approve of the check-off of dues used by FU cooperatives or of the practice of educational funds being paid by local cooperatives to the Farmers Union. Lassen also believed the Consumers

Cooperative Association to be a superior organization to the Central Exchange, — like a cat playing with a mouse.[23]

Union leaders became alarmed when George Anderson, a fieldman for the CCA, hosted a noon banquet in Sioux Falls on January 30, 1948, to work toward creating an educational council. In his speech following the meal, Anderson attacked the Farmers Union. He called the Union's leaders rabble-rousers and said that they fought against constructive measures. The leaders, he further charged, were sacrificing cooperatives to build up prejudiced groups that only wanted to perpetuate their own selfish interests. When he finished speaking, sixteen co-ops voted to join a CCA educational council; none of the cooperatives affiliated with the Central Exchange entered into this arrangement.[24]

The event in Sioux Falls generated a large amount of correspondence and conversation between the leaders of the State Union. Emil Loriks wrote a letter to Howard A. Cowden, the manager of the CCA, expressing his concern over the duplication and overlapping of certain cooperative activities. "For instance," Loriks wrote, "I am told that cooperatives that are contributing education funds to the Farmers Union as their educational organization, are now also being solicited by George Anderson to affiliate with and contribute to an educational organization he is setting up."[25] Loriks urged more cooperation at the top levels of the regional cooperatives to eliminate misunderstanding and competition.[26]

Former SDFU president John Skage attributed the controversy to personalities — Anderson versus Joe Fortner and Gust Moseson. Skage, who had not forgotten the role of the Central Exchange fieldmen in his defeat in 1944, believed that the SDFU's leaders should concentrate on building the State Union and let the fieldmen fend for themselves. Saying that the fieldmen were afraid of Cowden, he wrote to Loriks, "His [Cowden's] program became their program 2 or 3 years later — witness the ownership of refineries, building of fertilizer plants, feed mixing plants, LP gas program. They realized he is a little too fast for them. They always come to [the] FU asking for help."[27] The differences between Skage and Moseson continued until the latter's retirement.[28]

George Anderson held another banquet on February 18, 1948, at Brookings. There were several Farmers Union men among the approximately 100 guests. They were prepared to divert any discussion away from the Farmers Union and to direct it towards the future relationship between the CCA and the South Dakota Cooperative Association. One of the Farmers Union men asked if the creation of area councils would by-pass the state Cooperative Association. What would eventually become of the state association? In the spirited discussion that followed the question, the state association was recognized as the only overall organization and the only organization that would be neutral in sponsoring a comprehensive program. If the powerful regional cooperatives sponsored all inclusive area programs without coordination at a higher level, Union men argued that the result would be misunderstanding, controversy and conflict.[29] Elated over the discussion at Brookings, Loriks wrote Archie T. Erickson, "... we blew George's CCA mtg.[sic] clear into the stratosphere at Brookings Thursday."[30]

A special meeting of the South Dakota Association of Cooperatives soon followed the confrontation at Brookings. Agreeing there must be coordination at the highest level, the Association's representatives passed a resolution that denied endorsement of any area council not sponsored by or established by the state association. Anderson's effort, which struck at the heart of the Farmers Union and ignored all of the regional groups in South Dakota, had been thwarted. Loriks hoped that greater coordination might emerge from the affair. It might also bring some harmony to the Farmers Union and put a stop to the figurative "back-stabbing and knifing" that had been going on. Too much Union time had been spent on guerrilla warfare out in the country.[31]

An uneasy relationship continued to exist between the SDFU and the South Dakota Association of Cooperatives throughout the fifties. When the new secretary of the Association, Ralph Braun, began to organize councils and to plan for the local cooperatives' allocation of their 5 percent educational funds, his action was associated with the CCA. In a letter to M. W. Thatcher and Emil Syftestad on October 1, 1952, Loriks wrote, "The CCA has no educational organization — it is natural they want to take this over."[32]

The regional cooperatives believed they should play a larger role in the state Association of Cooperatives. M. W. Thatcher told Loriks he favored keeping the Association alive but with changes in its governance structure. He wanted the Central Exchange, GTA, and Great Plains Supply Company represented on the Association's board. "That's important," he said, "for the money we kick in."[33]

THE REGIONALS BUILD IN SOUTH DAKOTA

An agreement between Thatcher and Syftestad at the annual meeting of the FUGTA in December 1958 was good news for the SDFU. The two men had squabbled since the 1940s over the sale of feeds and fertilizers. At the December meeting, a joint-program for manufacture of feeds and fertilizer was announced. The GTA would market Central Exchange fertilizer at its outlets, and the Exchange would sell GTA feeds at its affiliated distributors.[34]

Within three years, 1959-1961, the two regional cooperatives made a major investment in South Dakota. For some time, the Grain Terminal Association had discussed where to locate a feed plant in the state.[35] M. W. Thatcher, the GTA's general manager, announced in early May 1959 that the firm would build the first of six modern feed plants on a twenty-acre industrial tract at Ellis, five miles west of Sioux Falls. The president of the GTA, Emil Loriks, and August Dahme of Mina, a GTA board member, approved of the site.[36] The grand opening of the twelve-story feed mill took place on May 3, 1961. The semi-automated facility produced thirty tons of feed an hour with a labor force of only twelve men. The mill cost approximately $1 million.

The FUGTA also purchased the Great Plains Seed Company on August 15, 1959. The twenty-five-year-old seed plant and warehouse were located across the railroad tracks from the new feed mill. The acquisition of the seed company enabled the GTA to expand its feed, seed and farm chemical service in South Dakota and western Minnesota.[37]

On December 4, 1959, the Farmers Union Central Exchange announced it would construct a million dollar soil service center at Ellis, the thirteenth center to be built but the first one in South Dakota. The soil service centers were bulk fertilizer plants that mixed fertilizer to fit specific soil needs. Personnel from the plant tested soil on an individual farm. After determining what plant food the soil lacked, the appropriate fertilizer was prepared and applied on the needed areas by spreader trucks owned and operated by nearby local cooperatives.[38] In the next two years, Central Exchange also erected fertilizer blending plants near Britton and Lake Preston. In each case, farmers were spared the task of soil sampling, choosing fertilizer ratios, mixing, loading and spreading the plant food.[39]

Lumberyard, Farmers Union Co-op, Nunda

THE PHYSICAL GROWTH OF THE FUGTA

The increasing number of regional cooperative facilities in South Dakota was only a minimal part of the expansion of the FUGTA and FU Central Exchange in the fifteen years following World War II. At the end of the fifties, FUGTA operated terminal elevators at Duluth-Superior, Minneapolis, St. Paul and Sioux City, with a total capacity of 26 million bushels. Subterminals were located at Rush City, Minnesota; Great Falls, Shelby and Lewiston, Montana.[40]

The Grain Terminal Association saved the Union's rendering plants in 1952. These plants had been purchased by the Farmers Union Livestock Company in 1947 and had brought the cooperative to the verge of bankruptcy. The GTA bought the firm and renamed it the Farmers Union Livestock Association. The GTA agreed to pay holders of stock and notes 40 percent of the amount due them in cash and the balance over a ten-year period. Because of the GTA's action, the livestock company remained in business.[41] Loyd Larson, the director for District III, was elected president of the Livestock Association for a three year term in December 1956. He had served on the board since 1948. The Association operated nine rendering plants in Minnesota and Iowa. Much of the grease was sold for export and to U.S. firms such as Lever Brothers.[42]

Great Plains Supply Company, a part of the GTA's business complex, purchased nine lumber yards in South Dakota in November 1955. Yards were bought at Alcester, Beresford, Bridgewater, Centerville, Chancellor, Elk Point, Hudson, Lennox and Wakonda. Great Plains Supply also had lumber yards at Arlington, Brookings, Canton, Elkton, Gary, Hayti, Hazel, Jefferson, Lake Norden and Ward. At the end of 1955, Great Plains operated 119 lumber yards on a cooperative basis.[43]

One of the GTA's major acquisitions occurred on October 1, 1958. The cooperative took over all of the physical assets of the seventy-two year old McCabe Company for $4.8 million dollars. An additional $3 million was paid for the firm's inventory and other assets. The purchase included fifty-seven grain elevators, four of them in South Dakota, and feed plants in North and South Dakota, Montana and Minnesota. In acquiring the McCabe elevators, the Grain Terminal Association added between five to six million bushels of extra storage space.[44] The Thatcher-directed GTA added thirty-eight wooden grain elevators and four concrete subterminals, located mostly in southern Minnesota. They were acquired from Archer-Daniels-Midland Company for an estimated $1.5 million on August 10, 1960.[45] With all of these acquisitions, the regional cooperative was marketing grain from its 196 line elevators and 450 cooperative elevator associations. Twelve percent of the marketed grain originated in South Dakota.[46]

Emil Loriks was president of the FUGTA when the elevators were purchased from McCabe and Archer-Daniels-Midland. An official member of the GTA board of directors since December 10, 1940, Loriks served several years as vice president before being elected president by a unanimous vote of the regional co-op's board of directors on December 11, 1957. He and Thatcher believed that farmers should engage in vertical integration through their own cooperatives. Vertical integration being the linking together of farming and marketing to avoid operating independent of one another. The farmer should take the initiative in processing his own products, eliminating the middleman. In his later years, Loriks argued that if the FUGTA had remained only a marketing organization, it would have declined.[47] He and August Dahme, a fellow board member, believed that the SDFU and the GTA grew stronger during the 1950s because they helped each other.[48]

Loriks and Dahme worked well with the sometimes difficult Thatcher. He was a frank, outspoken man with a strong personality. Born at Valparaiso, Indiana, on May 5, 1883, Myron William Thatcher first came to Minneapolis, Minnesota, in 1907, to audit the books of a milling firm. He moved permanently to Minneapolis in 1914. Before becoming general manager of the Grain Terminal Association in 1938, he had worked closely with President Roosevelt and Secretary of Agriculture Wallace in Washington, D.C. Thatcher demonstrated his mastery of cultivating powerful contacts and using them to improve the farmer's lot in life during his five years in the nation's capital.

Because of his dominant management style, some people swore at him while many swore by him. This great champion of the cooperative movement could be an individualist in his social relations. A mercurial figure, he could be affable and make friends easily when he wished to do so. He spoke in the old tradition of Simpson and Reno, which added to his personal magnetism. An admirer of his oratorical style said Thatcher was "sometimes peeling the hide with a six-oxen whip, sometimes applying salve to the open wound." Younger Union members admired him for his ability to make rousingly partisan points in a nonpartisan speech.[49] His effectiveness from the platform was demonstrated in his speech to the 1952 convention in defense of NFU President Patton's assertion that the farm economy was a legislated economy. Thatcher rhetorically questioned the convention delegates when he said:

> Since your price supports are given to you by legislation, by the Congress, and since your storage program and loans are given to you by the ... Congress — and since your credit system, the land bank, and the production credit and the banks for cooperatives, is given you by the Congress — do you agree with me that you live as farmers in a legislative economy?[50]

Thatcher continued to lead the FUGTA until his mid-eighties. The insertion of a retirement mandate in the cooperative's by-laws forced him to relinquish his position as general manager in 1968. He died on December 25, 1976, at ninety-three years of age.

In early August, 1960, the GTA moved into processing soybeans after purchasing the Honeymead Products' plant at Mankato, Minnesota for $6 million. The Honeymead plant could process 50,000 bushels of soybeans or flax daily. The beans were turned into oils, protein products, flours and flakes.[51] A little over a year later, the cooperative entered another processing venture. In early October 1961, M. W. Thatcher announced that the Grain Terminal Association had purchased the Minnesota Linseed Oil Company of Minneapolis.[52]

Loriks' old Union friend John Skage applauded the Grain Terminal Association and its new initiatives during the Loriks' presidency. In a letter to Loriks in January 1961, he said, "No wonder the GTA draws the largest number of farmers into its annual meetings. Here is an organization that is getting things done. How proud I am to be a part of this show."[53]

Other GTA physical assets in the late summer of 1960 included feed plants at Montevideo, Osakis and Perham, Minnesota; Hunter, Minot and Williston, North Dakota; and Chinook, Glendive and Great Falls, Montana. The cooperative also operated a durum wheat flour mill.

A significant change in the by-laws of the FUGTA was made at the end of September 1959. Stockholders voted overwhelmingly to change the GTA from a stock cooperative to a membership cooperative. Instead of individual voting stockholders, all farmers' elevators shipping more than 10 percent of grain to the cooperative, all GTA line elevator associations and farm suppliers, who bought supplies from GTA, automatically became members. Delegates to the annual stockholders' meetings would be elected by all of the cooperative association's members based on the number of patrons and the percentage of grain shipped to the Grain Terminal Association. Each association was entitled to one delegate for each 200 patrons if they shipped 100 percent to GTA.[54]

Twenty-two years after the Farmers Union Grain Terminal Association began with $30,000 of borrowed capital, it had a net worth exceeding $42 million and total assets of $73,200,000 in 1960. The GTA handled more than 150 million bushels of grain per year. It paid 20 percent of its annual savings to its patrons in cash, amounting to $690,000 for the 1957-1958 year. More than $181,000 in educational funds were paid out to Farmers Union organizations in the Dakotas, Minnesota and Montana.[55] During the 1950s, the GTA also made contributions to the non-tax supported colleges of South Dakota.[56]

THE CENTRAL EXCHANGE EXPANDS PETROLEUM PRODUCTION

The decline in farm prices during the fifties caused the Central Exchange to engage in a critical reevaluation of its physical assets. The Exchange liquidated or sold several failing enterprises, including the International Lumber Association, the National Farm Machinery Cooperative, the Illinois Mid-Continent Oil Company and the Premier Oil Refining Company. In 1951, the last CO-OP Tractor came off of the production line.[57] E. A. Syftestad provided an excellent summation of the state of the rural economy in an address to a NFU convention. He said:

> You can't have healthy cooperatives without a healthy farm community because the cooperatives are a part of the community. You can't have a healthy farm community without a healthy relationship between the price

Don Graff, Manager, Farmers Union Oil Co., Mitchell, standing by a Co-op tractor (1948).

farmers receive for their products and the prices they have to pay for their services and farm supplies.[58]

Although the Farmers Union Central Exchange had undertaken certain austerity measures, it continued to expand its key petroleum production and distribution facilities during the 1950s. In 1952, a $4.5 million expansion project at the Exchange's Laurel, Montana, refinery was completed. Production increased to 12,500 barrels per day, 7,500 barrels more each day than when the refinery was purchased in 1943.[59] A new oil field was opened in northeast Wyoming in the summer of 1960. The eight wells near Gillette, Wyoming, became a part of a series of 105 wells operated by Central Exchange in thirteen fields in Montana, North Dakota and Wyoming.[60] To handle the increased production, Central Exchange acquired total ownership of a 234 mile pipeline running between the refinery at Laurel to Glendive, Montana. An eight inch pipeline, costing $5 million, carrying refined oil from Glendive to Minot, North Dakota, was completed in the fall of 1960. It had a capacity of 15,000 barrels per day.[61] Central Exchange continued to own one-third of the $30 million refinery operated by the National Cooperative Refinery Association. Located at McPherson, Kansas, the world's largest coop oil refinery, refined 750,000 gallons of gasoline per day.[62]

During the decade, FU Central Exchange made a major effort to market liquefied petroleum gas. An LP gas line, the Mid-America pipeline, costing $75 million, was completed in the fall of 1960. The 2,000 mile pipeline ran from Texas to the Exchange's distribution centers at Sanborn, Iowa, Pine Bend, Minnesota and Janesville, Wisconsin. Central Exchange

The evolving logo of the Farmers Union Central Exchange.

distributed LP gas through twenty-three wholesale bulk storage plants and fifty-one retail plants operated by affiliated local cooperatives.[63]

Much of the Central Exchange's financial success resulted from vertically integrated operations through its ownership of oil refineries and fertilizer companies. The firm which began in 1931 with $525 in cash, a little borrowed capital, four employees and lots of heart, distributed over $91 million worth of farm supplies in 1959.[64] South Dakota's share of the sales volume was around 9 percent.[65] The state's co-ops received about $250,000 in patronage refunds.[66] In its first twenty-five years of operation, Central Exchange had paid-out over $200,000 in educational funds to its affiliates in an eight state area.[67]

Working hand-in-hand with the regional cooperatives in their expansion was the National Farmers Union. The NFU, looking to develop sources of fertilizer, acquired control of 17,000 acres of land near Artesia, New Mexico. Geologists found the land to be rich in potash, a mineral essential in the manufacture of commercial fertilizer. At the same time, Patton announced that the Union held an area in Utah that contained phosphate, also essential for making fertilizer.[68]

Emil A. Syftestad, the general manager of FU Central Exchange, died of a heart attack on July 16, 1957. Under his guidance, the cooperative's sales had grown from $1,678,000 in 1932 to $75 million at the time of his death. His successor was Thomas H. Steichen, a member of the Central Exchange since 1935 and assistant general manager of the cooperative at the time of his elevation to the presidency.[69]

Farmers Union Co-op Association Elevator, Mitchell (Circa. 1948).

Oscar Johnson of Oldham, South Dakota, first elected to the board of directors of the Exchange in 1944, continued to be reelected to the board throughout the fifties. He was joined on the board by Louis W. Vollmer of Mellette in 1948. South Dakota's second representative to the Central Exchange's governing body had been a member of the Farmers Union since 1926. Vollmer had considerable familiarity with the Exchange's oil activities, having been on the board of directors of the Farmers Union Oil Company at Redfield since it was organized in 1929.[70]

Gust Moseson, the state field supervisor for the Central Exchange, retired on July 1, 1960, after eighteen years of field work for the regional coop. The former manager of the FU Cooperative Brokerage started as a district representative for the Exchange in 1942 and became the state manager in 1950. Moseson was succeeded by Boyd Meyer, a member of the Exchange's field service staff. Previously, Meyer had been a vocational agriculture teacher and the manager of the Farmers Union Oil Company at Alexandria.[71]

The success of the cooperatives had not been reflected in the farmers' struggle to maintain parity for his products. After eight years of declining farm prices under the Benson-Eisenhower farm policy, Farmers Union members welcomed the election of John F. Kennedy in November 1960.[72] When the new president announced that Orville Freeman, governor of Minnesota, would be the new Secretary of Agriculture, Union members were elated. Freeman was considered a friend of the farmer, and the SDFU looked toward the 1960s with considerable optimism.

Chapter 15

THE POLITICAL STRUGGLE TO SAVE THE FAMILY FARM 1961 - 1981

"Keep the family in the country, Keep the country in the family."
— From a Farmers Union song.

"A few of us going broke each year is one of the healthy things there is. Bankruptcies and the unavailability of loans make other farmers more efficient."
— Kenneth Frick, ASCA Administrator, March 1975.

The highlight of the annual meeting of the Farmers Union Grain Terminal Association, in December 1961, was the announcement that Orville Freeman would be the new Secretary of Agriculture. "Farmers have a real friend in Governor Freeman," exclaimed an elated Glenn Talbott. Freeman's appointment, he said, "... is the best news we've had in eight years."[1] Freeman's role was to carry out the Kennedy administration's objectives in agriculture, including the reduction of storage subsidies.

GOOD NEWS ALONG THE POTOMAC

While keeping food prices relatively stable, steps were taken to raise net farm income per farm to a record high, a billion dollars a year over its 1960 level. New programs were introduced to reduce the farm surplus in storage, which during the Benson years had soared from $2.5 to $9 billion. Several hundred thousand dollars per day in storage costs were saved by expanding welfare food distribution at home, reducing wheat and grain acreage and by increasing farm exports by 70 percent. Domestic projects included a revival of food stamps for the needy, increased food distribution to the impoverished and an expanded school lunch and school milk distribution program. A new Rural Areas Development program that helped low income farmers to find new jobs and turn surplus cropland into recreation areas for fun and profit was also introduced.

Exports were increased through the Food-for-Peace program, the result of an amendment by Senator Hubert Humphrey of Minnesota. Food-for-Peace had limped along during the Eisenhower administration, caught in a crossfire between the Departments of Agriculture and State. Kennedy set up an expanded and independent Food-for-Peace office in the White House under a dedicated George McGovern of South Dakota. In eighteen months, more food was shipped abroad than Herbert Hoover and his associates shipped in ten years of relief to victims of World War I. Some of the food was donated, some delivered under long-term credit arrangements and some of it was paid for in local currencies.

Orville Freeman,
U.S. Secretary of Agriculture, 1961-1969.

Kennedy defied opponents on the far right by arranging the sale of 65 million bushels of wheat to the Soviet Union in 1963. Plagued by drought and crop failures, Soviet leaders had indicated an interest in purchasing American wheat at the world market price. The obstacle to the sale was Congress. At the height of the Berlin Crisis, Congress had added to the Agricultural Act of 1961 an amendment opposing the sale of subsidized agricultural commodities to unfriendly nations. Kennedy ignored the non-binding declaration of interest and sold the wheat. Although the sale hardly made a dent in the stored wheat, it added income to America's agriculture, benefitted the nation's balance of payments and slightly reduced federal storage costs.

The efforts of Kennedy and Freeman to fit food production to consumption encountered fierce opposition. It came from the larger and more prosperous farmers, who enjoyed being subsidized to produce crops for storage. It came from Congressmen opposed to the kinds of controls needed to make the farm program manageable. The Administration preferred to use the term "supply management" instead of controls.

The Farmers Union felt very comfortable with supply management. SDFU leaders praised the Food and Agricultural Bill of 1962. It included a new two-price system designed to keep wheat output permanently in line with market needs. The law required the Agriculture Department to support wheat prices between 75 and 90 percent of the "fair earning power parity level" if farmers accepted production controls.[2] Speaking to delegates at the 1962 annual convention, SDFU President Ben Radcliffe said that farm income increased by $1.1 billion from 1961 to 1962, a 10 percent increase over 1960. The farm surplus had been reduced by $1 billion, the first reduction in ten years.[3]

A DISASTROUS WHEAT REFERENDUM

Unfortunately, there were American farmers who were unhappy with price supports and acreage controls. Those who believed that farms should be family size and profitable for farm owners were opposed by those who wanted to see farms become big business enterprises. The battleground for the two sides was the annual wheat referendum. When farmers went to the polls in August 1961, a "yes" vote guaranteed $2.00 per bushel wheat, 84.4 percent of parity. It required a 10 percent reduction in acreage allotments.[4] For the ninth consecutive year, the proponents of price supports came out on top. Ninety-five percent of South Dakota's eligible farmers voted "yes." Nationally, 79.4 percent of the farmers favored the referendum.[5] Union leaders, however, became alarmed at the increasing number who opposed marketing quotas. In 1962, 68.4 percent of nation's farmers voted "yes," the margin of victory being slightly over one percent as a two-thirds vote was required to continue quotas.[6]

Poll workers supervising a wheat referendum (1950s).

The opponents of farm price-support programs, led by Charles Shuman, president of the National Farm Bureau Federation, launched an all-out assault to do away with them in 1963. The Farmers Union offered changes in the 1962 law. The Union proposed making direct payments at 100 percent of parity contingent on agreements to cut production but allowing support payments if farmers agreed to plant at the 1959 - 1960 level. The Union's plan was not accepted, and the referendum was based on the law passed by Congress in 1962.[7] Speaking at the National Farmers Union convention at New York City, President Kennedy said he viewed the wheat referendum as a test of policy for the wheat program. He said farmers should determine for themselves what role government programs should play in agriculture.[8]

"Parity Now" A wheat referendum in the 1950s.

With the referendum vote scheduled for May, the Farmers Union began to mobilize its forces in early February 1963. On February 7, 147 FU leaders from five states met at Huron to plan a campaign for a favorable vote. NFU President Patton told them that a failure to muster a favorable two-

The Wheat Referendum, Campaigning for a YES vote.

thirds vote might mean the "death knell" for further farm programs.[9] He later said, "We can either put permanent, practical farm programs on the books that will strengthen and stabilize the family farm, or we can do away with the farm programs altogether as some self-seeking groups would have us do...."[10]

Radcliffe was in the forefront in urging South Dakota farmers to sign up for the federal program and to vote "yes" on the supply management referendum. A South Dakota Wheat Committee was formed, and a giant rally was held at Aberdeen on April 4, with Patton as the featured speaker. The campaign for and against support prices and acreage controls took on the aspects of a political campaign with roadside billboards and advertising in every form of media. A large amount of money went into the campaign, especially on the side of a "no" vote. The Farm Bureau proclaimed that if you planted crop without restrictions you would get full price.[11]

When the votes were counted, a majority of the nation's wheat farmers, in a self-damaging assertion of independence, rejected the wheat program that required tight planting controls. Nationwide, only 47.9 percent of those eligible to vote had cast their ballots. There were 597,776 votes opposing price supports and acreage controls and 547,151 votes favoring continuation of the program. The Farm Bureau had mobilized the negative vote in the states of the Old Midwest, such as Michigan and Illinois. In South Dakota, 65.2 percent of the voting farmers favored the program, there were 21,771 "yes" votes" and 11,616 "no" ballots.[12]

In his report to the 1963 convention, President Radcliffe said the lack of unity among farmers had led Congress to consider farmers as a "babble of voters." He argued that farmers had the numbers to be heard, but they had "diluted [their] strength by dividing into three or four major farm organizations and dozens of commodity groups."[13]

JOHNSON RESURRECTS A FARM PROGRAM

Approximately a month after the convention, President John F. Kennedy was assassinated in Dallas, Texas. It was now up to his successor, Lyndon B. Johnson, to take up the cause of the family farmer. If there were no substitute proposal, there would be no program at all and a precipitous drop in wheat prices.

Farmer-oriented Hubert Humphrey, a native of Doland, South Dakota, pleaded with President Johnson to draft a new farm bill. The President backed a measure which bore considerable similarity to the Congressional Act of 1962. Knowing there was considerable opposition to a new farm bill in the House of Representatives, Johnson added cotton to the bill to secure Southern votes. He also emphasized food stamps for the needy to attract the votes of big city Democrats. The National Farmers Union led in the fight for the bill's passage. The House approved the farm act on April 8, 1964, 211 to 203. Johnson's wheat program was voluntary. The South Dakota Farmers Union urged farmers to sign up for the program despite its returning about twenty-five cents less to the producer than the wheat referendum which had been rejected in 1963. Radcliffe reported with considerable satisfaction, in mid-July 1964, that 84 percent of South Dakota's wheat acreage was signed up in the program. The 16 percent unsigned consisted primarily of small allotments.[14] The action taken by Johnson and Humphrey caused Radcliffe, in his "President's Column" in the Union Farmer, to urge Union members to support the Democratic ticket in the presidential election. There was no choice as the Republican candidate, Berry Goldwater, announced that he wanted to terminate the farm subsidy program. He also said that the REA had outlived its usefulness.[15]

RADCLIFFE MOBILIZES THE RANK AND FILE

By this time, Ben Radcliffe was putting, at least, 35,000 miles per year on one of the farm organization's cars, as he traversed the state fulfilling the Union's objectives.[16] A native of Wolsey, the tenth president of the SDFU graduated from Tulare High School in the depth of the Great Depression. From 1936 to 1942, he had worked in a gold mine, the Governor's Mine, south of Lancaster, California. He returned to South Dakota to farm in 1942. In 1944, Radcliffe joined the South Dakota Farmers Union, and he became a member of Crandon Local in Spink County. He soon became a director of the Spink County Federal Credit Union, an ASCS committeeman and Spink County FU president. Eleven years later, in 1955, he became the first chairman of the newly created District VII, entitling him to a position on the SDFU's board of directors. Radcliffe, in the same year, was appointed a director of the FU Livestock Commission at Sioux Falls. He was elected vice president of the state Union in 1958. Elected to the state House of Representatives in 1954, he was serving his fourth term in the Legislature when Harold Golseth succumbed to a heart attack in January 1961.

Because of his legislative responsibilities, Radcliffe did not officially assume his duties as acting president of the SDFU until early March. Almost immediately, he was named to the seven person executive committee of the NFU.[17] He was on the committee for twenty years, serving several years as chairman. In 1962, Radcliffe was named chairman of the Land and People Conference by Secretary of Agriculture Freeman. At the annual state convention in

**Ben Radcliffe,
President of the SDFU, 1961-1981.**

October 1961, he was formally elected president, beginning the longest presidential tenure, twenty years, in the history of the South Dakota Farmers Union. When he was in the Legislature, Radcliffe was known for his "honesty, ability, openmindedness and the fact that one always knew where he stood on an issue."[18] The same character traits applied to his leadership of the SDFU.

During the '60s and '70s, technology caused dramatic changes in American agriculture. Increasing mechanization led to a decline in the number of farms nationwide. From 1945 to 1970, the nation's farm population declined from 24.8 million to just under 10 million, or from 17.5 percent of the total population to 4.8 percent. From 1939 to 1964, the number of farms decreased by 2.4 million. One farmer lamented that "We lost country life when we moved to tractors." South Dakota, on average, lost between 1,000 and 1,200 farms per year during the 1960s.[19] South Dakotans witnessed the disappearance of farm families from the land equivalent to a rural county with all of the main street businesses, schools and churches that served a thousand families. Census figures indicated that for every six farms that disappeared, one main street business closed its doors.[20] Along with technological unemployment, inadequate farm income contributed to the declining number of family farms. During the 1960s, one-third of the farmers worked 100 days or more off the farm to meet their economic needs.[21]

The decline in the number of family farms meant a decline in the pool of potential members for the SDFU. In attempting to maintain membership and recruit new members, emphasis was placed on education, effective media, organization activities, young adult programs and cooperative relations. Upon the resignation of Edward Utecht in September 1962, Radcliffe appointed L. J. (Jim) Holwegner as organization director in 1962. Utecht had held the post since 1956. Usually two or three field men worked with the organization director. Fieldmen, during the 1960s, included Tom Terney, LeRoy Hardy, Daryl Walker, Hans Wagner and Maynard J. Whitmyre. Whitmyre assumed his duties on November 29, 1967. Wagner retired in June 1973. Walt Hofstetter of Wessington Springs was named acting fieldman. Donn H. Wright became the new director of Cooperative Relations for the SDFU in April 1961.

One of the novel approaches to recruiting new members occurred in 1969. Sixteen SDFU members flew to Kansas and Texas to do membership work while a similar number of North Dakota Union members flew to Huron to spend four days in South Dakota building membership. The latter group signed up more than 300 members.[22] Membership in the state Union totaled 18,678 in 1961. It reached a low point of 11,885 in 1975. Through hard work and aggressive recruiting, the membership had increased to 15,081 in 1981, the last year of Radcliffe's presidential tenure. Despite the drastic exodus from South Dakota's farms, the SDFU, relatively speaking, had done very well in maintaining its position as the state's largest farm organization. This feat was accomplished despite an assault upon a stable farm program throughout the 1970s.

EARL BUTZ: FAMILY FARMERS WILL HAVE TO TAKE THEIR LUMPS

Any hopes of a stable farm program vanished with the election of Richard M. Nixon as president of the United States in November 1968. Any hopes of maintaining the farm legislation of the Johnson administration were quickly dispelled. During the election campaign, Nixon had said that 74 percent of parity was not acceptable. Three years later, however, the parity ratio stood at 67 percent, the inflation rate was 6 percent and the United States experienced its first trade deficit since 1893.[23]

The Agricultural Act of 1970, designed by Nixon's White House staff and the Secretary of Agriculture, Clifford Hardin, was a disaster for the family farmer. It created a set aside program that encouraged more acreage planting and lower price support levels. Bountiful harvests flooded the grain market, forcing prices to drop. Farmers stored much of their grain, hoping that prices would rise.[24] When parity reached a post-Great Depression low, the Administration, through the Department of Agriculture, announced a change in reporting the parity ratio. A base period of 1967 was substituted for the 1910-1914 period as provided by federal law. Hardin, through this deceptive maneuver, set the new parity at 90 percent as of January 15, 1971. Under the old formula, parity was at 67 percent on December 15, 1970. Farmers Union officials protested against this self-serving manipulation of reality. Radcliffe noted that the old formula was basically sound. It had been continually updated to reflect new technology and the new kinds of expenditures required in agriculture.[25]

Farmers were not surprised when Hardin resigned in the fall of 1971. The SDFU considered the resignation a political ploy as Nixon began his campaign for reelection. Hardin's successor, Earl Butz, was a surprise. The former dean of agriculture at Purdue University had worked under Benson in the Eisenhower administration, and he went off the payroll of Ralston Purina to join the USDA in 1971. An indifferent President Nixon and an urban minded Congress let Butz have his way for five years. By late 1974, only sixty out of the 435 members of Congress had a definite commitment to agriculture.

The new Secretary of Agriculture was all things to all people in his first few months in office. An articulate man, backed by a group of skilled speech writers, he gained high marks from the national media. The Farmers Union saw through the media-created facade. Union leaders called him a showman. In his monthly "President's Column" in the *Union Farmer*, Ben Radcliffe called Butz a phony. The SDFU leader pointed out that during Butz's first few months in office, the Secretary had killed a bill that would have raised price supports on wheat and feed grains by 25 percent, had lowered the dairy support price and had undermined farm bargaining bills.[26]

When President Nixon signed the Agriculture and Consumer Protection Act on August 10, 1973, the government turned over the control of agriculture to the market place. Farmers had to depend entirely on market prices for their income. Under the new law, payments were made according to "target prices," established minimum prices. If the market price dropped beneath them, an eligible farmer received a government payment. Earl Butz, of course, set the target price for the first two years very low, eliminating any payment to farmers. Payments, if made, were limited to $20,000. Allotment and parity became obsolete as the law lifted acreage restrictions. The new farm bill, limited to four years, made it impossible to regain previous levels of government control.

It was Butz's good fortune that a worldwide food shortage began in 1973. Grain reserves fell sharply as the United States enjoyed an export boom. Between 1972 and 1981, American exports increased in value from $8 billion to over $44 billion. The price of wheat, over three years, increased 132 percent. It soared from $1.76 to $4.09 per bushel between 1972 and 1974. The price of corn rose 92 percent in the same time period. The Secretary of Agriculture became the cheerleader for the boom, and he exhorted farmers to plant every available acre. Between 1973 and 1981, the number of acres planted to crops increased by 12 percent, from 316 to 353 million acres. In October 1973, the parity ratio was 102 percent of the base period 1910-1914. The last time this level had been reached was between 1941-1952, when war-created shortages had brought parity and agrarian prosperity.[27]

There were several dark clouds that hovered over the resurgence in prices. Butz and President Nixon had killed the Strategic Reserve bill that would have established a strategic reserve for handling some of the surplus built up in 1971 under the disastrous set aside program.[28] Butz recklessly called in sealed grain in an effort to break the market price and push it below the world price.[29] In exporting wheat, he took advantage of omissions in the 1971 International Wheat Agreement. Because of objections from the Nixon administration, the wheat treaty had not provided for minimum and maximum prices, reciprocal rights or obligations of the participating countries.[30] In cleaning out government bins, subsidized U.S. wheat was exported below the International Grains Agreement minimums. The USDA hoped to outsell foreign farmers, driving them out of business.[31] In glutting the market, Butz caused American farmers to receive less per bushel than Australian and Canadian farmers.32 At the same time, American supplies of wheat were dangerously low at the end of 1973, and the Administration began to import wheat into the United States. Farmers Union officials suspected there had been collusion between America's giant grain corporations and the USDA in a secret grain deal with the Soviet Union.[33]

The price of farm commodities escalated in 1973-1974 but so did the cost of farming. The United States faced a serious fuel shortage in 1973, resulting in the importation of more petroleum — very expensive petroleum. War had broken out between Israel and the Arab states, and the Arab states cut off oil shipments to the United States, Japan and most of Western Europe in an attempt to compel Israel to withdraw from the war. The success of the Arab boycott led the Organization of Petroleum Exporting Countries (OPEC) to raise the price of a barrel of oil from $3.00 to $11.95. This action created a groundswell of inflation.

Testifying before a Congressional committee, SDFU Vice President Adam Seidel talked about inflationary farm costs. Since 1970, farm production costs had increased 30 percent, interest had risen 35 percent and taxes had gone up 25 percent. Farm debt at the beginning of January 1974 totaled nearly $70 million at a time when the interest rate was over 9 percent. Economically, farmers lost ground as fuel and fertilizer costs rose.[34]

By mid-summer of 1974, U. S. grain reserves were the smallest in twenty-five years. Senator George McGovern of South Dakota, a member of the Senate Agriculture and Forestry Committee, said an adequate grain reserve was essential, a grain reserve that would be protected from price manipulation by politicians and private traders.[35] Senator McGovern enthusiastically endorsed a Farmers Union proposal that a new farm bill must contain an escalator clause that would adjust the target price annually to reflect changes in the cost of production. The Administration turned a deaf ear to any action that was contrary to the 1973 Farm Act.[36]

Before his resignation on August 9, 1974, Richard Nixon's last official act was to veto a $13.5 billion agriculture appropriations bill for fiscal 1975. The bill included monies for rural

Sen. George McGovern.

**Robert Bergland,
U.S. Secretary of Agriculture, 1977-1981.**

industrial development, rural water and waste disposal and rural environmental conservtion.[37]

Earl Butz remained as head of the USDA under Nixon's successor, Gerald R. Ford. The Secretary of Agriculture continued to warn against higher loan rates and target prices. In an address to the Cotton and Grains National Advisory Committee, he said, "farmers will just have to take their lumps when market prices tumble."[38] It was before this same committee that Kenneth Frick, administrator of the ASCA, said that "Bankruptcies and the unavailability of loans make other farmers more efficient."[39] Butz's political demise came suddenly. The government career of the "fair-haired boy" of corporate agriculture and the American Farm Bureau ended abruptly in 1976. He was forced to resign on October 3, after a racist, obscene remark of his became public.

Butz's resignation was the good news for family farmers during 1976. The bad news for South Dakota farmers during the nation's bicentennial year was the severe drought in the central part of the state. It was the worst drought to strike the area since 1934. Temperatures in excess of 100 degrees with strong southwesterly winds eliminated a good portion of the state's corn crop. The Farmers Union made a list of available hay sources in an effort to preserve foundation livestock herds. Unfortunately, many farmers and ranchers were forced to liquidate livestock. One-third of the state's farmers suffered a severe economic setback from the drought. South Dakota incurred a loss of over $1 billion in farm income. While he remained in office, Secretary Butz turned a "deaf ear" to requests for federal economic assistance.[40]

All during Butz's tenure, President Radcliffe had wondered what would happen to the family farm system. It had received some hard knocks from the Secretary of Agriculture's so-called free market system. If Butz's policies continued, thousands more farmers and ranchers would be forced to leave the land, farmsteads would rot away and more small towns would die. The firm foundation provided by family farms would be gone. The result would be increased conglomerate activity and higher food prices.[41]

Ten days after Butz's resignation, the new Secretary of Agriculture, John A. Knebel, at President Ford's request, announced a 50 percent increase in the loan rate for wheat and a 20 percent increase for corn. Knebel helped reinstitute payments to farmers and government supported surpluses. His tenure was brief as Jimmy Carter triumphed over Gerald Ford in the presidential election of 1976. Carter chose Bob Bergland, a Farmers Union favorite, as Secretary of Agriculture. Bergland, whose mentor was Hubert Humphrey, was a Minnesota farmer who had been elected to Congress in 1970. Ray Fitzgerald of Plankinton, a former member of the State Union's board of directors, was named the new administrator of the Agricultural Stabilization and Conservation Service (ASCS). He had served for eight years as deputy administrator with the ASCS in the Kennedy and Johnson administrations.

In March 1977, the Carter administration began preparing a major agricultural initiative to present to Congress. Carter, deeply influenced by Chapter 41 of the Book of Genesis, which related the account of Joseph and the first ever-normal granary, wanted to establish an adequate grain reserve. Bergland proposed to do this by keeping target prices and loan rates at a level whereby farmers would not be encouraged to deluge government storage facilities with a huge surplus.

When President Carter signed the 1977 Food and Agricultural Act on September 29, 1977, the trend toward a market-oriented economy continued. The bill called for Humphrey's formula of $3.05 per bushel if the wheat crop was less than 1.8 billion bushels and $3.00 if the crop was larger. The Administration hoped that this formula would encourage farmers to grow less. The maximum for individual payments was raised to $50,000. By 1981, the last year of the four year act, Carter had allowed the target price for wheat to rise to $3.81 per bushel. The 1977 act authorized the Secretary of Agriculture to withhold land from production, but he had no power to compel farmers to comply. Bergland could ask for voluntary set asides, and if farmers ignored his request, they became ineligible for deficiency payments and loans. Although 13 million acres were taken out of production in 1978, farmers produced 300 million bushels more wheat than the year before.

A year earlier, the expected increase in income from exports did not materialize for farmers. Farm production had outstripped foreign demand, resulting in falling prices. Cash flow problems for heavily indebted farmers led to the American Agricultural Movement (AAM). Members of the AAM sent an ultimatum to Congress, demanding legislation that would ensure 100 percent of parity. If their demand was not met by December 14, 1977, they said that they would not plant a crop in 1978. Leaders of the AAM threatened to cancel membership in farm groups that did not support their proposal.

The Farmers Union did not join in the American Agricultural Movement's campaign. The AAM represented the fully commercial, well capitalized farmer, not the small farmer. Bob Bergland did not support the AAM "high rollers," and he was not intimidated by their angry "tractor march" on Washington, D. C., in February 1979. He and Jimmy Carter favored the family farm. Bergland's old political allies were members of the National Farmers Union.

Nevertheless, midway through the Carter presidency, the South Dakota Farmers Union became critical of the Administration's farm policy. In setting loan rates and target prices, the Carter government did not acknowledge the continued inflationary pressures on farmers. Ben Radcliffe said, "The Administration's program completely disregards the continuing increases in production costs and other inflationary costs that farmers must absorb."[42] The USDA's announced loan rate for 1979 was only 44 percent of parity while the target price was 64 percent of parity.[43] The Action Program of the sixty-fourth annual convention of the SDFU called for an incentive program that would guarantee 100 percent of parity prices for all production up to a family farm level. The delegates wanted a guaranteed annual net income equal to the average income of non-farm families. Farm production beyond the family farm level would be subject to market prices only.[44]

Senator George McGovern addressing an SDFU group.

When the Soviet Union invaded Afghanistan on December 26, 1979, President Carter announced an embargo on grain sales to the Russians. At that time $4.5 billion in grain had been purchased by the Soviets, but it still had not been delivered. In the months that followed, the Farmers Union accused the Carter administration of playing international politics with grain exports. Multinational grain traders had not suffered from the aborted grain deal, the victims of the embargo were family farmers.[45] Denying grain exports to punish political adversaries undermined the reputation of the U. S. as a reliable supplier among overseas customers and cost the U. S. the fatal loss of some markets.[46] The embargo was terminated in 1981, in the early months of the Ronald Reagan presidency.

In the closing months of his Union presidency, Ben Radcliffe renewed the call for 100 percent of parity to family-sized farms. Payments to farmers should equal the national median family income. The federal farm legislation for 1981 needed drastic revision. The open-ended payments of the past decade, he said, threatened the survival of the family farm. When the SDFU president delivered his final address at the 1981 annual convention, the farm parity index stood at its lowest level since 1933.

Low prices plus high interest rates threatened to put thousands of family farmers out of business. USDA statistics continued to indicate a declining number of farms in South Dakota throughout the 1970s. Acres were not taken out of production, they were incorporated into larger farms. Few of the departing farmers left because they disliked farming. They were forced to leave for the city because of a boom and bust farm economy that provided no guarantee of a decent living. While the Farmers Union had worked diligently to eliminate the causes for this rural exodus, engaging the anti-parity and corporate farm advocates in the nation's capital, the SDFU waged a successful fight to limit corporate farming in South Dakota.

THE SDFU LOBBY AT PIERRE

During the 1960s and 1970s, three men served as legislative counsels for the South Dakota Farmers Union during the sessions of the State Legislature. Roy Glover, who served as the counsel from 1951 until his death on December 11, 1963, was succeeded by Lafe Lunder of Canton. Lunder held the position through the legislative session of 1970. He was an experienced legislator who served four terms in the state Senate. He had vigorously promoted the enabling legislation which made it possible to organize, incorporate and operate rural electrical cooperatives under South Dakota law.[47] Lunder had been elected chairman of District I in 1961, and he became a member of the state Board of Directors the following year. He remained in these positions until his retirement in 1982.

Lafe Lunder, Canton. SDFU Legislative Counsel, 1963-1970; Member of the Board of Directors, District I, 1962-1982.

LeRoy Hardy of Sturgis became the Union's counselor in 1971. As chairman of District V, he was a member of the state Board of Directors from 1955 to 1967. After he resigned from the Board, Hardy became the SDFU fieldman for the West River area. He was a tour leader on a number of FU lobbying trips to Washington, D.C. in the late 1950s and early 1960s. Hardy was the Union's chief lobbyist until 1978.

During the legislative session, President Radcliffe would go to Pierre on Tuesdays and Thursdays to further strengthen the Union's voice in the state capital. With the appearance of Ag Unity, an organization which Radcliffe helped to found, the Farmers Union had a good ally in the halls of the Capitol building. It included several groups such as the South Dakota Beekeepers and Soil Conservation. Ag Unity worked hard to protect cooperatives.[48]

A Farmers Union Legislative Day was held during every session of the Legislature. Members were invited to the session from the various legislative districts, using individualized letters. They car-pooled or boarded the Union's buses at assigned locations for the journey to Pierre. At a morning session they reviewed the Farmers Union's legislative program. Following a noon luncheon, the SDFU faithful visited with their district representatives regarding the Union's legislative agenda.[49] In an effort to strengthen its political pressure, the SDFU, in 1979, designated ten people in each county as rural lobbyists. Well informed on specific issues, they were to respond to key legislation when the need arose.[50]

The Union needed articulate, knowledgable and energetic spokesmen in Pierre, because the opposition was well financed and experienced in legislative lobbying. In the opposition camp, year after year, were the lobbyists for the South Dakota Stockgrowers Association, Homestake Mining Company, the Farm Bureau Federation, John Morrell Packing Company and the "big oil" companies.[51] Despite the formidable opposition, the Farmers Union was able to pass significant pro-family farm legislation and to block the passage of several bills detrimental to the SDFU's principles and program.

UTILITY MATTERS

An amendment to the Consumer Power Law of 1950 became an irritant for Governor Archie Gubbrud (1961-1965). First introduced as the Fair Play Amendment, the legislation permitted a rural electrical cooperative to continue serving urban areas that might exceed 1,500 in population, if the cooperative received the approval of the governing body of the particular community.[52] Before the 1963 legislative session ended, the Fair Play REA bill became known as the Foul Play bill. In the final wording, the act authorized an electric utility serving the majority of customers within a city to acquire the electrical property and facilities of areas newly incorporated within the boundaries of a city or town. The acquisition took place after the utility paid a "fair and reasonable value" of the property and equipment. The amendment favored private power interests and drew the wrath of rural cooperatives.[53] REA proponents dubbed it the Foul Play bill because it forced rural cooperatives to sell out after an area it had served was annexed.[54]

The controversy persisted in subsequent administrations. Governor Nils Boe (1965-1969), immediately after his election, tried to resolve the jurisdictional differences between private and public power. He had representatives from public, private and municipal power companies meet for three months to work out their differences. At a final meeting in February 1966, they came to an accord. The Legislature created an Electric Mediation Board, but it was challenged in the courts and the state Supreme Court nullified the law.[55]

Boe's successor, Governor Frank L. Farrar (1969-1971), proposed a Gas and Electric Council to establish fair rates, settle territorial disputes between competing private utility companies and cooperatives and protect consumer rights. In House Bill 548, he also proposed that the utility industry be assessed the costs of the regulatory program. The Farmers Union opposed H. B. 548. The Union believed it would restrict the operation and expansion of rural electric cooperatives. Rural co-ops, argued the SDFU, did not need regulation or control.[56] Initially killed in the Senate, the measure passed on a 18 to 17 vote when a Republican senator changed his vote on reconsideration. Many lawmakers suspected that Governor Farrar had been unduly influential in the vote shift.

Believing HB 548 would not protect consumers but result in higher utility rates, the SDFU began a drive to collect signatures to refer the Gas and Electric Council law on April 5. By June 9, 1969, there were more than 50,000 signatures on referral positions, and the Secretary of State placed the measure on the 1970 general election ballot.[57] The contentious issue surfaced immediately during the Legislature's 1970 short session. House Bill 540, sponsored by twenty-four Republicans, called for repeal of H.B. 548. Governor Farrar agreed to the revocation if amendments were placed in the bill to provide protection for the consumer. The amended version resembled former Governor Boe's Mediation Board that had been subject to an unfavorable court decision. Repeal of the original Gas and Electric Council law eliminated the need to place the issue on the November general election ballot.[58]

TAXES: TRIUMPHS MIXED WITH FRUSTRATION

Given the Legislature's proclivity to increase the state sales tax when increased revenue was needed, any legislation to exempt items from the regressive tax was remarkable. The Farmers Union, arguing that none of the neighboring states imposed such a tax, convinced legislators to exempt fertilizer from the state sales tax in 1964.[59] In the following year, a bill was passed removing the sales tax from farm seeds.[60] Through the efforts of a "unity group," composed of all farm organizations with lobbyists in Pierre, the sales tax on agricultural chemicals was removed at the 1971 legislative session. It was estimated that elimination of the tax would save farmers and ranchers at least $125,000 a year.[61]

The Union's legislative efforts were directed mostly toward property tax relief. The SDFU believed that the best way to increase aid to public schools was through a graduated net income tax, based on a percentage of the individual's or corporation's federal income tax. As property tax levies increased in the late 1960s, the Farmers Union, through its twenty-five

Demonstrating for the repeal of the personal property tax (1978).

members in the State House of Representatives and Senate, introduced an income tax bill during the 1969 legislative session. The Union's proposal was backed by several other groups, including the South Dakota Consumers League, the State Federation of Labor, Lutheran Social Services of South Dakota, the South Dakota Education Association, Teamsters Union and the Independent Stockgrowers of America.[62] The proposed adjusted gross income tax bill received approval in the House but was rejected by the Senate.[63]

Approximately ten months later, at a meeting in Huron on December 12, the groups that had backed the ill-fated bill formed a coalition to initiate a state income tax. They called for repeal of the personal property tax over a three-year period, a homestead exemption to give needed real estate tax relief and an increase in state aid to education.[64] Throughout the winter of 1969-1970, members of the coalition for tax reform carried petitions across the state, securing 44,682 signatures.[65] In the summer of 1970, a fund drive to raise $45,000 for the success of the initiated measure in the fall election was conducted.[66] The initiated tax measure received only 39.2 percent of the vote in November 1970. The opposition argument that the proposal was an attempt to shift the tax load from the rural to the urban population took its toll. The "no" vote in urban centers ran between 70 and 85 percent. The farm vote favored the measure 60 to 40 percent.[67]

A positive aspect of the November ballot was the election of Governor Richard F. Kneip (1971-1978). He believed income and not property, to be a more accurate measure of wealth. Kneip immediately began a lengthy struggle to institute a tax structure based on ability to pay rather than on fixed assets. The new governor created a Council for Tax Decision to consider some kind of tax that was not regressive or on personal property. SDFU Vice President LeRoy Wiitala was named to the executive committee of the tax council.[68]

President Ben Radcliffe and the SDFU launched a different assault on the property tax. On October 11, 1971, he told delegates to the annual convention that attorneys were laying the groundwork for taking the state's property tax to court. The Union had decided to challenge the concept of financing public education primarily with local property taxes.[69] Through its attorney, Vincent J. Protsch of Howard, the SDFU filed its suit against the state's property tax in the Federal District Court at Sioux Falls on January 7, 1972. The Union argued that the property tax had forced farmers out of business and had compelled elderly people to work when they should have retired.[70] In the late winter of 1973, the U. S. Supreme Court rejected an appeal to hear the case, believing that the property tax as a main source of revenue for public schools was constitutional.[71]

For awhile, it appeared that Governor Kneip would have his way during the 1972 legislative session. His proposal for an adjusted gross income tax passed both the Senate and the House in different forms, but in attempting to resolve differences in the amendments attached to the bill by both houses, nothing satisfactory came out of the conference committee.[72] In 1973, the Democratic party controlled both houses of the Legislature for the first time since the New Deal, but the Governor's tax reform bill failed in the Senate by one vote.[73]

The Farmers Union had a political ally in Dick Kneip, and the organization, through the Ag Unity group, gave its wholehearted support to his tax reform proposals.[74] In the mid-1970s, farmers made up 25 percent of the state's population, but they earned only 24 percent of the income in South Dakota. At the same time, they paid 42 percent of all property taxes and 55 percent of personal property taxes, including 70 percent of all taxes paid to support public schools.[75] The Union cause was joined by two other groups in 1977. The SDFU, the School Administrators of South Dakota and the South Dakota Education Association decided to bring a lawsuit challenging the constitutionality of current funding programs for education in the state. With the reliance on property taxes, the prevailing system did not provide for uniform and equal educational opportunities for all children.[76]

The twenty year effort by the Farmers Union to change the tax system ended when the 1978 Legislature agreed to repeal the personal property tax or regressive "liar's tax." The tax had developed a bad reputation through minimal validation of self-assessment forms, and it had not been uniformly applied throughout South Dakota. A $16 million surplus in the state treasury also made the lawmakers more amenable.[77] In the years that followed, the South Dakota Farmers Union continued to call for an income tax or, at least, a corporate income tax. In the final year of Radcliffe's administration, the SDFU backed the tax cutting Dakota Proposition, which voters rejected at the ballot box in November 1980.[78]

THE FAMILY FARM ACT

The South Dakota Farmers Union launched a full-scale effort to curb corporate farming and its food chain monopolization of agriculture in the mid-1960s. In his message to the annual convention in October 1967, Ben Radcliffe said that corporations found farm land and agri-business to be "cozy havens for tax savings." His audience directed the state organization's leaders to initiate legislation to limit the growth of corporate farming in South Dakota.[79] The 1968 Legislature referred Senate Bill 92, designed to regulate corporate farming, to the Legislative Research Council for further study.

President Radcliffe and Governor Richard Kneip looking at a petition for the repeal of the personal property tax (1978).

President Radcliffe renewed his efforts to alert the state's populace to the threat of corporate farming. He warned that if the corporate movement into agriculture was not halted, it would "seal the doom and write the epitaph for family farming in agriculture." Many of the corporate acquisitions had taken place since 1960. An SDFU survey, in January 1968, indicated that one acre in twenty-seven in South Dakota was owned by corporations for a total of 1.6 million acres.[80] South Dakota was losing an average of 1,200 farms per year.

The foremost opposition to an anti-corporate farming bill came from the State Senate. A House-supported bill which prohibited corporations from engaging in farming in South Dakota, unless 51 percent of the stock in the corporation was owned by residents of the state, passed the House by twenty votes but was killed in the Senate State Affairs Committee in 1970. Governor Farrar was a primary opponent of the bill.[81] His opponent in the 1970 gubernatorial election, Richard Kneip, had introduced a similar measure in 1968 as a member of the State Senate.

Vested interests continued to defeat the pro-family farm legislation during the legislative sessions of 1971-1973. The key opposition came from the South Dakota Stockgrowers Association and the banks. An anti-corporate farm law would not permit banks to farm in South Dakota.[82] The family farm advocates, however, would not be denied, and the Family Farm Act was herded through the 1974 Legislature by Governor Kneip. This crowning piece of Farmers Union sponsored legislation protected family farmers against non-farm corporate giants. It explicitly prohibited conglomerates or other non-farm corporations from engaging in agriculture.[83] Concerned about foreign investments in rural America, the 1979

Governor Kneip signing Family Farm Act (1974).
At Kneip's left is SDFU President Ben Radcliffe. SDFU lobbyist, LeRoy Hardy is at right front.

Legislature voted to limit the purchase of farm land by non-resident aliens to no more than 160 acres.[84]

THE SOUTH DAKOTA RAIL AUTHORITY

The changes in technology following World War II had a devastating impact on railroads in South Dakota. The miles lost to rail service rapidly increased in the mid-1960s. In the decade following 1965, 540 miles of rail service were lost and another 400 miles were pending abandonment.[85] New Interstate Commerce Commission regulations allowed railroads to abandon any line that did not average at least thirty-eight carloads per mile per year. The railroad had to prove it had lost money on the particular section of track.[86] The Milwaukee Road, which operated over 1,424 miles of the 2,988 miles of railroad track in the state, verged on bankruptcy. The result was that South Dakota had one of the worst rail service and abandonment rates in the nation.

Radcliffe, as early as 1973, called for state and federal action to save the railroads. He asked for a state transportation task force to study South Dakota's transportation system. South Dakota's entire rail system was rapidly deteriorating to the point where it could no longer handle the food crop. Trains could only move at five to ten miles per hour on branch lines because of the extremely poor condition of the roadbed and track.[87] Concerning the railroads' competitor, the trucking industry, it took six times as much fuel to move freight on the highways as on steel rails. Economically, the cost of transporting wheat by rail was up to twenty-five cents per bushel lower than shipping grain by truck.[88]

Radcliffe and Emil Loriks were appointed to the South Dakota Task Force on Railroad Abandonment Policy. The group was asked to analyze abandonments with a view toward their economic impact upon farmers and communities. The task force also was to explore alternatives to abandoning branch lines.[89] Emil Loriks contacted an acquaintance in Minnesota's House of Representatives as to how that state was dealing with the crisis. He was told that the State of Minnesota had authority to purchase or possibly condemn branch lines scheduled for discontinuation. The primary purpose of any lines acquired by the state was that they should serve the public and not necessarily make a profit.[90]

South Dakota was confronted with the wholesale collapse of its railroad system by 1978. Fifty-one percent of the system was being abandoned. The largest percentage of any state in the nation. The shortage of rail hoppers and box cars meant that shipping costs to farmers would increase.[91]

The deteriorating situation alarmed Governor Harvey Wollman (1978). He was completing the unexpired term of Richard Kneip, who had been appointed ambassador to Singapore by President Carter. Wollman held a series of unique hearings to call public attention to the problem. A study prepared by economists at the University of South Dakota indicated that abandonment of Milwaukee Road tracks would cost farmers more than $30 million in lost grain sales.[92] The Farmers Union demanded a moratorium on railroad abandonments. Time was needed for either governmental or cooperative acquisition of railroads and for planning future operations.[93]

"Without available rail transportation," Ben Radcliffe told the annual meeting of the GTA in 1979, "I don't see how South Dakota farmers will be able to compete with grain producers located nearer to terminal markets or on good rail systems. The price spread will be too great."[94] He went on to say that saving the rail system might mean public ownership of railroads or a combination of public and private ownership. The 1978 Legislature had passed a bill allowing cooperatives to get into the railroad business, and the SDFU president was encouraged by the growing interest in the idea of cooperative ownership of railroads.[95]

When Governor William Janklow (1979-1987) brought his proposal for a South Dakota rail authority to the 1980 Legislature, 150 SDFU members were in Pierre to urge passage of the legislation.[96] Radcliffe argued that state and local ownership was necessary for the rail network to continue.[97] Despite opposition from the Farm Bureau and the Stockgrowers Association, the lawmakers agreed to establish and maintain a railroad core system. An extra penny was added to the state's sales tax to purchase track and to pursue future operations of a rail system. Using the $25 million raised through the tax increase, the administration purchased 765 miles of track for immediate use and an additional 219 miles for possible future operation. The 1981 Legislature added a penny to the state's gas tax to help fund operation of the state system.[98] Leaders of the SDFU were among the first to make the state's populace aware that there was a railroad crisis, and the Union played a very significant role in getting Governor Janklow to salvage essential rail service in South Dakota.

One reason for the SDFU's successes in the Legislature was Radcliffe's ability to build political coalitions. The Farmers Union worked with labor organizations in the state, and the National Farm Organization (NFO) moved from competition to cooperation.[99] Other allies included the National Consumer Federation, the South Dakota Consumers Federation and public power groups. These groups, working together, launched a petition drive for single

SDFU President Ben Radcliffe delivers petitions to Secretary of State Alice Kundert in 1981. The petitions called for a constitutional amendment to establish single-member state senate districts.

member state Senate districts. Dr. William Farber, professor emeritus at the University of South Dakota and a leading critic of multimember districts, was a guiding force in placing the issue on the 1982 general election ballot.[100] Prior to the election, three districts, Brown, Minnehaha and Pennington, elected 30 of 105 state legislators on an at-large basis. Voters approved the constitutional amendment requiring senators to represent and to be accountable to a certain district.[101]

In the legislative halls at the national and state levels, the Farmers Union worked diligently to save the family farm. Hostile administrations and congressmen in the nation's capital contributed significantly to the abandonment of many South Dakota farms in the sixties and seventies. In South Dakota, where there was still a substantial rural population, the SDFU labored effectively to have legislation approved which contributed to the survival of the small farmer. Removal of the sales tax on fertilizer and seeds, elimination of the personal property tax, passage of the Family Farm Act and the saving of a large part of the state's rail system were legislative triumphs for the Union and victories for the rural residents of South Dakota. In the same two decades, the South Dakota Farmers Union sought to build new bridges of knowledge and understanding within the state organization, with other groups within South Dakota and with people from overseas nations.

Chapter 16

BUILDING BRIDGES 1961 - 1981

The Farmers Union, a truly grassroots organization,
has been the greatest force for the welfare of agriculture in our time.
It has taken an active interest in legislation and the building
of farm cooperatives, but has made education a primary concern
preliminary to its many undertakings.
I am happy that education is the base of the Farmers Union triangle.
— Emil Loriks, 1981.

In past decades, different farm organizations had come together to form coalitions. In the 1960s and 1970s, they were unwilling or unable to come together. With the decline in the farming population, it became necessary for the Farmers Union to form coalitions with consumer groups to develop the necessary strength to win legislative battles. Family farmers, for their own well being, had to think of these consumer groups not as adversaries but as allies.

The key to building bridges with urbanites was education. It was also the key to ensuring future membership in the South Dakota Farmers Union. The youth program of the Union grew considerably during the Radcliffe years. Using a mix of traditional and new programs, South Dakota's Union youth and young adults learned the merits of Farmers Union membership through a variety of learning experiences.

Any pretexts of being isolationist or parochial were abandoned as the SDFU sought a closer relationship with farmers in other countries. Peace could best be achieved by reaching out to those people having a common vocation in distant lands. Once again, education was the key in building new bridges, bridges of understanding and unity.

THE NEW INTERNATIONALISM

The National Farmers Union, in cooperation with the International Cooperation Administration, had begun an international exchange program with several nations of Western Europe in 1957. Eight farm trainees from Western Germany, England, Austria and Ireland were placed in South Dakota in 1959. In turn, young men and women from rural America received a similar experience in Europe.

In the second year of the Kennedy administration, a new program was introduced. Under the auspices of the Agency for International Development (AID), a part of the Alliance for Progress, the Farmers Union established the Farm Leadership Program with five countries of Latin America in 1962. Dr. John Eklund, Director of International Projects for the NFU, was overall administrator for the Latin American program. As part of a technical assistance program, the Farmers Unions provided in-country training of selected Latin American leaders in rural youth activities, farm association programming, membership and organizational techniques, cooperative education and rural farm communications.[1]

Arlene Schley, Director of Education for the SDFU, received a ten week leave of absence in late 1962 to study in Caracas, Venezuela. She and five other Union personnel had an intimate exposure to Latin American culture, and they became familiar with the basics of the Farm Leadership Program.[2] Upon her return to South Dakota, Schley took charge of the state program, preparing for the arrival of the first group of young farm leaders in the spring of 1963.

Fourteen young farmers from Chile, Peru and Venezuela arrived in the state on May 10, 1963, for a four month stay. All of them were from the lower economic class, and they had very little formal education.[3] Prior to their arrival, they received four weeks of intensive English language training at Jamestown College, Jamestown, North Dakota. For four of the next five summers, South Dakota farm families played host to a different group each year. South Dakota did not participate in the program in 1967. The majority of the Latin Americans were placed with farm families in Davison, Hanson, Hutchinson and Miner counties. All of these young men and women had demonstrated leadership ability in farm and community development. Each person was affiliated in some way with either a *campesino* organization, cooperative or some municipal function in his home community.[4]

Although the month of language training proved inadequate, the Spanish speaking visitors were able to learn basic farm methodology and organizational skills from their Union hosts. A Spanish speaking coordinator was available at all times in the state to surmount major language barriers. Russell and Florence Tilberg of rural Ethan said their two guests were very interested in what took place on an American farm. One of them even got down on his knees to take a picture of a boar feeding at a trough. The guests, in turn, were very interesting people. One of the Tilberg guests liked to play the ukulele and sing; the strains of "Cielito Lindo" were often heard at the Tilberg farm. Their young man from Chile would eat chicken down to the bone, and he returned to his native land very well fed.[5] In 1965, the entire Latin American contingent took a two week short course in community development at South Dakota State University.

"Bien Venido Amigos" ("Welcome Friends") was the sign that greeted campers and eight Latin American guests when they arrived for the first Senior state camp at Camp Placerville in June 1965. At the second camp at Camp Wytamca in July, all sixteen Latin Americans and their host families were present. Arlene Schley arranged for each country's group to present music and dances of their native lands. Under the direction of John Eklund,the Education Department of the NFU filmed the social, farm, and cooperative experiences of the Latin Americans to recount the story of the Farm Leadership Training Program. This significant cultural and learning experience came to an end in 1968 as newly elected President Nixon abandoned what remained of Kennedy's Alliance for Progress and introduced his own policy toward Latin America.

The international theme continued to be emphasized at youth camps and in Junior and Senior study topics. Campers focused on the newly established Peace Corps in 1962. In ensuing years, world hunger, as the single most important threat to world peace, human rights, Latin America, the threat of nuclear war and the military industrial complex were subjects for discussion. In 1965, campers celebrated twenty years of the various programs of the United Nations.[6]

The South Dakota Farmers Union continued its strong support of the United Nations. A United Nations luncheon, with a guest speaker, continued to be one of the special features of the annual convention. In August 1966, President Radcliffe was appointed to the 100 member U. S. National Commission on UNESCO matters. He became the official representative of the National Farmers Union on the Commission; he was reappointed in 1969.[7]

The SDFU President became a global traveler as one of the NFU's four delegates to the general conferences of the International Federation of Agricultural Producers (IFAP), which had been formed in the late 1950s. The only world-wide association of professional farm groups, the IFAP included representatives from more than sixty countries. Farmers of different cultures and different races came together to discuss grain agreements, tariffs, soil conservation and the application of modern technology to agriculture. During his presidency, Radcliffe attended IFAP sessions in various national capitals, including Paris, France, Helsinki, Finland, Wellington, New Zealand, and Buenos Aires, Argentina.[8]

BARRIERS TO PEACE AND INTERNATIONAL UNDERSTANDING

American involvement in the Vietnam War (1964-1975) was always a subject for discussions at FU conventions in the late sixties and early seventies. The Union's opposition to American military involvement in Southeast Asia was exemplified through some of the guest speakers. A longtime critic of the war, Senator George McGovern, addressed the 1968 convention. The delegates, in their action program, asked for a halt to the bombing of North Vietnam.[9] In 1969, the Reverend James Armstrong, Resident Bishop of the United Methodist Church for the Dakotas, delivered an address which was very critical of the U. S. role in Vietnam. He had just returned from Paris, France, and a meeting of the World Council of Churches, at which the conflict was the center of discussion.[10] Before the convention adjourned, delegates asked for American withdrawal from Vietnam, recommending that American responsibility for the conduct of the war be handed over to the South Vietnamese government. The resolution stated:

> Though we have spent billions of dollars and the lives of 60,000 of our young men, the Saigon government has not implemented any meaningful social reforms. This, coupled with corruption among officials of that government, has made it impossible for them to get support among their own people.
>
> The waste of human and natural resources in Indo- China will continue so long as our military presence is there and the bombing continues — We believe it is not only in our interest but in the interest of the people of Indochina that we withdraw all military presence and aid immediately.[11]

Union members were in general agreement that the war did not truly represent the national interests of the United States. In addition, the large consumption of funds by the

Former SDFU president John Skage, NFU President James Patton, SDFU President Ben Radcliffe and former SDFU presidents Paul Opsahl and Emil Loriks

military-industrial alliance caused huge cuts in monies needed to solve domestic problems.[12] The Farmers Union also saw the hand of the military- industrial complex in the construction of the $5.5 billion anti-missile complex (ABM) in northeastern North Dakota in 1969. Believing that the United States had an abundance of overkill, the Union viewed the program as unnecessary and wasteful. Once again, the dollars spent on the missile complex could be better applied to solving domestic problems. Rural common sense proved correct, but only after the $5.5 billion had been expended. On October 1, 1975, the missile complex was declared fully operational. A week later, the project was declared obsolete and, within a few months, was completely closed down.[13] The Farmers Union believed that the Vietnam War and reckless military spending were unnecessary obstacles to building bridges of international understanding.

A STEADFAST ALLY

In his final convention address, in October 1981, Ben Radcliffe said that if farmers and ranchers were to survive, they had to forge coalitions with nonfarm groups. He urged the State Union "to look for new friends for the family farm."[14] New friends were found, including Ag Unity, churches and consumer groups. In the search for new allies, Radcliffe was not discounting a longtime friend, the urban laborer. For over fifty years, the Farmers Union had had a working relationship with organized labor.

Tony Dechant, President of the NFU, 1966-1980.

The bond between the two was annually reaffirmed at the Farmer-Worker Conference, first held in 1954. A primary purpose of each conference was to maintain an open dialogue between farmers and urban workers. Both groups had to contend with inflation, energy costs, high interest rates, inadequate income and adverse legislation. Each year a theme was developed for the conference which related to common interests and problems. The theme of the 1969 conference, for example, was "Are There Dangers in a Corporate Society?" One of the speakers was Victor K. Ray, Director of Public Relations for the National Farmers Union. He had recently authored a book entitled *The Corporate Invasion of American Agriculture.*[15] Discussion at the conference focused on three issues: 1) the corporate threat to society, 2) corporations run by managers, not stockholders and 3) a society endangered when the incentive for profit and desire for money overrode human values.[16]

About 200 farmers and workers would come together to hear NFU President James G. Patton and his successor, Tony T. Dechant, and a representative of organized labor such as Jack McCoy, the director of the AFL-CIO Committee on Political Education. Organizers of the conference were the president of the SDFU, the State Union's director of education, Cliff Schrader, president of the South Dakota Federation of Labor and Francis McDonald, secretary of the State Federation of Labor.

The Farmers Union always supported the efforts of the South Dakota AFL-CIO to repeal the "right-to-work" provision in the state constitution. When an effort to initiate a referendum to this effect failed to come out of a state legislative committee in 1973, members of the Union assisted in circulating petitions to have the repeal of the amendment referred to voters in the 1974 general election.[17] Voters rejected the repeal proposal, and the income of South Dakota workers continued to be limited by the "right-to-work" clause.

KEEPING THE RANK AND FILE INFORMED

An informed membership is an active and committed membership. Operating on that premise, the SDFU continued to enlarge its program to bring significant information to the rank and file. Each January and February during the 1960s and 1970s, mid-winter conferences were held at different locations across the state. Always on the agenda were the major issues facing the family farmer. Conference topics ranged from "The Farmers' Share in the National Economy" to "Who Will Control American Agriculture — Corporations or Independent Farmers" to "The Family Farm — Ghost of the Past or Reality of the Future?" In January 1968, the Union staged a series of town and country meetings, seeking to build a solid rural town and country alliance in the fight for equitable taxation, better farm prices and the regulation of corporate farming.[18] The SDFU tried a different format in 1972. One day Mid-Winter Seminars were held on four college campuses around the state. Young people and senior citizens were invited to these seminars to meet and talk with family farmers.[19]

Music at State Office meeting in the early 1960s.
***Union Farmer* editor John Holum and SDFU President Ben Radcliffe play a duet.**

During the 1970s, the SDFU scheduled a series of legislative action conferences each December, usually a week before Christmas. Staff members of the South Dakota Farmers Union joined prominent South Dakota legislators in leading discussions. Consideration was given to federal farm legislation and to crucial state issues, such as tax reform, land tenure and railroad abandonment.

By the 1960s and 1970s, a new generation of farmers was tilling the soil of South Dakota. They were the children of the cooperators of the 1930s and 1940s. Knowing only of the hardships of the Great Depression through textbooks, they had little awareness of the necessity of the cooperative movement. It was necessary for the Farmers Union and established cooperatives to acquaint these new farmers with the benefits and the values that co-ops could bring to their lives.

Radcliffe and his staff sought ways to keep the post-high school to age thirty group in the Farmers Union and to encourage their participation in cooperatives. The first Young Adult Seminar was held in the state office building auditorium in late December 1969. Approximately twenty young adults participated in a discussion of "The Corporate Invasion of American Agriculture."[20] The seminar format was continued in ensuing years, but participation was broadened to include the wives of the young farmers. At a Young Couples Weekend at Huron in March 1974, farm families learned about cooperatives and farm credit.[21] Six couples were selected from the weekend event to attend a National Young Couple's Camp at Bailey, Colorado, during the summer of 1974.[22] They joined other young adults and their children to discuss a proper land policy for the 1970s.[23]

By this time, David Wentzlaff had joined the Farmers Union staff as director of the SDFU Young Farmers' Program. In January 1975, he instituted a series of banquets built around the cooperatives in each area. By May, over 800 young farm couples had attended banquets in six different communities. After the meal, the diners participated in a question and answer session and formed into discussion groups.[24] These programs helped bridge the generation gap. In the post-banquet discussions, the young farmers and ranchers were encouraged to become involved in cooperatives and assume leadership roles in their communities.[25]

David Wentzlaff,
Director, SDFU Young Farmers' Program.

Wentzlaff, who had joined the staff in May 1974, resigned effective July 1, 1976, to become education director for the Rocky Mountain Farmers Union, headquartered at Denver, Colorado.[26] His replacement was James Eichstadt of Wolsey. He was a 1976 graduate of South Dakota State University. Under Eichstadt's direction, the Young Farmer program continued to gain momentum. By the spring of 1971, he was coordinating a series of county Young Farmer banquets, the SDFU Young Farmer seminars and Farmers Union participation in the Central Exchange-sponsored Young Couples conferences in South Dakota.[27] At the co-op appreciation banquets, the young farmers came together to visit with Central Exchange, GTA or Farmers Union representatives on a variety of farm issues vital to beginning farmers.[28]

Wentzlaff rejoined the SDFU staff in December 1978 following Eichstadt's resignation. Wentzlaff's responsibilities were enlarged upon his return to Huron. He continued to direct the Young Farmer program, but he also became a full time lobbyist for the Union. In the latter role, he replaced LeRoy Hardy, who continued as the SDFU's West River fieldman.[29]

Mrs. Wilma Putzier, SDFU Director of Education, joined with her counterpart in North Dakota to stage the first bi-state seminar for 18 to 25 year olds at Aberdeen on December 29-30, 1971. Forty youths from the two states came together to examine health care problems, possible national health insurance and the importance of young voters in the political process.[30] The event proved successful, leading to an expanded seminar in 1972, which included young adults from Minnesota and Wisconsin as well as the two Dakotas. Sherry Leach, NFU Director of Education, assisted in planning the second Young Adult Seminar on the campus of the University of Minnesota.[31]

New bridges of understanding were also built for the women of the South Dakota Farmers Union. Upon announcing a special Women's Conference at Sioux Falls, on February 14-16, 1977, Ben Radcliffe stated: "If we are to preserve family agriculture, we need the strong efforts of the entire family." The goal became to increase the involvement of farm women in the Farmers Union, in cooperatives and in other state and community organizations.[32]

In addition to an in-depth examination of the problems of family agriculture, women were urged to be involved politically. At a Women's Action Conference at Sioux Falls, in early May 1980, the women considered a variety of ways in which they could affect the polit-

President Ben Radcliffe and David Wentzlaff entertaining at the annual SDFU picnic.

ical process, everything from running for office to being a lobbyist to educating urban consumers regarding the problems of family farmers.[33]

The adult education program begun during Radcliffe's administration was comprehensive and effective. Thousands of adults became aware of rural economic conditions, of what the Farmers Union could do for its members and what they could do for the Union.

AS THE TWIG IS BENT

One dimension of the education of the pre-adult membership, the camping program, continued to flourish while new and innovative programs were being developed for their elders. Because of modern technology, particularly the automobile and television, there were greater distractions for young people in their home communities. By the end of the 1960s, there were a number of locals that did not have an active Junior or Senior group. A decade later, NFU headquarters in Denver believed it was necessary to establish a committee to get more of the youth involved in the family farm organization, affirming the East Asian adage that "as the twig is bent, so grows the tree."[34]

The future of the Farmers Union depended on knowledgeable and committed young people.

Leadership of the SDFU's education program remained strong through the two decades. Radcliffe was a staunch adherent of a vital youth program and several very competent people served as the Director of Education. Arlene Schley was the director for seven years, from 1959-1966. In the first year of her tenure, a state Senior Camp was held at Placerville for the first time. In 1964, two state Senior camps were scheduled for the first time, one at Camp Placerville and the other at Camp Watymca, on Lake Kampeska, near Watertown.[35]

SDFU Torchbearers, early 1950s. Back row: Harlyn Hellen, Jasper, MN.; Erlin Ellison, Vermillion; Glen Amdahl, Jasper, MN; Robert Perkins, Miller; Wayne Leuning, Jasper, MN; and Dale Braatin, Bryant. Second row: Shirley Fedt, Bryant; Ardella Van Rosendale, Montrose; Marilyn Fedt, Bryant; Ludelle Fedt, Bryant; Betty Porath, Ortley; Virginia Mielke, Elkton; Arlene Johnson, Winfred; Norma Goettsch, Ramona; Dora Carstensen, Bryant; and Betty Stienocker, Summit. Front row: Verla Coffee Barber, Wessington Springs; Josephine Atkins, Wessington Springs; Jeanette Johnson, Winifred; Patricia Viere, Elkton; Bonnie Sprong, Henry; Connie Christianson, Bryant; and Darles Winters, Huron.

Beginning in 1968, both senior summer camps were held at Camp Placerville. It had been Schley who capably directed the in-state training program for Latin American leaders. She resigned in July 1966 to accept a position with the General Services Administration.[36] During the following eleven months, Robin Ellison, the assistant education director, had charge of the program. On June 1, 1967, Joan Olsen was appointed Director of Education. A graduate of Hayti High School, she received her Bachelor of Science degree from Augustana College in 1967, with a major in sociology and minors in social science and education.[37] She resigned in July 1969, announcing her intent to marry in August and to move to Minnesota.[38]

Barbara Heidenreich of Northville was appointed acting director for one year. When Mrs. Wilma Putzier was appointed Director of Education on September 1, 1970, Heidenreich returned to St. Paul, Minnesota, to complete her college degree at Concordia College.[39] Mrs. Putzier became the director after having taught for eleven years in the rural schools of Hanson and Miner counties. She had been director of education for Hanson County and District II for the five years preceding her appointment. She received her education at Alexandria High School and Dakota State College.[40]

Putzier brought new stability to the position. It was during her tenure that the Young Adult program began to flourish. When she stepped down from her post at the beginning of 1979, she remained on the state staff as coordinator of the SDFU Travel Program. Her replacement, Kenneth Birgen, was a former Torchbearer. A graduate of Beresford High School and South Dakota State University, he had been teaching physical education at

Central Lee Community Schools, near Argyle, Iowa, before becoming Director of Education.[41]

At the national level, one of the major forces in shaping the Farmers Union education program, Mrs. Gladys Talbott Edwards, died in mid-November 1975. She had been education director for the NFU until her resignation in 1953 because of ill-health. The first Director of Junior Education for the North Dakota Farmers Union, beginning in 1929, she had been appointed National Director of Education for the Farmers Union in 1937.[42] Her successor, Mrs. Flossie Harris Nikkel, retired in 1972, and she was replaced by Sherry Leach.[43]

The age and grade categories for the Junior program were revised in 1963. Under the new Youth Education program, youngsters were grouped according to the grade division they were familiar with in school. A comparison of the old and new categories follows:

OLD		
Ages	6 - 13	Reserves
	14 - 21	Juniors

NEW		
Grades	1 - 7	Juniors
	8 - 12	Seniors
Ages	18 - 21+	Young Adults.

Under the new structure, members began their Senior work in the eighth grade. After five years of work, they could become a Torchbearer while in school and still at home. Under the old program, they completed their five years of work one year after high school graduation. The requirements for becoming a Torchbearer remained the same.[44]

One of the historic highlights for Torchbearers occurred in March 1971. Thirty-one South Dakota Torchbearers attended the 1971 NFU convention in Washington, D.C. This was the largest number of young people from any state that had ever attended a national convention. For many of them it was their first plane ride, an unforgettable experience. During their four days in Washington, they dined at the home of Senator and Mrs. George McGovern, visited foreign embassies, where presentations were made concerning agriculture in their respective countries, enjoyed breakfasts at Capitol Hill and toured the Capitol building. In time away from convention headquarters, they visited several of the nation's historic sites, including Mount Vernon, the Lincoln Memorial and the Smithsonian.[45]

Young people, who had participated in the Senior program for three years, were awarded a bus trip to sites in Minnesota. Education was combined with recreation and fellowship as the Seniors spent three days touring the Union's regional cooperative headquarters, the Central Exchange and the Grain Terminal Association, at St. Paul. They also visited the GTA's grain-loading facilities and storage sites on the shores of Lake Superior at Duluth, Minnesota, and Superior Wisconsin.[46]

CAMP ACTIVITIES

John, John the farmer's son,
Never did have any fun
Till he became a Farmers Union lad
Now he says he's very glad.
He goes to meetings

Farmers Union State Camp (1962).

And camps you see
And says, "The Union's the place for me![47] — Mildred Nielson

A number of camp sessions were given over to developing leadership and examining career opportunities. Through journalism, music and recreation workshops, campers developed skills that could be applied in their locals when they returned home. Camp themes for 1965 and 1966, respectively, were "Your Days of Decision" and "Your World of Work." Participating youths were made aware of the vocational opportunities available after they completed their formal education. In considering career choices, campers in 1977 visited a number of Rapid City businesses, including a veterinary and farm supply center, hospital, beauty shop, trucking firm, television station and the School of Mines and Technology.[48] Camp attendance reached a new high in June 1975 when 196 campers participated in the annual event.[49]

As in past decades, campers in the sixties and seventies listened to speakers, organized a co-op store, performed KP duty, attended classes and enjoyed many recreational opportunities. Hiking, performing skits, dancing and singing favorite tunes from the Farmers Union song book around a roaring camp fire became camp traditions. The president of the South Dakota Farmers Union would be present for at least one day of the camp, and during his presidency, Ben Radcliffe came to camp with his guitar to sing traditional and modern ballads plus an occasional tax reform song.

POLITICS AND PICNICS

The South Dakota Farmers Union had begun its Washington "Bus-Ins" in 1949. Committed Union members spent three or four days in the nation's capital lobbying mem-

Farmers Union Fly-In, Destination, Washington, D.C.

Farmers Union Fly-In (1996), National Farmers Union participants.

State Fair 4-H Barbeque (1977 or 1978).

bers of Congress for favorable farm legislation. In the 1960s, the SDFU began the Washington "Fly-Ins." On June 10, 1965, over one hundred South Dakotans flew to Washington, D.C., aboard a chartered Constellation aircraft. Fly-Ins replaced the bus trips. The Farmers Union began to use its two busses for summer tours. In 1966, approximately 2,300 members used the busses to travel to the Twin Cities to tour the GTA and Central Exchange facilities and to see the Minnesota Twins play baseball. Other tours were taken to the American Southwest, touring the Grand Canyon, Carlsbad Caverns, Dodge City, Kansas, and the Farmers Union headquarters in Denver. In the years that followed, bus tours became a regular service of the SDFU.[50]

The Fly-Ins to the nation's capitol expanded by the late sixties. There were annual Ladies Fly-Ins with the women lobbyists trying to persuade Congress to retain farm parity and improve net farm income.[51] In 1973, three Fly-Ins were scheduled, one each for Young Couples, farm program advocates and Women-In-Action.[52]

Special picnics for participants in the Washington lobbying effort were held each summer at the State Fairgrounds in Huron. The first annual state picnic, which was open to all Union members, was held on July 18, 1965. After a potluck noon luncheon, with coffee and cold drinks provided by the State Office, those present listened to a speaker discuss some aspect of farm legislation. The picnic site was moved to the lawn on the south side of the state office building in June 1974, where it continued to be held until the end of the 1980s.[53] Special guests each year were the people who had participated in the Washington, D.C., Bus-In or Fly-In.

Beginning in September 1975, the SDFU, GTA, and Central Exchange began to sponsor the South Dakota State Fair 4-H beef barbecue. John Morrell and Company, which had sponsored the barbecue since 1949, had withdrawn its support of this state fair event. The Union and its cooperative affiliates took on a formidable enterprise. Eighty to one hundred people were needed to cook, prepare food and serve the 6,000 to 6,300 people who came to the barbecue. Supplies included 2,500 pounds of beef, 6,000 buns and packages of potato chips, 7,000 paper plates and thousands of cartons of milk, chocolate milk and orange juice.[54] This good will gesture cost the State Union and the regional cooperatives up to $10,000 per year.[55]

Charles (Chuck) Groth, Director of Communication, SDFU, 1974-present.

PUBLIC RELATIONS AND THE PRESS

Very few of the State Union's members were not reached in one way or another by the education program. Pivotal in providing information on available programs plus reports and analysis on rural and Unions affairs at the local, state and national levels, was the *Union Farmer*. Several young men served as editor of the paper as well as director of public relations in the 1960s and 1970s. In charge of media relations during the two decades were:

Roland Olson	1958-1961
Richard Gaikowski	1962-1963
John Holum	1963-1965
Dick Ricci	1965-1970
William Wagner	1970-1971
Allan Burke	1971-1974
Charles Groth	1974-to present.

Charles "Chuck" Groth, the current editor and director of public relations, grew up on a farm near Inkster, North Dakota. He received his bachelor's degree in journalism (1970) from the University of North Dakota and a master's degree in political science from the same school in 1972. Prior to joining the SDFU staff, he was director of communications for the North Dakota Regional Medical Program at Grand Forks.[56]

In 1980, the South Dakota Farmers Union observed the fiftieth anniversary of the formal establishment of its Junior local program. Its founders would have marveled at the magnitude of the program five decades after its inception. The State's Union leaders adopted, changed and enlarged the program to meet the needs of each new generation while never abandoning the basic principles of the Farmers Union. No organization or structure can survive without a solid foundation. Education continued to provide that base for the South Dakota Farmers Union.

Other important matters must be considered before moving on from the '60s and '70s. Chapter Seventeen describes the state Union's administrative structure and staff, new programs adopted because of the rapidly changing social and technological climate and the continuing important economic role of the local and regional cooperatives.

Chapter 17

TORCHES ARE PASSED 1961 - 1981

"Farm supply co-ops are the farmers 'Ace in the hole'"
— John C. McKay, 1973.

"The cooperative's greatest value is not in the immediate savings on its daily transactions but in the service it performs to secure better and more stable prices over a long period of time."
— M. W. Thatcher.

At the annual convention of the National Farmers Union in March 1966, President James G. Patton and Vice President Glenn Talbott announced their retirements. Tony Dechant and Ed Christianson were elected to replace them. The fifty-one year old Dechant had been secretary-treasurer of the NFU for twenty years, and Christianson had presided over the Minnesota Farmers Union since 1950. Dechant, a coalition builder, retired in March 1980. His successor and the eighth president of the National Farmers Union was George W. Stone of Oklahoma.

An era ended for the national organization with Patton's retirement.[1] The NFU had grown and developed substantially during his twenty-six year presidency. He had served the interests of what he called the "farmers with income in the lower two-thirds" of the national rural economy. Patton made a household word of the term "family farmer."

He brought unity to the national body. No one opposed him for the national presidency after he was elected for the first time in 1940. This was in stark contrast to the power struggles within the leadership prior to that date. Ben Radcliffe, too, faced minimal opposition during his two decades as leader of the South Dakota Farmers Union. He was challenged only once, in 1971, in a presidential election. Retiring in 1981, he passed the torch of leadership to Leland Swenson.

The regional cooperatives also experienced significant changes in leadership. Emil Loriks retired as president of the Grain Terminal Association board in 1967. The next year, M. W. Thatcher, who had guided the GTA for thirty years, announced his retirement. Thomas W. Steichen, general manager of the Central Exchange for fifteen years, died in 1972. In the next eight years, the Exchange had three changes in leadership, with Darrel Moseson, a native South Dakotan, assuming the helm on January 1, 1981.

Despite the changes in leadership, the cooperatives, local and regional, flourished financially during the 1960s and 1970s. Co-ops remained the best method for farmers to pool their

NFU President James Patton and President Lyndon B. Johnson at the White House (1964).

buying and selling power to protect themselves from those in the economy who would take advantage of their individual economic weakness. Cooperatives continued to help farmers increase their net income. In retrospect, one of the most significant policy decisions made by the South Dakota Farmers Union during the 1930s was the decision to embark upon a program of building and promoting cooperatives.[2]

DECISIONS AND DECISION MAKERS

Three different men served as vice president during Radcliffe's twenty years as president of the South Dakota Farmers Union. One of three contested elections from 1961 through 1981 occurred in 1969 after Herman Pietz of Parkston announced his retirement as vice president. Lafe Lunder of Canton and LeRoy Wiitala of Frederick competed for the vacancy. Both men sought the support of veteran FU leaders. Loriks backed Lunder, whereas August Dahme supported Wiitala, a colleague and friend in the Brown County Farmers Union.[3] At the 54th annual convention of the SDFU, Wiitala emerged victorious, beating Lunder and W. E. (Bill) Holloway of Hartford.[4] The defeated Lunder expressed his bitterness in a letter to Loriks on December 16, 1969. He believed that lack of support from Radcliffe and the state office contributed to his defeat. Lunder was also told by two fellow members of the Board of Directors that he was considered too conservative and that he would be replaced as lobbyist at the 1971 legislative session. Lunder told Loriks that his energies would now be directed to representing District I and supporting the REA and credit unions.[5]

A gathering of SDFU presidents at the 1976 state convention.
Seated: Paul Opsahl and Emil Loriks; Standing: Ben Radcliffe, Oscar Fosheim and John Skage.

Two years later, Wiitala challenged Radcliffe for the presidency. It was the only time in his twenty years in office that Radcliffe was opposed in a presidential election. Wiitala's candidacy had been advanced by August Dahme of Aberdeen.[6] Supporters of both sides canvassed the membership to gain last-minute support. Radcliffe easily won reelection by a 9,218 to 2,340 vote. The vice presidential office opened up with Wiitala's announced candidacy for the presidency. Adam Seidel of Menno, a supporter of Radcliffe, defeated George Kane of Lemmon, a backer of Wiitala, by a 3,000 vote margin.[7]

The third contested election occurred in October 1981, following Radcliffe's announced retirement. Lee Swenson, who returned to Huron to be Director of Field Services for the SDFU in 1980, became the eleventh president of the state Union. He defeated Paul Symens of Amherst, the chairman for District III. Radcliffe did not endorse either candidate.[8]

Several people served as secretary-treasurer of the Union during the 1960s and 1970s. Holders of the position included:

Mrs. Joy L. (Johnson) McGregor	1959-1963
Mrs. Illene M. Harrison	1963-1967
Mrs. D. Maxine Peterson	1967-1970
Leland (Lee) Swenson	1970-1978
Norman Paul Tofflemire	1978-to the present.

In searching for a field director, Radcliffe had contacted the president of the Minnesota Farmers Union. He recommended Lee Swenson, a twenty-three year old graduate of

Leland (Lee) H. Swenson, President of the SDFU 1981-1988. President of the NFU 1988 to present.

Dorothy Authier, Vivian, Member of the Board of Directors from District IV (1969-1973) with SDFU President Dallas Tonsager.

Mankato State University. Swenson was employed in Tracy, Minnesota, at the time of his appointment to the SDFU post. He was a very successful organizer, and the membership of the South Dakota Farmers Union increased in six of the eight years of his tenure. Swenson resigned in late December 1978 to accept a position in the Field Services Department of the National Farmers Union. Norman P. Tofflemire, a graduate of South Dakota State University in agricultural education, taught vocational agriculture in public schools between 1973-1976. The Wessington Springs native was supervisor of the Feed and Farm Supply Store at the Selby Equity Union Exchange when he joined the State Union office staff.[9]

The first woman to join the state Board of Directors was Mrs. Oris (Dorothy) Authier of Vivian. She represented District IV from 1969 to 1973. The second female to be elected to the Board was Mrs. Carroll McKenna of Zeona, representing District VI as of 1975. The third woman to be elected to the SDFU's executive board was Mary Jo Nemec, representing District VII after November 1994. Two board members with remarkable terms of service died in the seventies. Harry E. Caslin of Kennebec died on April 13, 1979. He had been elected to the Board of Directors of the SDFU in 1922. He served until 1925 and was reelected, nineteen years later, in 1944. Caslin continued as District IV chairman for twenty-five years, retiring in 1969. He was the first president of the board of the Farmers Union Co-op Elevator of Kennebec in 1947, and he was board president in 1971, at the age of eighty.[10] The other venerable board member was Ewald Osterberg of Salem, who died on January 10, 1975. Osterberg had served as director from District II since 1959. An insurance agent in the Salem area since 1949, he was well known for his work with electric power cooperatives. Osterberg was one of the prime movers in starting the McCook Cooperative Telephone Company in the early 1950s.[11]

Because of the growth of Farmers Union insurances, the Board of Directors, in April 1975, developed plans for an addition to the state office building. The 3,950 square foot addition included a basement and a ground-level floor. It would allow for a small meeting room

A new addition to the SDFU State Office, Huron, July 1975.

for the Board of Directors and quarters for a research library. In June, the Board accepted bids totaling $159,686 for the construction of the 42 foot by 47 foot addition, plus remodeling existing offices. Picek Construction Company of Huron began working on the project on June 6, 1975. In February 1976, several of the office staff were able to move into the new addition.[12]

The auditorium of the state office building was brightened by the addition of several rural scene murals in the fall of 1971. Mrs. Edith Willie of Frankfort had presented the murals in memory of her husband, Leo. A. D. Clark of South Bend, Indiana, had been commissioned by Mrs. Willie to do the art work.[13]

Added to the Union's physical assets was a new thirty-nine passenger bus in September 1971. "There will be no more traveling in a hot, often broken-down vehicle," Radcliffe joked. Actually, it showed the growing maturity of the SDFU travel program. With two modern buses at its disposal, tours and visits to the regional cooperatives increased. To pay for the new vehicle and its maintenance, Radcliffe asked for contributions from cooperatives, county and local Farmers Union groups and the several hundred people who had ridden the FU buses in the past.[14]

CAUSES AND CONCERNS

The Farmers Union became involved in programs for the rural elderly in the mid-1960s. As a part of President Johnson's "Great Society," the Green Thumb horizon, which was

Paul Symens, Amherst, member of the Board of Directors from District III (1979-1984) with Dr. Paul Middaugh, pioneer in ethanol fuel at SDFU State Convention, circa. 1979.

funded by the Department of Labor through Title V of the Older Americans Act, was launched by the National Farmers Union in 1965 to provide employment and community service for rural elderly men and women. Through the efforts of President Radcliffe, Green Thumb started in South Dakota on March 1, 1968. Gerry R. Eisenbraun became the first state director, the Farmers Union (Green Thumb, Incorporated) being the grantee of the Senior Community Service Employment Program.[15]

When the program began, the Green Thumb enrollees were all-male crews working for government entities in construction, maintenance and weatherization. The state program advanced from seventy-two workers in five counties to more than 367 workers statewide in 1991. There was a shift in the type of employment for these low income senior citizens. More women than men became Green Thumb participants. Many held skilled clerical and paraprofessional jobs. Others were cooks at senior centers and schools or park employees.[16]

The National Farmers Union provided another service for retired and semiretired people. The FU Senior Member Council, in the spring of 1965, began providing a low-cost drug service for members of the organization. Paul Opsahl, one of the first to use the service, said he saved about one-half of the money he normally spent for pharmaceuticals. The Farmers Union cooperated with Greenbelt Consumers Cooperative through which the direct drug service was established.[17]

Political instability in the Middle East and threats of boycotts by the OPEC nations caused Americans to look for alternative energy sources. Discussion ensued on the production of alcohol fuels and the positive impact it would have on family agriculture in South Dakota. The Union staged a workshop at Huron on October 26, 1979, devoted to an in-depth examination of the methods and finances needed to distill alcohol. The 150 farmers who attended the meeting were made aware of the new grain and other farm products market that would emerge if alcohol fuels became a major source of energy.[18] In the early winter of 1979, the state Union appointed an Alcohol Fuels Advisory Committee.[19]

One cause that the South Dakota Farmers Union pursued but never satisfactorily resolved was the Oahe Irrigation Project. Governor Ralph Herseth (1959-1961) believed that "water was more precious than oil" and that extensive conservation practices must be introduced. As a result, he encouraged the passage of the South Dakota Conservancy Law, the first step in the proposed Oahe Irrigation Project.[20] Delegates to the annual conventions of the SDFU urged support of the Oahe Irrigation Project and the development of the James River Basin. Over 800 people attended a meeting at Huron in May 1971 to express their concern over the Nixon administration's impounding of funds for the Oahe Conservancy Subdistrict. South Dakota needed $1.5 million to start the project. A project which would provide economic growth, increase the tax base and improve the quality of life.[21]

Governor Richard Kneip tied the state's future in water development to the controversial Oahe project. The Flood Control Act of 1944 (Pick-Sloan Plan) authorized this project in the James River Valley as a replacement for the almost one million acres of land lost when the four mainstem dams were built on the Missouri River. Not everyone in the area affected favored the irrigation plan. In 1974, an opposition group, the United Family Farmers, challenged the project in federal court. Although the court rejected their suit, United Family Farmers continued to organize opposition to the plan.

The emotional debate reached a peak in 1977. The project's antagonists demanded that construction cease; Kneip would not support this. After the fall elections of 1976, a majority of the members on the Oahe Conservancy Subdistrict board either opposed or questioned the project. President Jimmy Carter agreed and requested that the program not be funded. When Kneip left office in 1978, federal funding for the project had been terminated, and efforts had been undertaken to deauthorize the Oahe plan and ultimately to kill it.[22]

THE TORCH IS PASSED

In September 1981, President Radcliffe made public his plan to retire. Earlier, he had notified the two men interested in succeeding him, Leland Swenson and Paul Symens, of his intention.[23] Swenson was elected president of the SDFU at the convention at Huron, beginning October 25. Delegates reelected Adam Seidel to his sixth two year term as vice president. He overcame a challenge by Henry Elsing of Chelsea. Radcliffe administered the oath of office to Swenson, who promised to "preserve, protect and defend the family farm system of agriculture."[24]

In his final "President's Column" in the October 1981 issue of the *Union Farmer*, Ben Radcliffe reflected on the many changes on the farm during his administration. Changes that had taken place because of new technology and the decline of the farm bloc in Congress. Technology had determined the transition from a two-cylinder John Deere to a four-wheel drive supertractor that cost more than a family farm did in 1961. American agriculture began to be driven by machines. These machines made machines out of the people who ran them and the land they were used on. Farmers' lives were driven by the need to pay for these machines.

In the same two decades, a farm program was lost as well as 12,000 family farms in South Dakota. Radcliffe said the defeat of the wheat referendum in 1963 was a turning point in American agriculture. It marked the beginning of the end of a stable farm program that had provided a loan program of nearly 90 percent of parity since the 1940s.[25]

The Sioux Falls Argus Leader paid tribute to Radcliffe in an October 1981 editorial. He was described as "a forthright champion of agriculture and the family farmer." The newspaper cited his legislative accomplishments and the State Union's ability to maintain its membership at about 15,000 despite the steady decline in the number of farms. *The Argus Leader* echoed Radcliffe's belief that saving the family farm was the biggest battle facing South Dakota agriculture. The editorial concluded by saying that Radcliffe ranked "among the state's finest citizens."[26]

THE STATUS OF SOUTH DAKOTA'S COOPERATIVES

Most cooperatives had been organized in South Dakota during the late 1930s and the 1940s. Many of them celebrated their 25th anniversary in 1965. People were reminded that their presence was due to the efforts of the Farmers Union, the Farm Security Administration, which provided the financing, and affiliation with the Union's two regional co-ops, the GTA and Central Exchange. Some of the state's cooperatives fell by the wayside following World War II. Nationwide, the number of farmer-owned cooperatives declined 9 percent from 1950 to 1961. Ironically, the surviving cooperatives did 50 percent more business in 1961 than they had done eleven years earlier.[27]

With the exodus of people from rural South Dakota, multi-purpose cooperatives appeared to be the best hope of sustaining small towns. When the population of a community declined, absentee-owned businesses left, reducing the services available. Population loss and reduction of services were a vicious circle. A cooperative, however, owned by the people it served, remained in the community, and it could be expanded to provide the services that were lacking. Furthermore, savings made on the co-op's business operations stayed in the community as patronage refunds to those people who patronized the firm. The governance of Farmers Union cooperatives remained as democratic as the traditional town meeting. FU co-ops continued to adhere to the principle of the Rochdale pioneers of "one member, one vote."

The National Tax Equality Association (NTEA) continued its assault on cooperative enterprises. An article in the *Reader's Digest* magazine, entitled "Why Should These Co-ops Get Tax Privileges?" argued that because cooperatives did not pay taxes on patronage dividends, they could use tax-exempt money to buy out their private enterprise competitors. This was the standard fare of the NTEA. Two co-op tax bills were considered by the Congressional House Ways and Means Committee at the urging of the NTEA. One would have required cooperatives to withhold taxes on dividends when the dividends were not paid out but held in reserve. The second bill required patrons to annually authorize in writing the reinvestment of patronage dividends in their own cooperative enterprises. Neither bill became law.

The Eighty-Seventh Congress finally clarified the income tax on patron refunds. The Federal Revenue Act of 1962 stated that: 1) patronage refunds belong to the patron, 2) taxes on refunds are paid by the patron, 3) a cooperative must pay the patron 20 percent of the annual refund in cash to help him pay the tax, and 4) the cooperative reports refunds of ten dollars or more to the United States Treasury. The new law further stipulated that when the patron does not elect to pay the tax, the cooperative must.[28] The NTEA also pressed its case in the State Legislature. Authored by John Buehler of Emery, House Bill No. 714, under the guise of an income tax, would have imposed a state tax on regional cooperatives. Fortunately, Boyd Meyer, fieldman for Central Exchange, discovered the subterfuge, alerted the Farmers Union leadership, and the bill was killed.[29]

The SDFU, in addition to being the leading advocate for cooperatives, maintained formal contacts with the major cooperative organizations. President Radcliffe named Gerry Eisenbraun as the Director of Cooperative Relations in the spring of 1965. The former manager of the Farmers Union Oil Company at Sioux Falls, through formulating and directing programs, was to build a closer working relationship between the SDFU and the Farmers Union regional cooperatives and their local affiliates.[30] Lafe A. Lunder, a member of the FU Board of Directors, had been appointed to the board of directors of the South Dakota Association of Cooperatives in June 1962.

For eighteen consecutive years, through 1965, the South Dakota Farmers Union staged an annual Livestock Market Day in Sioux Falls. The emphasis continued to be on youth, with young Union members invited to consign at least one head of livestock to the firm on Market Day. The day's schedule remained basically the same. The participants went through the steps for marketing cattle, toured the stockyards, the GTA feed mill at Ellis and enjoyed entertainment and a free noon meal in the Homemakers Center at the Sioux Empire Fairground.

In February 1965, Loyd B. Larson, cattle and hog farmer near Brookings, became the new manager of the Farmers Union Livestock Commission. Larson had been a member of the state Board of Directors for seventeen years. At the time, the FU Livestock Commission was one of nine commission firms operating on the Sioux Falls public market.[31] Larson's position was transitory as the Sioux Falls Livestock Commission was merged with the Farmers Union Marketing Association (FUMA) in December 1965.[32] The merger strengthened the financial structure of the Sioux Falls firm, and it was a step toward bringing the livestock marketing firms "under one roof." The Marketing Association, based in St. Paul, Minnesota, would maintain an office in Sioux Falls, as well as in West Fargo, North Dakota, and South St. Paul, Minnesota. The St. Paul Livestock Market was the second largest livestock terminal in the world, and it was considered the price-setting market for the northern Prairie Plains livestock industry.[33] Family farmers were more likely to receive competitive prices at the St. Paul livestock terminal. Ewald Osterberg of Salem was named to a two year term on the FUMA board of directors. David Hanson, another member of the SDFU Board of Directors, was appointed to a one year term on the same board.[34] Eighteen months after the Livestock Commission merger, on July 1, 1967, the Farmers Union Marketing Association merged with the Farmers Union Processing Association, which operated nine rendering plants in Minnesota and one at DeSmet, South Dakota. Headquarters for the former Processing Association remained at Redwood Falls, Minnesota, but the new "umbrella name" for the two organizations became the Farmers Union Marketing and Processing Association.[35] Frank Butler of Bison, SDFU Board representative from District VI, represented South Dakota on the board of the Marketing and Processing Association until his death in a plane accident in northwest Canada in 1975.[36]

Clifford Ott, State Manager, Farmers Union Insurance (1963).

Marvin Winter, Farmers Union Insurance.

The Sioux Falls Livestock Commission had been in operation for thirty-six years at the time of the merger. When it opened in February 1929, there were two employees — the manager who was also the hog salesman and an office supervisor. In 1965, the firm employed fourteen people plus the manager. Since its inception, the Commission had paid out well over a half-million dollars in dividends to its farmer-owners.[37]

The diminutive Paul G. Erickson, who developed and built the South Dakota Service Association (Farmers Union Insurances), died on May 20, 1963, twenty-one months after he retired as the first manager of the state Union's insurance division. President Radcliffe described him as a determined man who fought for his convictions whether popular or not. He "put principles ahead of every expedience and followed his convictions to the point of doggedness."[38]

Robert Kudrna, Erickson's successor, resigned after two years as state manager to rejoin his Farmers Union agency in North Dakota. Clifford Ott, a Verdon farmer and an FU insurance agent, succeeded Kudrna on September 1, 1963. For the next twenty-one years, Ott played an important role in establishing South Dakota as one of the top producing states for Farmers Union Insurance. Within two years after he assumed the post as SDFU Insurance Manager, agents submitted $1,500,000 more life insurance than in 1964. Marvin Winter of Huron became the first Farmers Union agent in the history of the company to top the $2 million mark in life insurance in force.[39]

Ott became involved in the Farmers Union after listening to a speech by August Dahme, whom Ott described as one of the most persuasive public speakers he ever encountered. The Verdon farmer was elected chairman of the Mud Creek Local, and in 1945, he became Brown County Farmers Union president. In July 1948, he attended a meeting at Mitchell, called to help recruit FU leaders as part-time agents for the fledgling NFU Insurance Company. Three years later, Ott was the Farmers Union State Insurance Manager for Oregon. He returned from Salem, Oregon, to his farm at Verdon after five months. He continued to farm and operate his local Farmers Union insurance agency until 1961, when he became a district manager. Significant changes took place in Farmers Union Insurance in the two decades that followed his accession to State Manager. Many agents shifted from being part-time to full-time agents. There was an increase in the variety of policies offered, making the insurance business more complex.[40]

THE REGIONAL COOPERATIVES: CENTRAL EXCHANGE

Farmers Union Central Exchange, along with its local cooperative owners, continued to pursue vertical integration, as producers, refiners, manufacturers, wholesalers and distributors of products. By the end of the 1960s, Central Exchange owned oil wells, refineries, pipelines, terminals, an oil blending plant, LP gas bulk plants, fertilizer manufacturing and blending plants, seed and feed plants and warehouses.

The National Farmers Union and Kerr-McGee oil industries began to jointly develop a potash project near Carlsbad, New Mexico, in 1962.[41] Within three years, production was underway at the Kermac potash plant, a $30 million mine and mill development. Annually, the plant could produce one-half million tons of premium grade muriate of potash, a basic fertilizer ingredient.[42]

Central Exchange increased its refinery capacity in March 1976. The regional cooperative joined with eight other cooperatives to purchase Atlantic Richfield Company's (ARCO) East Chicago, Indiana, refinery. Fed by a direct pipeline from the Gulf of Mexico, the refinery had a rated capacity of 126,000 barrels per day.[43] With the production of oil in northern Alaska, FU Central Exchange began to look for a reliable delivery system to convey some of this crude oil to the Prairie Plains states. In the summer of 1979, FUCE made an equity investment in Northern Tier Pipeline. As a first owner, the firm would be a shipper and a receiver of crude oil from the pipeline, which would run from Port Angeles, Washington, to Clearbrook, Minnesota. Northern Tier Pipeline would supply the Central Exchange refinery at Laurel, Montana. In 1979, the Laurel refinery had a daily capacity of 40,000 barrels per day with half of the daily output coming from Canadian crude oil.[44] Unfortunately, the governor of Washington, John Spellman, refused to grant Northern Tier the necessary construction permits, and the pipeline plans were abandoned in April 1983.[45]

Believing that the regional cooperative's survival depended on more than production and supply, Central Exchange (CENEX) emphasized performance and service. To enhance performance, CENEX sent cooperative specialists into the field to promote new business and marketing procedures. They also conducted employee training schools and clinics for directors and managers.[46]

Management emphasized that service was coequal to supply. Service ranged from on-the-farm tire service to leasing transport vehicles to local cooperatives. In extending its service capabilities, CENEX purchased Northern Cooperatives, Incorporated, of Wadena, Minnesota. The Wadena firm was a distribution cooperative, owning a sizable number of truck tractors and general merchandise and petroleum transport trailers. Supplies could now be delivered in vehicles owned by farmers. A few months later, CENEX and four other regional cooperatives created the Service Leasing Corporation. This joint venture made it possible for the regional and local cooperatives to lease trucks, cars and trailers.[47]

The sales volume for Central Exchange products soared during the two decades. The sales volume in the early 1960s had been around $104.5 million. By 1977, it had surpassed $645 million. Even with the inflation factor included, the increase was astounding. South Dakota's share of the sales each year usually ranged between 9 and 12 percent. Annually, local co-ops saved between seven and eight dollars on every $100 worth of supplies they purchased. In 1977, South Dakota had sales of $56.9 million, and the state's cooperatives shared over $4 million in savings.[48] The cooperatives in Toronto, Britton, Watertown, Lake Preston, Gregory, Corsica, Salem, Menno, Sioux Falls and Huron were consistently among the top ten in receiving the largest share of the savings. All of these co-ops paid educational funds and checked-off Farmers Union dues.[49]

Beginning in 1965, Central Exchange, in pursuit of excellence, began presenting Sparkle Awards. As part of its regional customer relations program, a "Mr. Sparkle," unknown to dealers, checked and numerically graded Central Exchange served companies on employee appearance, service awareness, attitudes and the general appearance of facilities. The Farmers Union Oil Company of DeSmet was a three year winner in 1967.[50] Only forty-one of all Central Exchange service stations, throughout the region, qualified the first year, 5 percent of the total. By 1974, the Farmers Union Oil Company at Chelsea had become a six time winner. Five time winners included the co-op oil companies at Alcester, Columbia and Wecota.[51]

FUCE helped farmers in northeastern South Dakota enter the "Information Age" during the summer of 1981. A cooperative, computer farm management system was made available to tillers of the soil. Nine cooperatives from Sisseton west to Columbia and south to Watertown founded the new co-op, which was headquartered at Webster. CENTROL, designed by Central Exchange in cooperation with Control Data Corporation, provided farmers with advisory services in crop and livestock management, soil testing, crop monitoring, irrigation scheduling and record keeping. Each of the nine participating cooperatives invested $7,500 in the system.[52]

During the seventies, CENEX broadened its service and supply west of the Rocky Mountains. In 1971, the Grange Cooperative Wholesale, with headquarters at Spokane, Washington, allied its sixty affiliates with CENEX. Five years later, in 1976, CENEX purchased the Utah Cooperative Association, further expanding its patron base in the western United States. The acquisition of Pacific Supply Cooperative in the following year brought an additional sixty co-op stations into the CENEX organization.

As of 1980, Farmers Union Central Exchange served over one-half million farm families in thirteen Prairie Plains and Pacific Northwest states. More than 1,000 local co-ops were affiliated with what had become popularly known as CENEX. Eight years earlier, in April 1972, the decision was made that CENEX would be the trade name and general corporate communication for the regional cooperative. The official corporate name remained Farmers Union Central Exchange. Central Exchange President John McKay said, "This is the first stage of a comprehensive reidentification program for cooperatives served by Farmers Union Central Exchange."[53] Three months later, Norval Ellefson, chairman of the board of Central

Sparkle Awards. Al Oukrop, CENEX vice president (mid-1970s).

Cenex Board Members. Doane Mortenson and Roy Kopperud.

Exchange announced the selection of the CENEX logo.[54] The gold co-op shield which had been the cooperative's insignia would be replaced by a new red and white CENEX sign.

Oscar C. Johnson of Oldham, first elected to represent South Dakota on the South St. Paul cooperative in 1944, retired in March 1962. He was replaced by Doane Mortenson. A former chairman of the Farmers Union Oil Company at Howard, Mortenson owned and operated a 640 acre grain and livestock farm near Howard.[55] Orrin Korth of Watertown was South Dakota's other representative on the thirteen member CENEX board until his resignation on March 8, 1974. His successor was the former chairman of the board of the Lake Preston Farmers Union Oil Company, Roy Kopperud. Kopperud, who also served on the board of the South Dakota Association of Cooperatives, farmed 1,600 acres in the Lake Preston area.[56] Boyd Meyer continued to be the South Dakota state manager for CENEX.

The president and general manager of the Farmers Union Central Exchange for fifteen years, Thomas H. Steichen, died of a heart attack on August 30, 1972. Steichen had been raised on a farm near Watertown and educated in the Watertown public school system.[57] His successor, John C. McKay, a native of Eddy, Montana, served for only sixteen months, announcing his retirement on December 31, 1973. Jerry G. Tvedt, a Wisconsin native and a former senior vice president of CENEX, presided over the firm until 1980. A native South Dakotan, Darrell Moseson, was chosen to lead the Central Exchange during the 1980s. Moseson was a lifelong member of the Farmers Union. His father, Gustav L. Moseson, had headed the Brokerage until the cooperative merged with Central Exchange in 1942. The elder Moseson then served as senior field representative in South Dakota for Central Exchange until his retirement on July 1, 1960. Following his retirement, Gustav Moseson started the South Dakota Fertilizer Association. In addition to promoting the fertilizer industry, he collaborated with his former neighbor, Oscar Fosheim, to get the sales tax removed from farm fertilizer.[58] Darrell Moseson, a 1950 graduate of Augustana College, had joined Central Exchange as a member of the Credit Department in 1951. In 1959, he became credit manager, and he was the Senior Vice President of Finance at the time of his elevation to the presidency.[59]

Thomas H. Steichen, President and General Manager of the Farmers Union Central Exchange.

Darrell Moseson, Sioux Falls. President and General Manager of the Farmers Union Central Exchange, 1980-1987

THE REGIONAL COOPERATIVES: GRAIN TERMINAL ASSOCIATION

Four years after the Grain Terminal Association was established, the regional cooperative began a methodical program to integrate production and processing. The process began with a durum mill in 1942, followed by the construction of feed plants in the 1950s. GTA began processing soybeans with the purchase of Honeymead Products of Mankato in 1960. Two years later, the processing of flax seed was added with the acquisition of Minnesota Linseed Oil Company. In November 1965, having declared that the barley business must be integrated, M. W. Thatcher, the GTA's general manager, announced that the cooperative had purchased the Froedtert Malting Corporation of Milwaukee, Wisconsin, the largest malting company in the world. The purchase included the company's subsidiary, the Red River Grain Company of Moorhead, Minnesota.[60] Froedtert, with plants in Milwaukee and Winona, Minnesota, processed 18 million bushels of barley per year, producing malt for the brewing, distilling and food industries. GTA members marketed between 30 and 40 million bushels of barley per year.[61]

Through its Honeymead Products oilseed processing division, GTA purchased a food processing firm, Holsum Foods, in July 1977. Holsum packed a variety of products, including salad dressing, peanut butter, jelly, olives, honey, syrup and processed mustard.[62]

As the Butz-dominated Department of Agriculture continued its sale of grain overseas in the early 1970s, including grain sales to Communist bloc countries, the GTA decided to increase its share of the export market. The FUGTA added facilities for loading grain into barges on the Mississippi River. In 1975, the cooperative leased a one million bushel terminal on the river at St. Paul from the St. Paul Port Authority and completed construction of a barge-loading facility at Winona, Minnesota.[63] The GTA also moved to share ownership of an export elevator at the mouth of the Mississippi River near New Orleans with five other regional grain marketing cooperatives. The group planned another export elevator to be located on the Texas Gulf Coast.[64] Construction began on another new barge-loading terminal at Savage, Minnesota in 1981. This new facility was also equipped to "load-out" unit trains.[65]

The Farmers Union Grain Terminal Association handled between 260 and 280 million member bushels of grain and oil seeds in the late 1970s. After taxes, earnings were almost $7.4 million at the close of the fiscal year on May 31, 1978. The GTA had earned a total of $203 million over its forty year existence.[66] The physical assets of the firm approached $450 million. Among these holdings were 118 building supply yards, twenty-one poultry and livestock feed manufacturing plants, three edible bean processing facilities, three barley malting plants and a mill for grinding durum, a hard wheat used in the manufacture of spaghetti and macaroni products. In 1977, Great Plains Supply Company opened a plant in Lennox, South Dakota, to provide farmers with fence gates and farm needs. The plant supplied local supply yards with jigs, frames and other equipment for efficient low cost building construction.[67]

Other physical assets included terminal and country elevators. Thirteen of GTA's 143 line elevators and eighty-eight of the firm's affiliated elevators were located in South Dakota in 1978. Bushels of small grains and corn received in 1978 from South Dakota members were 25,751,409.[68] The GTA had a feed service center at Chancellor, a regional marketing office at Aberdeen and the feed plant at Ellis. A two story office building to serve the GTA's feed division was built next to the division's feed manufacturing plant at Ellis in the early 1970s.[69]

Emil Loriks, a GTA board member for twenty-seven years, retired at the regional co-op's 1967 convention. He had been president of the board between 1957-1967. The transformation of the GTA from a grain commission and storage operation into an integrated cooperative marketing and processing enterprise began during his presidency. There were many times when Loriks, as the GTA president, had to play the role of diplomat. M. W. Thatcher was a dominating personality, and often, there were misunderstandings between the general manager and other GTA personnel. In his eighties and still wanting things to always go his way, GTA board members sought means by which Thatcher could be relieved of his duties. Working with Loriks, the board implemented a retirement at age seventy policy in 1967. Since Loriks was seventy-two in the summer of 1967, he was also obligated to step down. The "Co-op Warrior," M. W. (Bill) Thatcher, finally announced his retirement in June 1968. He was eighty-five years old. His successor was B. J. (Barney) Malusky of White Bear Lake, Minnesota. Thatcher died eight years after his retirement.

Before his retirement, Thatcher attempted to interfere in the election of Loriks' successor to the GTA board. Thatcher's efforts to work through SDFU board members Frank Butler and David Hanson were stymied when Loriks got Lafe Lunder and Harry Caslin to nominate and make speeches for Norman Olsen of Canton. Loriks seconded the nomination and Olsen was elected.[70] Olsen, a member of the SDFU board for five years, had been a delegate from the GTA affiliated elevator at Canton to the GTA convention for eight years. He was present at the GTA board meeting when the decision was made to ask for Thatcher's retirement.

Olsen became first vice chairman of the regional's board in 1978. South Dakota's other representative was Philip Testerman of Wessington, South Dakota. In 1972, he replaced August Dahme, who retired after twenty-six consecutive years on the GTA board. The amount of grain received from a state determined the number of representatives it might have on the board. Meeting once each month, the GTA board usually met in the Twin Cities, but at least once a year, they held meetings in Washington, D.C. and Milwaukee, Wisconsin.[71]

As the regional co-ops gained strength, the rights guaranteed to farmers under the Capper-Volstead Act came under increasing attack. In the fall of 1978, a national commission for the Review of Anti-Trust Laws and Procedures concerning agricultural anti-trust immunities made a special study of these rights. There was a general feeling that co-op farmers would have to meet new challenges.[72] Agriculture, in general, would be faced with new challenges during the 1980s. Farm exports declined dramatically, beginning in 1980, surpluses mounted, and the American farmer faced his greatest economic crisis since the Great Depression.

Chapter 18

LEGISLATION — FEDERAL ADMINISTRATORS VS. THE FAMILY FARMER 1981 - 1999

"If a person stands up for what he believes in, it is a true exercise in democracy, not one in futility."
— Frank Kloucek, 1988.

"When someone tells you that they understand the politics of the Farmers Union, please remind them that our politics are progressive. We are made up of both Democrats and Republicans with a vision for opportunity."
— Dennis Wiese, 1994.

In his valedictory address in October 1981, Ben Radcliffe said that the nation's farm economy was in the worst shape it had been in fifty years.[1] It became worse by the middle of the decade. The causes of the farm crisis of the mid-1980's can be traced back to 1973. The Golden Age of American Capitalism came to an end in the early 70s. The post-World War II economic boom ended with the Arab oil embargo, beginning on October 6, 1973. The Organization of Petroleum Exporting Countries (OPEC) enacted the embargo to pressure Americans into taking a pro-Arab stance as Egypt and Syria engaged in a war against Israel. At the time, the United States was importing one-third of its oil supplies. Although most of OPEC's members lifted the embargo in March 1974, oil prices remained high. The price of a barrel of oil almost quadrupled because of the success of the boycott. The radical price increase created a groundswell of inflation, increasing to a frightening 14.4 percent in 1979. The embargo also drove up unemployment, reaching 10.2 percent in the early 80s. The inflationary trend coincided with an increased demand for American farm products overseas. By the mid-seventies, one-third of all the farmers' production went to foreign customers. Producers were urged to plant from fence row to fence row as high prices brought land, some of it marginal, into production. During the 1970s, the acres cropped in the United States increased from 294 million to 363 million.

FROM BOOM TO BUST

The resulting rise in agricultural prices caused many farmers to borrow heavily to expand their output. Borrowing was advantageous in a time of high inflation and low interest rates. Between 1975 and 1983, the national farm mortgage debt soared from less than $50 billion to more than $112. In 1985, about half of the farmers in the country were in debt and the interest payments on their debt came to $21 billion.

The economic boom of the seventies became a "bust" beginning in 1979. The Federal Reserve System decided to fight runaway inflation with a tight-money policy. Interest rates rose and inflation fell. The rate of inflation declined from 12 percent at the beginning of the 1980s to 4 percent by 1984. Simultaneously, the demand for American grain in foreign markets declined radically. U. S. farm exports decreased in value by 21 percent in 1980 and 1981. In 1981-1982, the nation experienced the most severe of nine recessions since World War II.

There were three primary reasons for the decline in exports. First, the dollar rose in value against other currencies, resulting in an increased cost of U. S. exports to other nations. Second, other world food producers worked to capture export markets through increased subsidies. Finally, world production of grains increased significantly. Many third world countries moved toward self-sufficiency because of better seed, new plant varieties and new farming technology. The use of grain as a diplomatic weapon also contributed to the decline of exports. The Carter Administration imposed an embargo on grain to the USSR after the Soviets invaded Afghanistan. The declaration of the embargo, on January 4, 1980, cancelled sales of 17 million tons of grain.

Without a supply management program, U. S. farmers continued to produce for an export market that no longer existed. By the end of 1982, the United States had acquired one of the largest grain surpluses in history. From 1973 to 1982, corn in storage increased from 484 to 3,120 million bushels. The amount of wheat stored increased from 340 to 1,515 million bushels and soybeans, in storage, rose from under 60 to over 344 million bushels. In the early 1980s, the United States was selling its grain on the world market for less than production costs and subsidizing the sales with tax dollars.

In confronting a deteriorating situation, the Farmers Union and the Republican administration, in Washington, D.C., presided over by Ronald Reagan, took radically different approaches. The new president of the South Dakota Farmers Union, Leland Swenson, became a leading spokesman for the Union's position. Within seven years, the thirty-three year old native of Westbrook, Minnesota, went on to become the president of the National Farmers Union.

He grew up on a diversified farm in Minnesota. His family received a National Farmers Union Farm Family Living Award in 1953. Active in the farm organization's youth program, Lee Swenson earned Torchbearer honors in 1965. He graduated from Mankato State University and served in the U. S. Air Force prior to becoming secretary-treasurer of the SDFU in 1971. As Field Service Coordinator for the National Farmers Union from 1978-1980, Swenson directed the nationwide "Save Our Co-ops" campaign.

Adam Seidel continued on as SDFU vice president until his sudden death on August 23, 1985. Ironically, he had announced two days before his death that he would not seek a record eighth consecutive two year term. At the annual convention in October, the delegates selected Michael O'Connor of Alcester as his successor in a three way race with Dallas Tonsager of Oldham and Richard Moser of Wessington. Two years later, in 1987, O'Connor said that he would not be a candidate for reelection. Tonsager, in an uncontested election, became the vice president.

Swenson had called for a positive supply management farm program and substantial increases in commodity loans as the best way to meet the farm crisis. President Reagan's response was found in the four year farm bill which he signed on December 22, 1981. SDFU President Swenson described the legislation as the worst farm bill in fifty years. The law became effective as the farm parity level plunged to 57 percent at the end of the year.[2] The new bill set a minimum loan rate for wheat at $3.55 per bushel and a target price of $4.05. The law expanded the grain reserve program. In the years that immediately followed enactment of the farm legislation, the amount of money spent on farm programs increased from $4 billion to close to $12 billion, but prices fell to the disadvantage of the family farmer.

Farm prices dropped by 12 percent in 1981, while the prices farmers paid for consumer goods rose from 5 to 15 percent. A Statewide Farm Crisis Committee, which included President Swenson, called for the establishment of an international Mercy Food Program to avert a total collapse of the agricultural economy. Providing food to the hungry millions in developing countries would reduce the commodity surplus and increase prices.[3] Radical solutions were needed as there was a 400 percent increase in farm and ranch bankruptcies in South Dakota in 1982, a result of high interest rates and the lowest farm prices since the 1930s.

John Block, Reagan's Secretary of Agriculture, implemented the "Payment in Kind" (PIK) program, in early 1983, in another attempt to reduce the crop surplus. Farmers agreeing to withhold additional land from production would not be paid in money but in corn, sorghum or wheat from the farmer-owned reserve, grain under loan or Commodity Credit Corporation stocks. The SDFU board of directors gave qualified endorsement to the program. Eighty-six percent of South Dakota's wheat farmers signed up for PIK in 1983. Over 80 percent of the corn and sorghum farmers enrolled in the program. Farm land taken out of production in the state totaled 3.8 million acres.[4] Nationally, 80 million acres were removed from production in 1983.

Additional acres were removed from production in 1985 through the Conservation Reserve Program (CRP). While aimed at reducing excess production, the CRP also preserved land resources. Nationwide, thirty seven million acres of erodible land were removed from cultivation for ten years. More than a half-million acres of South Dakota farmland were idled under CRP. The enrolled land was seeded to permanent grasses and converted into windbreaks or wildlife cover. Many older farmers used the program to reduce their farming operations or they rented their remaining cropland to their rural neighbors.

Leaders of three of the state's farm and ranch organizations met in Sioux Falls in early 1985 to formulate a unified approach to national farm policies. The three presidents, Leland Swenson of the SDFU, Ken Eckman of the National Farmers Organization and LuVerne Jensen of the South Dakota American Agricultural Movement called for:

1. an immediate moratorium on farm foreclosures.
2. the establishment of a program to restructure farm debt, reducing interest rates to a serviceable level.
3. a substantial increase in commodity loan rates and target prices.
4. an increase in loan rates to at least the cost of production.[5]

Following President Reagan's veto of an emergency credit bill, more than 8,000 farmers and ranchers, members of the three organization, gathered for a Farm Alliance Rally at Pierre on February 12, 1985. The Union's state office staff was responsible for organizing the rally. Lowell Hanson, Jr., had provided twenty-five Jackrabbit Line busses that brought many of the farmers and ranchers to the State Capitol. It was the second largest gathering of farmers in South Dakota's history. The first, also organized by the Farmers Union, brought 10,000 people to a park in Huron in 1932 to hear John A. Simpson.

After listening to speeches for three hours at Riggs High School, the assembly marched to the Capitol, According to Charles Groth, SDFU Communications Director, the march was an awesome sight. Streaming out of the school building, the farmers marched ten to twenty abreast toward their meeting with the Legislature.[6] Deeply impressed by the size of the throng, the House of Representatives responded to their beleaguered agrarian constituents by appropriating the funds necessary to send the 105 member legislative body, plus Governor William Janklow, to Washington, D.C., to lobby for the restructuring of farm and ranch debt at serviceable interest rates. Ron Larson of Hetland presented Governor Janklow with a check for $20,000. Through the "Give-A-Buck" campaign, promoted by Larson and KWAT-radio, Watertown, the money had been raised to support the lobbying effort.[7] While in the nation's capital, the legislators made Congress and the media aware of the severity of the farm crisis. President Reagan said that he did not have time to meet with the delegation.

Unable to confront reality, the Reagan administration continued to argue that lower market prices would make the United States more competitive in world markets. Even though parity dipped below 50 percent in the spring of 1986, the Department of Agriculture froze target prices at the 1985 level. Loan rates were reduced with wheat being cut to $2.40, corn to $1.92 and oats to 99 cents per bushel. After acreage reductions were announced, the Administration's overall objective was clear — farmers were being asked to produce less for less.[8]

The Farmers Union's pleas for mandatory production limits were given little attention by the new Secretary of Agriculture, Richard Lyng. When only one-fifth of the ballots were returned in a national poll, asking producers if they supported production limits on wheat, Lyng disregarded the poll. Over 65 percent of South Dakota's farmers had favored controls.[9] With little assistance from Washington, the South Dakota Farmers Union did what it could to assist financially hard-pressed farm families and ranchers. Using a grant of $10,000 from country-western singer Willie Nelson's Farm Aid, Incorporated, a credit counseling service was established, called Farm Help. The service emphasized counseling and advice for Farmers Home Administration (FmHA) borrowers who were threatened with foreclosure. By calling a toll free Farm Help Hotline, farmers and ranchers were made fully aware of their rights, of rules and regulations and the FmHA appeals process.[10]

An ironic tragedy was taking place. The United States had a farm program costing more than any other in its history while small and medium sized farm operators were going broke. Rural businesses were selling out and small town schools were closing. Land prices in South Dakota dropped 45 percent between 1982 and 1986, the sharpest decline in the 20th century. The sale price for the state's farm land in mid-1986 had fallen to 1970-1972 levels when adjusted for inflation.[11] From 1986 to 1987, the average land price fell 17 percent, to $178 per acre. It was the first time in a decade that farmland values had fallen below $200.[12] By October 1986, there had been over 500 bankruptcies in South Dakota, one hundred more than the record set in 1985.[13] In the following year, South Dakota's 5.2 per thousand rate of

Chapter 12 bankruptcy filings was the highest among surrounding states. The FmHA turned 353 state farmers and ranchers, who had already voluntarily liquidated, over to a collection agency.[14] Twenty-six percent of the more than 12,000 FmHA borrowers in South Dakota were delinquent in their payments.[15]

FARMS IN SOUTH DAKOTA[16]

YEAR	NO. OF FARMS	AVERAGE ACRES
1931*	84,300	
1978	38,741	1,147
1987	36,376	1,214
1992	34,057	1,316
1995	33,000	1,333
1996	32,500	1,354

** Record high of farms in South Dakota*

President Ronald Reagan spoke to a largely Republican audience at the Sioux Falls Arena on Monday, September 29, 1986. Outside of the Arena, approximately 500 farmers and union laborers staged a three hour farm rally, protesting against the Administration's farm policies. Following his speech, Reagan invited two of the protest leaders, Charlie Johnson of Madison and Richard McBrayer of Ethan, to ride with him to the airport. In the ensuing dialogue, Reagan advanced the Administration's standard arguments regarding farm policy. He said the Administration helped farmers by lowering interest rates and inflation. He went on to say that the Administration had provided more money for farmers than the last five previous administrations, ie. $26 billion in 1986. The President argued that expanding world markets, not government subsidies, would bring farmers a fair price. McBrayer told Reagan about farmers' credit problems, particularly with the Farmers Home Administration (FmHA). The Ethan farmer said the FmHA in South Dakota was not using all of the options available, such as deferring loans and rescheduling payments. South Dakota's farmers did not need more loans, they needed proper service on the ones they already had. Johnson and McBrayer told the President that money should go to small, family producers.[17]

The Reagan administration's expected increase in agricultural exports never materialized. Exports declined by more than 37 percent between 1981 and 1986. In 1986, for the first time in fifteen years, the United States incurred monthly agricultural trade deficits.[18] Members of the SDFU continued to call for a revised farm program that emphasized supply management. Mandatory control measures accompanied by increased commodity loan rates would more effectively reduce supply than the diversion programs devised by the federal government. A primary objective of the Farmers Union was to repeal the Findley Amendment to the 1985 farm bill. The amendment allowed the Secretary of Agriculture to reduce loan rates by 5 percent per year.[19]

In the spring of 1987, farmers in South Dakota could anticipate spending $168 to produce an acre of corn and selling it for only $107, based on 1985 production costs and March 1987 corn prices. The farmers' costs included fertilizer, repairs, miscellaneous expenses, seed costs, fuel, chemical and drying corn. Fixed costs included overhead, taxes, insurance and interest payments. Corn, which sold for $3.06 a bushel in April 1981, was placed on the market for approximately $1.30 per bushel in 1987. It cost farmers $103 to produce an acre of wheat. Their crop from the same acre sold for $67.

CORN PRICES PER BUSHEL IN SOUTH DAKOTA
1981 - 1996 *

1981	$3.03	1989	$2.45
1982	2.35	1990	2.17
1983	2.53	1991	2.23
1984	3.05	1992	2.31
1985	2.49	1993	1.86
1986	2.15	1994	2.48
1987	1.30	1995	2.09
1988	1.78	1996	3.46

* *The statewide monthly per average for March.*

The Administration's continued emphasis on exports with low market prices and large federal subsidies that flowed to corporate farms was devastating to rural South Dakota. The number of people living on farms and ranches declined from 112,854 in 1980 to 76,170 in 1990, a decline of 32.5 percent.[20] Addressing delegates to the SDFU's 71st annual convention, held at Aberdeen, President Swenson said that the problems of rural America could be overcome. "It won't be easy," he said, "It will take involvement and the support of those who care about the future of our state and nation."[21]

He, personally, carried the struggle to a higher level when he successfully ran for president of the National Farmers Union in March 1988. The former SDFU president began his campaign in August 1987, not long after NFU President Cy Carpenter announced that he was retiring.[22] At the national organization's annual convention at Albuquerque, New Mexico, Swenson emerged victorious in a close contest with John Stencel, the president of the Rocky Mountain Farmers Union. Swenson had 126,801 votes to Stencel's 118,505. South Dakotans played key roles in Swenson's election. Tom Neuberger of Canistota and board member Carroll McKenna of Zeona delivered the nominating and seconding speeches. More than one hundred other SDFU members attended as delegates, distributed caps and campaign buttons, put up posters and balloons and acted as hosts in the Swenson for president hospitality room.[23]

The South Dakota Farmers Union Board of Directors named Vice President Dallas Tonsager of Oldham as the acting president of the SDFU on March 15, 1988. Tonsager, a 1976 graduate of South Dakota State University, and his wife, Sharon, operated a dairy farm and raised small grains and corn south of Oldham. The thirty-three year old dairy farmer had been active in the Farmers Union for several years. He became chairman of the FU Dairy Task Force in 1982, and he continued to serve as chairman of the SDFU Dairy Council. The Oldham farmer was also president of the Kingsbury County Farmers Union and District III cooperative director. Very involved in the cooperative movement, Tonsager was a member of the board of directors of the Oldham Farmers Elevator for seven years, and he was one of the Land O'Lakes Volga Area Voting Unit board.[24]

Dallas Tonsager,
President of the SDFU, 1988-1993.

Another dairy farmer, Clifford Gascoigne of Yale, was Tonsager's opponent when delegates to the 73rd annual convention, meeting at the Mitchell Corn Palace on November 4-5, voted to decide on Swenson's successor. Dallas Tonsager overcame the challenge of the former Beadle County FU president, and current member of the Board of Directors from District III, to become the twelfth president of the South Dakota Farmers Union. Cecil Hall, the legislative director for the Brookings County Farmers Union, defeated Tom Neuberger of Canistota in the vice presidential contest.

Federal farm policy changed very little when George Bush succeeded Ronald Reagan as president in January 1989. Clayton Yeutter, the new Secretary of Agriculture, recommended "more of the same" when formulating the 1990 farm bill. Yeutter wanted a continuation of the target price trends established under the 1985 act. He also favored the Findley Amendment, giving him authority to lower loan rates to "Market Cleaning Levels."[25] The Farmers Union, through its Fly-In program, lobbied for a "bushel-based" supply management farm bill for 1990. It would tie farm program bases to actual bushel or pound production instead of the more traditional acreage bases. The Union's plan called for farm supports to be paid on a certain level of crop production beyond which farmers would be at the mercy of the market.[26] Yeutter got his way and small farmers continued to face threats to their net income.

THE UNION'S POSITION ON INTERNATIONAL TRADE AGREEMENTS

During the early 1990s, the Union's leadership at the national and state levels was preoccupied with the international trade agreements which were being negotiated in Washington D.C. The Farmers Union, as early as 1989, took a stance against the General Agreement on Tariffs and Trade (GATT), realizing that it could create export dependency, destabilize domestic farm programs and lower commodity prices. The agreement called for percentage reductions in export subsidy levels. By reducing all subsidies at the same level, American farmers were the losers because it locked in the advantages some countries already possessed. Congress would be restricted in the use of the domestic tools used to foster the farm and rural economy. Overall, the agreement had a negative impact on family farm producers because it would lead to domestic program cuts and increased imports. At the end of 1993, the FU lost a four year struggle as the final GATT agreement was accepted. The only winners were the multi-national purveyors of farm products.[27]

The Farmers Union hoped that the farm economy would improve after William J. Clinton defeated George Bush in the 1992 presidential campaign. NFU President Lee Swenson believed that the Clinton administration would be more farmer friendly. Swenson did not foresee any changes in the 1990 Farm Bill, but he believed it would be administered in a more pro-farmer manner.[28] Union members were not disappointed. In March 1993, Mike Espy, President Clinton's Secretary of Agriculture, announced suspension of FmHA foreclosure actions to provide borrowers with the opportunity for a new review of their cases by an independent panel. Insisting that the FmHA wanted to be a friend in need, Espy said borrowers, who were in foreclosure but who had not been referred to a court, would have thirty days to request that their case be reviewed by an independent review board.[29]

The South Dakota Farmers Union, for the third time, played host to the NFU convention when the delegates gathered for their annual meeting at Sioux Falls on March 5-8, 1993. The

Parade float during the NFU national convention in Sioux Falls (1934).

national body had met in Mitchell in 1925 and in Sioux Falls on November 20-21, 1934. The changes wrought by sixty-years of technological change were apparent in examining the two national conventions in Sioux Falls.

In 1934, over 5,000 Farmers Union members from thirty states had met at the Coliseum for their official sessions. Public address systems were in their infancy. Many of the delegates came to Sioux Falls by train. Surviving South Dakota delegates fondly remembered the two mile long parade through downtown Sioux Falls. Floats and banners emphasized the New Deal's AAA and the Farmers Union's program for cost of production, farm refinancing and monetary reform. Many floats were horse drawn. Several State Union members went to Sioux Falls two days prior to the parade to build their floats.[30] In 1993, the delegates came to Sioux Falls either by airplane or automobile. There was no parade, and their headquarters were the convention facilities at a motel, the Ramkota Inn. No horses were visible, there was no passenger train service to Sioux Falls and the Coliseum was a condemned building, awaiting the wrecking ball. President Clinton spoke to delegates via video tape in March 1993. Clinton said he would like the upcoming 1995 farm law to better serve the needs of family farmers. He proposed to put family farm agriculture in the forefront of his plans for rebuilding the nation.

Dennis Wiese, Flandreau, President of the SDFU, 1993-present.

An action of the new Democratic administration changed the leadership of the South Dakota Farmers Union. Dallas Tonsager was appointed as the new administrator for the Farmers Home Administration effective May 3, 1993. The SDFU board of directors selected thirty-two year old Dennis Wiese of Flandreau to serve as acting president until the state convention at Pierre on November 4-6. Wiese, and his wife Julie, operated a diversified 960 acre farm near Flandreau. He had served as Moody County Farmers Union president, and he was an organizer and member of the planning committee for the Farmers Union Families and Friends (the Farmers Union young farmers' group). He and his wife were serving as co-directors of the Farmers Union education program for Moody County at the time of his appointment to the presidency.[31]

Delegates to the 78th annual convention elected Wiese to a full two year term as president when they assembled in Pierre in early November. He was the only candidate to file petitions for the position. Less than five months later, he was selected to fill a vacancy on the National Farmers Union Executive Committee.[32] Cecil Hall, vice president since 1989, overcame a challenge from Jeff Moser, the former president of the Hand County Farmers Union, to retain his position.

At the same convention, the delegates rejected another international trade agreement, the North American Free Trade Agreement (NAFTA). They believed the proposed agreement did not provide proper protection for agriculture. Insufficient consideration had been given to NAFTA's potential impact on agricultural commodity prices, livestock health and environmental issues. NFU President Swenson said NAFTA would hurt South Dakota's economy. He estimated that the state could experience a drop of over six percent in gross agricultural income.

An earlier Canadian Free Trade Agreement had resulted in huge increases in agricultural imports at the expense of U. S. producers. By March 1994, Canadian grain was flowing into the United States in ever increasing amounts. The amount of spring wheat entering the U. S. had increased fivefold since the trade agreement was ratified in 1988. Canada's share of the U. S. durum market had gone from zero to 22 percent in less than six years. The Union called on the Clinton administration to impose Section 22 of the agreement, which permitted the USDA to impose tariffs and/or quotas whenever such imports threatened or undermined U. S. farm programs.[33] The belief that international free trade would mean higher prices was an illusion. Commodity prices decreased as the trade in grain increased. With bi-partisan support, the Congress, nevertheless, ratified NAFTA. The approval of GATT and NAFTA reflected the loss of agricultural power on Capitol Hill. The number of Congressional districts dependent on agriculture had decreased by 50 percent between 1984 and 1994.[34]

The Farmers Union viewed President Clinton's agricultural program with mixed feelings. On the negative side, the Administration had supported both GATT and NAFTA. The REA and the hunger and food assistance programs had been subject to budget reductions. Likewise, funding for mediation programs for farmers needing assistance in dealing with the federal credit system had taken a blow from the budgetary axe. On the plus side, loan rates began to go back up in 1994, the Administration had moved to check the influx of Canadian wheat and new ten year contracts had been issued for continuation of the Conservation Reserve Program (CRP).

In the fall of 1994, preliminary discussion began on the content of the 1995 farm bill. An agricultural summit conference was held at Huron as a preliminary step to a National Rural Summit at Des Moines, Iowa, on December 1. The 220 South Dakota farmers, ranchers and commodity people who gathered at Huron agreed that there should be a substantial increase in commodity loan rates. They asked for a shift to a supply management oriented farm program based upon bushels rather than acres. The assembly supported continuance of the Conservation Reserve Program, providing that the program was fiscally responsible and targeted at highly erodible lands. NFU President Swenson argued that CRP should not be funded by reductions in the commodity program. Congress should provide separate funding because the program benefitted all Americans with its emphasis on wildlife habitat, water quality and the reduction of soil erosion.[35]

South Dakota's family farmers had weathered the most severe agricultural crisis since the Great Depression. The costs of the struggle had been high. Many people had been forced from the land because of the misguided and misdirected policies emanating from the nation's capital. Significant challenges lay ahead for the South Dakota Farmers Union as the debate continued over the 1995 farm bill. There were philosophical differences within the Department of Agriculture. Changes in the farm program would come slowly, however, because of the entrenched bureaucracy in the Department of Agriculture. The bureaucrats persevered despite changes in presidential administrations. Urban Congressmen wanted a greater share of the budgetary pie at the expense of the rural economy, and with a Congressional majority obsessed with budget reductions, the Union faced a formidable task in fulfilling its policy objectives.

Chapter 19

LEGISLATION — HOGS, TAXES AND IDENTITY 1981 - 1999

"The democratic principle is that there never was, never has been, never will be, room for the ruthless exploitation of the many for the benefit of the few."
— President Harry Truman

Tax reform, anticorporate farm legislation, rural economic development, building alliances and strengthening identity through the increased involvement of its membership were primary goals of the SDFU from 1981-1999. The increasing burden of real estate taxes, possible large-scale hog and poultry confinement production facilities and the need for programs to counter the assault upon rural life in the 1980s, caused the SDFU to strengthen its grassroots lobbying efforts in Pierre. President Swenson scheduled the first annual SDFU Co-op Legislative Day at Pierre for February 9, 1982. He wanted cooperative managers and directors to be more involved in the legislative process. Their presence during the annual legislative session would help counter the impact of the anti-farmer owned cooperative lobbyists who were present whenever the lawmakers assembled in Pierre.[1]

Dennis Wiese revised the longstanding format for the annual Farmers Union Legislative Day program, beginning with the 1995 legislative session. Union members would be in the halls of the Capitol Building on four different days rather than one. Wanting the legislators "to know that farmers and ranchers are in town," Wiese scheduled three separate Legislative Day program with members from the seven districts assigned to particular days. The Board of Directors would also meet in Pierre when the legislators were in the city. The traditional morning meeting at the King's Inn was eliminated. The grassroots lobbyists would go directly to the Capitol Building at 10:00 A.M. for a short briefing session and attend the second round of morning hearings.[2]

COUNTERING CORPORATE HOG FACTORIES

Corporation farming, once again, became a major issue in South Dakota, eleven years after the SDFU secured passage of the original anticorporation farming law in 1974. An amendment to the Family Farm Act, Senate Bill 131, vigorously supported by Governor Janklow, was approved by the House of Representatives, 48-20. The bill allowed corporations that built or considered building agricultural processing plants in South Dakota to own land sufficient to produce 30 percent of the commodities needed by the plant.[3]

By May 1985, the Union had begun a petition drive to submit Senate Bill 131 to a vote of the people. The proposed exemption to the anticorporation farming law never was placed on the election ballot. Responding to pressure from the Farmers Union and the South Dakota Pork Producers, the 1986 Legislature voted to repeal Senate Bill 131. The Governor had favored the establishment of a corporate potato company near Pierre. When the company went bankrupt, Janklow seemed to lose interest in Senate Bill 131.[4]

Within eighteen months, the corporate farming issue was once again a major concern of the Farmers Union. National Farmers, Incorporated, a Kansas City based corporation, planned to build major hog confinement operations in Hughes and Brule counties. It was estimated that the two operations would each market 300,000 hogs per year. The 1974 Family Farm Act did not limit ownership of agricultural feeding operations; it did prohibit corporate ownership of agricultural land. In 1986, 2,644,000 hogs had been marketed in South Dakota. The output of one of National Farmers' confinement operations would be more than 11 percent of that total. The Farmers Union feared that National Farmers' proposal, if implemented, would displace thousands of South Dakota livestock producers.[5] Homer Ayres was inspired to write a few lines of protest poetry. The poem, entitled "Corporate Ham and Bacon," included the following verses:

Ten thousand pigs up north of Pierre
That ought to make one quiver
The drainings of that mess of stink
Will get down to the river.

The smell will rise high in the sky,
and float for miles around
Repulsion of the Sunshine State
Will grow by leaps and bounds!

From Maine to California
The state will then be known;
As the stink capital of the U.S.A.
Where "all them pigs is grown."[6]

The Farmers Union brought a proposal to the 1988 State Legislature to amend the 1974 Family Farm Act by prohibiting non-family farm corporate ownership of hog production facilities. The Legislature concurred with the Union's position, but Governor George S. Mickelson (1987-1993), a strong advocate of corporate hog production, vetoed House Bill 1130. When the Legislature was unable to override the Governor's veto, the SDFU, in cooperation with the South Dakota Pork Producers Council, began circulating petitions to put the measure on the 1988 election ballot.[7] Mickelson's efforts to form a coalition to defeat Dakota Proposition II and the amendment to the Family Farm Act, Initiative Measure No. 3, failed. The only group to openly express opposition to the anticorporate hog farm amendment, prior to the November election, was the South Dakota Farm Bureau.[8]

Voters approved the amendment but the issue refused to go away. In the spring of 1992, a large-scale hog confinement facility was proposed for Tripp County. It appeared that its promoters were trying to circumvent the 1988 amendment to the Family Farm Act.[9] They attempted to use the ruse that the corporation president was the actual owner of the proposed 2,000 sow confinement center near Winner.[10] The pressure to bring giant, corporate hog farms into South Dakota came from both out-of-state and in-state interests. Certain in-state farm leaders, and members of the state government, called for the repeal of the 1988 amendment. SDFU President Dennis Wiese said these people "apparently believe that it is more important to have corporate hogs in South Dakota than it is to have family farmers."[11] The pressures for the corporate enterprise came at a time when family farmers were considering cutting back on hog production. Hogs, in the fall of 1994, sold for $34 per hundredweight, and the establishment of large-scale production facilities would destroy the family farm market. When the South Dakota Pork Producers resoundingly rejected a Governor's task force proposal to repeal the 1988 amendment, the contentious issue was either dropped or shelved for the time being.[12] During the 1995 legislative session, no attacks were mounted on the twenty-one year old anticorporation farming law.[13]

The specter of corporate hog farms in South Dakota raised its head once again during the summer of 1996. Out-of-state conglomerates such as Murphy Farms, Tyson Foods and DeCoster Farms proposed to negotiate contracts with South Dakota farmers whereby the corporate firms would supply the hogs, feed and veterinary services and farmers would provide the buildings and labor. The proposal immediately raised major economic and environmental questions. Were such operations permitted under the Family Farm Act? The Attorney General, Mark Barnett, believed that the Act allowed the development of conglomerate-contract farms. He called on the Legislature to clarify the law.

Would such contracts be beneficial to South Dakota farmers? In-state proponents of corporate hog farms argued that South Dakota farmers needed to raise more hogs to meet the demands of the state's two major East River meat packers, John Morrell and Company, Sioux Falls, and Dakota Pork Industries in Huron and Mitchell.[14] Only 35-40 percent of the hogs slaughtered by the meat packers came from South Dakota. Members of the South Dakota Pork Council said contracts would increase slaughter hog numbers while helping young farmers finance expensive hog confinement operations and would offer all farmers a steadier way of living.[15]

Others cited the economic disadvantages of the contract program. Speaking for the SDFU, Charles Groth expressed concern about conglomerate profits leaving the state, thus evading tax payments on their South Dakota earnings.[16] Farmers who signed contracts would be obligated to pay for building costs, bonding and insurance, increased property taxes and possible damages or pollution cleanup.[17] Farmers currently engaged in hog production had the potential of making $29,000 per year on 2,500 hogs. The conglomerates offered farmers $25,000 per year to produce 10,000 hogs per annum. Fewer farmers would produce more hogs for less money. The decline in the number of farmers would further shrink rural communities and the tax base for public schools.[18]

Contract farming would contribute to the further vertical integration of the hog industry. A few companies would control the many phases of hog production — "from breeding to feeding to processing to selling."[19] Independent farmers might lose access to competitive markets if contract farming occurred on a large scale. SDFU President Wiese said the program would make it possible for corporations to control hog prices as they did in the beef industry. Ironically, the attempt to further vertically integrate the hog industry occurred as attempts were being made to deconcentrate what had happened in the beef industry.

What would be the impact of hog contract farming on the state's natural environment? In responding to this question, Governor William J. Janklow immediately entered the controversy. Large-scale confinement production of hogs in other states had caused both air and water pollution. Hogs produce a large amount of manure and its stench could be inhaled by people over a wide area. Lagoons would be constructed to store the waste. If the containment areas seeped or the walls broke, nearby sources of drinking water would be contaminated. In mid-August, 1996, Janklow proposed new regulations to govern hog farms. He recommended the following new rules:

1. New swine units with 2,500 pigs or more must apply for a general permit.
2. Lagoons and waste disposal sites must not be closer than 1,000 feet from public drinking water or 250 feet from private drinking water.
3. Earth lagoons must be lined to minimize seepage.
4. Containment structures must meet minimum designs and be able to store at least 180 days of manure.
5. A producer violating the new regulations could be fined $10,000 a day and face criminal penalties.
6. Local governments should control zoning, air quality and distance regulations.
7. Large hog farms that existed before August 14, 1996, would not need a permit unless complaints were filed against them and state officials decided that a permit was needed.[20]

Upon securing the permit, large-scale hog farmers would be informed that they must adhere to existing rules protecting lakes, streams and rivers from pollution.

Janklow had proposed that the rule changes be enacted by October 1. Many of the state's residents, including Union President Wiese, called for a delay in implementing the Governor's recommendations. Wiese, along with several of the state's leading newspapers, asked for more time for consideration of the issues. They proposed that the 1997 Legislature make the necessary decisions. Wiese pointed out that about one-half of the state's counties had no zoning regulations.[21] President Wiese went on to say that if the lawmakers failed to provide adequate protection for the environment and independent producers, the issue would probably end up on a general election ballot. "I don't want this state to be known as the `Dances With Pigs' state," Wiese stated.[22] Frank Kloucek, a legislator and a hog and cattle farmer from the Scotland area, stated, "We're at the crossroads in family farm operations. We can give in and allow conglomerate agriculture to take over or we can take a stand and work for solutions that benefit us all."

With the continued threat of the "Tysonization" of agriculture in South Dakota, a coalition, including the Farmers Union, was formed to revise and strengthen the 1974 Family Farm Act in May 1997. The coalition formulated Amendment E which prohibited non-family farms, corporate ownership of agricultural land and livestock or livestock production facilities. The coalition, consisting of the SDFU, Dakota Rural Action, the SD National Farmers Organization (NFO), SD Resources Coalition, the SD Independent Pork Producers, the SD Livestock Marketing Association, the Northern Plains Sustainable Agricultural Society and the Peace and Justice Center, secured 31,162 certified petition signatures and delivered them to the Secretary of State on November 3, 1997. Amendment E was scheduled to be voted on by the general public in November 1998.

Supporters of Amendment E argued that corporate intrusion into South Dakota agriculture would eliminate a good farmer to land ratio. They believed that a farmer knows his land

and what is good for it. In addition, a good animal to land ratio required that wastes be recycled to the land, returning them to the soil rather than becoming a hazard to the natural environment. Corporate farms, a modernized version of the feudal system, would force individual producers from the land. Like the company towns in 19th century America's coal fields, South Dakota's farmers would become wage employees of the corporate giants. As proven in North Carolina. which has large corporate farms, property values and population decreased dramatically in the affected rural areas. Factory farms in South Dakota would only increase the threat to rural areas. Towns detached from urban centers were losing population. Approximately 200 communities dependent solely on agriculture had experienced a population decline during the last three decades of the 20th century.

As the date of the general election drew near, the debate over Amendment E became heated. A strong proponent of the amendment, Phil Cyre, a Hazel farmer and vice president of the SDFU, said "We are being overwhelmed by corporate megafarms whose sole intent is to control food production in the United States. Nothing looming on the ballot today is more important than Amendment E to family farmers and the agricultural structure that we have today." Opponents questioned the legality of the amendment, saying that it would lead to lawsuits. SDFU president Dennis Wiese responded that Amendment E was not that complicated. He reiterated that it would prohibit a corporation from owning agricultural land or engaging in farming, unless a majority of the ownership interests were held by family members. One of these members would have to reside on the farm or be engaged in its day-to-day labor and management. The proposal also exempted farmer-owned cooperatives that met certain criteria. It would not allow, however, nonfamily members to form limited liability partnerships or companies to farm or feed livestock. The amendment was patterned after a similar measure approved in Nebraska that had withstood legal challenges on constitutional grounds. With the assistance of urban voters, Amendment E was approved by the electorate at the general election on November 3, 1998. The vote was decisive, 149,470 (58.67%) favored the measure, 105,282 (41.33%) were opposed.

In late June 1999, owners of five livestock operations plus the South Dakota Farm Bureau and the South Dakota Sheep Growers Association filed a lawsuit in U.S. District Court in Pierre, challenging Amendment E. President Wiese, speaking for the Union, commented that "There are certain individuals out there who want to take from the land, from the people, and not live up to their responsibilities. This [Amendment E] is about protecting a way of life, and in many cases, there are basic health issues involved."

TAXES: GIVING AND TAKING AWAY

Tax reform became an overriding issue in the mid-1980s. At the annual state convention of the SDFU in 1985, the assembled members called for the Legislature to place a cap on annual property tax increases. Their motion read: "The confiscatory level of property taxation has contributed to the economic crisis now facing rural communities. In many areas, property tax delinquency rates are increasing rapidly and threaten the stability of funding for education and county government."[23] A proposal by the State Union to use some of the state's budget to fund an $18 million property tax rebate was rejected by the Legislature.[24]

In the early fall of 1987, the decision was made to let the people decide on tax reform in a general election. The SDFU board of directors gave qualified endorsement to Dakota Proposition II, a proposal to reduce property taxes and to establish a limit on future increases. Lee Swenson indicated that rural property taxes had increased by almost $20 million a year since 1981. At the same time, agricultural land values had declined by over 50 percent in many areas of the state.[25] Union members joined in the successful effort to secure enough signatures to put the measure on the 1988 election ballot. The SDFU became part of a speaker's bureau favoring the issue. Union members traveled to various parts of the state in a spirited campaign for Dakota Proposition II. Insufficient numbers at the ballot box resulted in the defeat of Constitutional Amendment "C," and it was "back to the drawing board" in the struggle for tax reform.

When the Legislature met in January 1989, a Farmers Union bill to roll back property taxes by 25 percent, if the Legislature failed to enact property tax relief by 1991, never made it out of the Senate Tax Committee.[26] The growing resentment against the property tax did motivate Governor Mickelson to place a two year freeze on property tax increments. The tax freeze did nothing to alleviate the tax burden on property that had given rise to the tax reform movement. Property taxes would remain at the 1989 level through 1991.[27]

Anxious to continue the assault on the property tax that had begun in 1987, the South Dakota Farmers Union, in the early spring of 1990, joined with the South Dakota Education Association, the South Dakota Association of Senior Centers, organized labor and the Voice of South Dakota Taxpayers to form a coalition for constructive tax reform. The People for Tax Reform circulated petitions to put their proposal, Initiative 2000, on the November ballot. The coalition called for a 25 percent reduction in property taxes, elimination of the state sales tax on food and the contractors' excise tax. State coffers would be replenished through a graduated corporate and personal income tax based on taxable income, after all deductions had been made.[28]

Two years after the failure of Initiative 2000, a new coalition, Citizens for Property and Sales Tax Relief, turned in petitions to get property and sales tax reductions on the general election ballot. The proposal was very similar to Initiative 2000, with the sales tax being eliminated on clothing and utilities. If successful, property taxes would have been reduced by 20 percent.[29] The reform measure, Initiative No. 3, fared poorly in the November election.

Persistence had its virtues. During the 1994 legislative session, an attempt to raise the sales tax as a means for property tax relief was voted down. The defeat of this regressive legislation was heartening to Union members as farmers paid almost 30 percent of all sales taxes in South Dakota — about $3.00 for every dollar of tax paid by the average urban taxpayer.[30] Within a month after the lawmakers returned home, a Beresford housewife, Barbara Halweg, initiated a measure to limit property taxes to one percent of assessed evaluation. She had the immediate support of the Farmers Union.[31]

With the strong possibility that Initiative I, the Halweg proposal, might win in November, the candidates for governors, Democrat James Beddow and Republican William Janklow, promised to reduce property taxes if they were elected. Opponents of Initiative I used costly ads and scare tactics in their efforts to encourage a "no" vote. They said that if Initiative I was implemented, public schools and state government would be crippled for lack of funds. Initiative I lost by the narrowest of margins but Governor Janklow, who initially called for a 30 percent property tax reduction, did implement a 20 percent reduction beginning in 1996. To make up for a possible shortfall in revenue, he proposed many new taxes, several of which affected farmers. Included were 4 percent taxes on veterinary services and personal property (excluding livestock) sold at auctions, and taxes on cattle semen and intrastate trucking.[32] The Governor backed down on a 2 percent sales tax on agricultural feed, seed, chemicals and fertilizer. The proposal had aroused the ire of farmers, ranchers, cooperatives and other agricultural suppliers.[33]

In the fall of 1995, the South Dakota Farmers Union decided to mount a petition drive to force a public vote on the new taxes in the 1996 general election. The SDFU needed to file 15,581 signatures with the Secretary of State by May 7, 1996, to place the issue on the ballot. Opponents criticized the Union for having no plan to replace the expected revenue, approximately $17.7 million. Union leaders were accused of wanting an income tax substituted for the Janklow proposal. Charles Groth, SDFU Communications Director, replied that "If you specify a replacement source, that becomes a debate, and we want the debate to be these taxes passed last session."[34] He denied that the Union's motive was an income tax, stating that farmers and ranchers were discovering that the sales tax they would pay on services would equal or surpass their property tax break. In many cases, rural residents would break even, in other cases, they would experience a loss.[35]

The Union's efforts to repeal the trucking tax and other new sales taxes were stymied during the summer of 1996. In July, Attorney General Mark Barnett advised the Secretary of State's office not to place the SDFU measure on the November ballot. Barnett indicated that the SDFU could carry the issue to the State Supreme Court. Commenting on Barnett's action, President Wiese said, "Obviously our alternative is to take it to the Supreme Court and do it in a very direct ... fashion. Frankly, what they're doing here is stalling on an issue they've lost not once but actually twice and they are making a mockery of our judicial system and the people's right to initiate."[36] In mid-August, the South Dakota Supreme Court ruled that the sales taxes on trucking and other services were necessary for the support of state government and therefore were not subject to a referendum vote. Responding to the Court's decision, Wiese commented, "It's the people who lost. The Constitution has been turned on its head." Looking to the future, the Union would seek to elect people who would ultimately overturn the tax in the Legislature. Wiese said voters should examine the voting record of current legislators and the views of their challengers on the tax issue.[37]

RURAL ECONOMIC DEVELOPMENT

Throughout much of the 1980s, the Legislature proposed few remedies for the farm depression. The SDFU, meeting in convention in 1987, urged the enactment of a Rural Economic Development Authority to aid farmers, ranchers and small businessmen. Governor Mickelson proposed an economic development program to be funded through a one cent increase in the sales tax. With the Farmers Union being the only organization to oppose an increase in this regressive tax, the Rural Economic Development Initiative (REDI) came into being. The first proposals by the Farmers Union to use REDI funds to finance low interest livestock production loans and rural business loans in small communities were tabled by the Legislature.[38] When members of the SDFU gathered in Pierre for the annual Legislative Day, on February 9, 1989, President Dallas Tonsager said, "We feel that agriculture should be a partner in economic development and share in REDI funds. The money should be used to promote agriculture more than it is."[39] The economic development money was not being distributed equally. Only two of the loans, to date, had gone to agriculture.[40] The larger cities were the recipients of the bulk of REDI funds.

Five years later, in 1994, the Legislature did pass several rural development bills. The pro-farmer legislation included: 1) loans for beginning farmers, 2) a microenterprise loan

Demonstrating the process for making ethanol.

program, 3) tax incentives for livestock feeders, 4) scholarships for an Adult Farm Management Program and 5) the creation of a council to explore ways to add value to grains and livestock before they were exported from the state.[41] The favorable legislation can be attributed to two factors — first, the creation of a bipartisan Rural Renewal Coalition in the State Legislature. The coalition, led by Bernie Hunhoff of Yankton, was composed of six Democrats and six Republicans, all Farmers Union members, from the House and the Senate.[42] The second factor involved changes in the composition of the Legislature. Following the 1992 general election, South Dakota had a divided government for the first time in sixteen years. The Republicans had a 41-29 plurality in the House while the Democratics controlled the Senate, 20-15. Twenty-eight members of the SDFU had been elected to the Legislature.[43]

Efforts were made to have the state government invest in industries using products grown in South Dakota. The Union did what it could to encourage the production of ethanol. The Farmers Union urged the required use of ethanol in state-owned vehicles, and it helped force the Governor to abandon his plan to eliminate the two cents per gallon tax on ethanol blend fuels.[44] As a part of its efforts to promote alcohol fuel development in South Dakota, the South Dakota Farmers Union converted two cars to operate on a blend of up to 85 percent ethanol. The E-85 conversion kits installed in the vehicles were provided by the South Dakota Corn Utilization Council. The two autos became traveling advertisements for the efficiency and clean-burning qualities of ethanol blend fuels.[45] In 1992, the Legislature approved providing ethanol producers with motor fuel tax credits, amounting to twenty cents per gallon of ethanol produced. Ethanol producers had received twenty cents per gallon from the state general fund under the previous system.[46] As of September 1996, South Dakota had two ethanol plants. Broin Enterprises of Scotland, established in 1988, produced seven million gallons of ethanol per year. Heartland Grain Fuels of Aberdeen, begun in 1993, produced six million gallons of ethanol fuel from 2.3 million bushels of corn, virtually all from Brown County. Because of demand, Heartland Fuels planned to boost its production to eight million gallons of ethanol per year. South Dakota drivers were using between 16 and 18 million gallons of ethanol annually.[47]

By the fall of 1996, the two ethanol plants had received $8,000,000 in state subsidies since incentive subsidies began in 1988. State subsidies to ethanol plants were limited to a lifetime payment of $10 million. The Scotland plant, having received $4.5 million, could have its subsidy terminated at the beginning of the 21st century. The production center in Aberdeen, recipient of $3.5 million in subsidies, expected it state assistance to end in five years. Both plants prepared to outlive the production payments.[48]

About 8 percent of all gasoline in the United States was an ethanol blend. The cleaner burning fuel was composed of nine parts gasoline to one part ethanol. In adding ethanol to gasoline, the octane improved by three points. In addition, 30 percent less carbon monoxide was produced. Because ethanol is produced from grain, it does not deplete the world's petroleum reserves.

Despite opposition from Governor Mickelson, state lawmakers, in March 1990, agreed to financial assistance and possible partial state ownership of a major agricultural processing plant for the wet-milling of corn. The legislation had the support of the South Dakota Corn Growers, who hoped to entice AGP, an Omaha cooperative, to locate a $300 million corn wet-milling plant in South Dakota.[49] In the following year, the Legislature extended producer incentives to 2003, increasing the amount of annual funding available for payments from $1.4 to $2.5 million but limiting payments to any individual producer to ten years.[50]

During the fifteen year period, the Union experienced other wins and losses in the legislative arena. In 1981, the SDFU, and other organizations, had secured 10,000 more signatures than needed to place a measure on the general election ballot requiring single-member state Senate districts. Amendment A, mandating such districts, was approved by the voters by a margin of 52-48 percent in November 1982.[51] The Farmers Union was one of the losers in November 1992, when voters rejected Initiative No. 4, which would have repealed video lottery. The SDFU believed that South Dakota should not be built on gambling receipts.[52]

Tax reform remained a frustrating issue. Although some relief from property tax was on the horizon, the Farmers Union remained committed to a tax system based on the ability to pay. This would only be accomplished through the enactment of a corporate profits tax and a low rate personal income tax.

THE GREEN THUMB PROGRAM

The Farmers Union sponsored Green Thumb program almost succumbed to the budget scalpel of the Reagan administration in 1982. President Reagan vetoed a supplemental appropriations bill that included $210 million to be used for developing part-time employment opportunities for older economically disadvantaged men and women and other such programs authorized under Title V of the Older Americans Act. Fortunately, a bipartisan vote in the Congress in October overrode the President's veto. Under Gerry R. Eisenbraun's

John Riedy, Thunder Hawk, being inducted into the S.D. Co-op Hall of Fame, 1987.

leadership, the program provided employment for 425 elderly, low income, rural South Dakotans in the mid-1980s.[53] Program participants served as teachers' aides, bus drivers, fork-lift operators, computer technicians and as support staff for women's shelters, food distribution projects and centers for the homeless.[54] The Green Thumb project was conducted in conjunction with the Job Training Partnership Act.

Eisenbraun retired on December 31, 1992, having directed the program since its inception in South Dakota on March 1, 1968. His successor was Jean Bennett of Sioux Falls. She had been office manager and executive support administrator for the South Dakota Green Thumb since 1981.

RECOGNIZING LEADERSHIP AND SERVICE

Beginning in 1980, the South Dakota Farmers Union began a program, the Meritorious Service Awards, to recognize those individuals who had given distinguished service to the organization. Memorial awards for Meritorious Service were added to the program as of 1987. Many of the recipients had first joined the Farmers Union in the latter half of the 1920s. In 1992, Mrs. David Lovinger of Ethan and Dorothy Authier of Vivian were the first women to be recognized. The men and women recognized for Meritorious Service follows:

1980	Emil Loriks, Oldham
1981	Ben Radcliffe, Hitchcock

1982	Lafe Lunder, Canton
1983	Herman J. Pietz, Parkston
1984	Clifford Ott, Huron
1985	Paul W. Opsahl, Huron
1986	John C. Skage, Sioux Falls
1987	John Riedy, Thunder Hawk
	Frank Butler, Bison
1988	LeRoy Hardy, Sturgis
	Oscar Fosheim, Howard
1989	Harvey Faulstich, Orient
	Oscar Johnson, Oldham
1990	Jens Christiansen, Wessington Springs
	Harry Caslin, Kennebec
1991	Homer Ayres, Sturgis
	Richard Pastian, Lake Preston
1992	Dorothy Authier, Vivian
	Mrs. David Lovinger, Ethan
1993	George Levin, Hereford
	Karl Pete Jensen, Bruce
1994	Robert Fortner, Mt. Vernon
	Harold Golseth, Lake Preston
1995	Norman Olsen, Canton
	J. E. Krull, Sr., Watertown
	Ewald Osterberg, Salem
1996	Cecil Hall, Bruce
	Oris Authier, Vivian
1997	Carroll McKenna, Zeona
	Henry Bell, Flandreau
1998	Loyd Larson, Brookings
	Lewis Tofflemire, Wessington Springs

A South Dakotan, Dwyte Wilson, received the NFU "Award for Meritorious Service to Farmers Union and to American Agriculture" in March 1992. In South Dakota, Wilson managed FU livestock and creamery cooperatives and served as camping director for the SDFU before becoming an organizer for the National Farmers Union. He conducted FU membership drives, and he helped to form cooperatives from the Dakotas to Texas and from Oregon to Pennsylvania.[55]

PICNICS AND CONVENTIONS

The locale of the Union's annual picnic, since its inception on June 18, 1965, had always been Huron, either on the State Fairgrounds or on the lawn of the SDFU state office. President Tonsager moved away from this longstanding practice in 1987, when the traditional summer event was staged on the banks of Lake Herman, just west of Madison. The picnickers had the opportunity to tour Prairie Village and to see a musical production at the Prairie Village opera house.[56] Since 1987, SDFU picnics have been held in various parts of the state, including Wylie Park at Aberdeen, the shores of Lake Randall near Yankton, and on the

David Wentzlaff and Maynard Whitmyre at the annual SDFU picnic (1975).

grounds of the Custer State Game Lodge. These assemblies at different sites were a part of "the Union's policy of moving programs like the picnic and state conventions around the state to maximize the opportunity for members to participate in the activities of their organization."[57] The new policy had the desired result as there was a steady rise in attendance after the annual summer event was held at different locations.

The 71st annual convention of the SDFU was held in Aberdeen in early November 1986. It was the first time a state convention had been held outside of Huron since the early 1950s. Aberdeen, Mitchell, Yankton, Brookings, Rapid City, Pierre, Huron, Sioux Falls and Watertown have all hosted the Union's delegates since 1986.

Within a decade, the convention delegates approved two increases in membership dues. In 1983, annual dues were doubled from $10 to $20 per year. It was the first dues increase since 1967.[58] Eight years later, in 1991, dues were increased to $25.00 per year, as a result of an increase in NFU dues.

BUILDING ALLIANCES

The South Dakota Farmers Union continued its efforts to secure alliances in order to secure better market prices and save the family farm. The SDFU came together with the state's larger Christian denominations to discuss the "Future of Rural America" in a conference at Mitchell on May 21, 1982. The meeting was organized by the Farmers Union, with assistance from the South Dakota Association of Christian Churches and the Catholic Rural

Life program. The conferees discussed the moral and philosophical questions involved in the current trends in farm land ownership and taxation.[59] The Roman Catholic Church said that the concentration of landholding was a scandal, depriving a large part of humanity of the fruits of the earth. The church called for a moratorium on the construction of large-scale livestock production facilities.[60] In the following year, 250 people gathered for another Farm-Church-Land Conference at Huron, again on May 21. Using the theme "Chaos in the Face of Abundance," the participants considered soil conservation and domestic and international food aid programs.[61]

Although no longer held on an annual basis, Farmer-Worker conferences were held during the 1980s. Emphasis was placed on how farmers and workers could do a better job of communicating their views to the public, given the decreasing influence of both groups in the U. S. Congress. As noted earlier, the bipartisan Rural Renewal Coalition was very successful in securing approval of a package of rural development bills in the 1994 legislative session. Five years earlier, nine farm, church and labor groups in South Dakota had begun a campaign to make the public aware of the growing economic stranglehold of three major meatpackers, Con-Agra, Excel (Cargill) and IBP on the beefpacking market. In 1989, the three packers controlled over 70 percent of the market, giving them considerable control over the price paid to farmers and ranchers for their livestock and the prices that consumers paid at the meat counter. Dakota Rural Action, through its public education program, wanted pressure put on state leaders and the congressional delegation to seek enforcement and strengthening of existing anti-trust legislation to eliminate a possible monopoly of the meatpacking industry by the three giant firms.[62]

At a December 19, 1990, news conference, an announcement was made of a joint marketing agreement between the SDFU and the National Farmers Organization (NFO). The livestock marketing agreement allowed Farmers Union members to market livestock through the NFO's eight South Dakota collection points without paying NFO membership dues. The agreement did not constitute a merger of the two farm organizations; its objective was to provide an additional marketing option for South Dakota Farmers Union members.[63]

INVOLVEMENT OF THE MEMBERSHIP

Efforts continued to keep all members of the State's Union informed. The Legislative Action conferences, usually held each December, were increased in number to encourage greater attendance and to reduce the distance traveled. On December 16, 1994, the Farmers Union launched a radical change in the conference format. The annual series of legislative conferences moved to the statewide Rural Development Telecommunications (RDT) network. In this move onto the "information highway," FU members and other interested rural South Dakotans gathered at nine locations and were linked by the RDT interactive system. The state operated system permitted any participant at any one of the nine locations to go on-screen with a question or comment that could be simultaneously viewed and heard by participants at all of the other locations. The traditional Legislative Conference publication, which contained a review of the national and state issues affecting the state's farmers and ranchers, was distributed to participants at each of the meeting sites.[64]

Young Farmers' Conferences were scheduled to enable young farm and ranch families to participate in the FU Legislative Day program at Pierre. Women's conferences emphasized study of the food marketing system, who controls it, what is the role of advertising and what women could do to make a statement with their food dollars.[65] Carroll McKenna of Zeona, a member of the SDFU Board of Directors from 1975 to 1993, was appointed the NFU's representative to the Agricultural Women's Leadership Network (AWLN) in March 1990. The AWLN, which began in 1988, had as its purpose "to advance rural viewpoints on issues such as food research and safety, animal welfare and the image of agriculture."[66] The group met with Clayton Yeutter, President Bush's Secretary of Agriculture to voice their concerns.

Carroll McKenna, Zeona, member of the Board of Directors from District VI, 1976-1994.

THE FARMERS UNION TRAVEL ASSOCIATION

The Union's bus program came of age during the 1980s. With the addition of a forty-seven passenger bus in the spring of 1990, the SDFU owned three busses. The Farmers Union Travel Association, a wholly owned subsidiary of the South Dakota Farmers Union, offered tours to East Coast metropolitan centers, Branson, Missouri, Las Vegas, Nevada, the Rose Bowl Parade, Disneyland and to a host of other places. The Travel Association also scheduled overseas trips to Scandinavia.[67] The success of the Travel Association was enhanced by Union members contributing $150,000 to a Bus Fund during the 1980s.[68]

STAFF CHANGES

When Leland Swenson assumed the presidency of the SDFU in 1981, he restructured the administrative staff. He added an associate education director and a membership coordinator. The duties of Education Director Ken Birgen were expanded. He became the Director of Education and Cooperative Services. A half year later, Birgen resigned as he planned to marry and move to Colorado. Pamela G. Lubinus of Salem replaced Birgen as Director of Education on July 1, 1982. She was a 1981 graduate of South Dakota State University with a bachelor of science degree in Child Development, Family Relations and Psychology.

David Wentzlaff left the staff, on May 6, 1983, to become Director of Admissions at Huron College. As the SDFU Director of Legislative Services, he had been an effective lobbyist at Pierre. Wentzlaff also worked as field coordinator for Districts I and II at the time of his resignation.

Maynard Whitmyre of Webster, who had joined the SDFU staff in 1967, in late August, 1988, announced his plans to pursue a career in photography in the Twin Cities area. In his twenty-one years on the staff, Whitmyre had been a fieldman, education director, cooperative membership services director and bus driver. His replacement, Randy Brown, a native of Harlan, Iowa, was named Director of Cooperative and Membership Services. Brown returned to Iowa in early 1992 to rejoin the staff of the Iowa Farmers Union, where he had been employed for three years before coming to South Dakota. Jeff Willer, a graduate of North Dakota State University, accepted the position of director of field services in March

1992. He became responsible for coordinating membership and development activities and promoting farmer-owned cooperatives. The Dickinson, North Dakota, native had been a Union Torchbearer in 1982. Prior to joining the Farmers Union staff, he had spent four years as an extension agent in Haakon and Beadle counties.[69] In January 1995, the field staff included: Jeff Willer - Field Services Director, Susan Dufek - Membership Coordinator, Gene Ellingson - Fieldman, and Don Falla - part-time Fieldman with emphasis on membership development. Willer resigned in July 1995 after accepting a position as Transportation Coordinator for the North Dakota Farmers Union.[70]

The South Dakota Farmers Union built for the future during the most severe farm depression since the 1930s. Out-of-state corporate firms were held at bay and some tax relief seemed in sight for the family farmer. The Union, to further strengthen the organization, consolidated some of its activities in the last two decades of the 20th century. The SDFU also reached out, more than ever, to other groups and allied with them in solving rural economic, environmental and social problems. The same policies would be followed by local and regional cooperatives. Mergers took place, and the cooperatives sought ways to add value to farm products before they were exported from South Dakota.

Chapter 20

COOPERATION AND EDUCATION — PREPARING FOR FUTURE CHALLENGES 1981-1999

In every new crisis it seems to be forgotten that the demand for freedom means a reaching out for mental activity, for greater scope of thought.
— John Dewey, 1920

The theme for the 1994 annual convention of the South Dakota Farmers Union was "Renew the Spirit," recognizing the 150th anniversary of the founding of the modern cooperative at Rochdale, England, in 1844. A century and a half after the co-op was established, on August 15, 1844, 700 million people around the world shared the legacy of the twenty-eight founders of the Rochdale Equitable Pioneers Society. In the United States, 120 million people were members of a network of 47,000 cooperatives.[1] To further commemorate the event in South Dakota, a regional co-op village was created at Camp Lakodia, near Madison, between June 26 -29, 1994. Youths, ages 14 to 16, from Minnesota, Iowa and Wisconsin were invited to join South Dakota youngsters at the camp site. They were taught the mechanics involved in creating and operating a cooperative.

SOUTH DAKOTA'S COOPERATIVES

As indicated in Chapter Eight, most cooperatives on the northern Prairie Plains were started in the 1930s and 1940s, during a time of great economic stress. They flourished during the fifties and sixties, but their numbers decreased nationwide, during the 1970s, when monetary inflation and the agricultural policies of Earl Butz took their toll. The USDA's Agricultural Cooperative Service reported that more than 1,300 farmer-owned co-ops, nationwide, disappeared during the '70s. Only 6,445 of the 7,790 farm co-ops that existed in 1970 remained in 1979. Although 31 percent of the 1,345 co-op casualties merged with or were acquired by another co-op, fully 50 percent simply went out of business.[2] South Dakota was a part of the national trend. With the decline in farm population, the smaller cooperatives were forced to terminate their operations or merge with another co-op.

October became Co-op Month in South Dakota. As part of its promotional activities for Co-op Month, the SDFU held the first of its annual co-op banquets in October 1982. The venture was accompanied by a major radio advertising campaign. At the fourth annual banquet, two of the state's co-op pioneers, Emil Loriks and Arthur Jones, were inducted into the

South Dakota Co-op Hall of Fame. The Hall of Fame honored those individuals who had made important contributions to cooperative development in South Dakota through:

1. organizing local or regional cooperatives
2. serving effectively on one or more cooperative boards
3. providing outstanding work in the area of cooperative management or government or other public service on behalf of cooperatives and their member owners.[3]

Emil Loriks, the first member of the Hall of Fame, compared building a co-op to that of the construction of a great cathedral.[4] On September 28, 1991, President Dallas Tonsager presided over a dedication ceremony for the South Dakota Co-op Hall of Fame. It was to be permanently located in the SDFU auditorium at Huron. Containing records and photographs, the Hall of Fame preserved the memories and stories of the people of South Dakota's cooperatives. Tonsager said that the Union needed "to revitalize and rebuild the co-op pioneer spirit." He went on to say that "Cooperatives were the first great economic developers in South Dakota," and they currently possessed $1 billion in equity.[5] The roster of Hall of Fame members from 1985 to 1994 included:

1985	Emil Loriks, Oldham
	Arthur Jones, Britton
1986	Herman D. Holt, Olivet
	Albert Schramm, Winner
1987	August Dahme, Mina
	John Riedy, Thunderhawk
1988	John C. Skage, Sioux Falls
	J. Evan Rasmussen, Beresford
1989	Darrell Moseson, Sioux Falls
	Warren G. Grebner, Aberdeen
1990	J. E. Krull, Sr., Watertown
	Lawrence Stoley, Highmore
1991	Clifford G. Bell, Doland
	Warren "Bud" Moe, St. Onge
1992	Kenneth Holum, Sr., Bethesda, Maryland
	John Cink, Parker
1993	Douglas Murphy, Woonsocket
	Clement Kloucek, Scotland
1994	Phil Testerman, Wessington
	Max Farrar, Hurley.
1995	Quentin Louden, Martin
	Oscar Johnson, Oldham
1996	Leo P. Flynn, Milbank
	Vernon Berg, Woonsocket
1997	Ben Radcliffe, Hitchcock
	G. L. (Gus) Moseson. Howard/Sioux Falls.
1998	George A. Hargens, Orient
	J. D. Lynd, Huron

Groton native, Kenneth Holum, Sr., was inducted into the National Cooperative Hall of Fame on April 28, 1993, in Washington, D.C. A recipient of the NFU Meritorious Service Award in 1983, Holum was a board member of Northern Electric Co-op and an original organizer of East River Electric Power Cooperative.[6]

A co-op pioneer, John Riedy of Lemmon, received the national Cooperative Spirit award in December 1991. His work on behalf of the state's cooperatives spanned almost fifty years. He was one of the original incorporators of the Grand Electric Cooperative at Bison in 1946, and the chairman of the board for thirty-nine years. He served on the boards of the South Dakota Rural Electric Association, the Farmers Cooperative Association at Thunder Hawk and the former Equity Union Creameries of Aberdeen. Riedy helped organize the Dakota Propane Gas Cooperative, the Federal Credit Union and the Cooperative Food Store, all located in Lemmon. Serving in the State Senate from 1973 to 1976, he was a prime sponsor of SB 261 — legislation establishing the territorial integrity of South Dakota's electric utilities.[7]

During the 1990s, local cooperatives in South Dakota tended to merge and to demonstrate co-op unity. As with rural school districts, mergers were deemed necessary because of the decline in population in the rural areas of the state. The new associations created increased diversity and additional working capital, which, for several cooperatives, meant increased purchasing power. Among the co-ops merging since 1991 were:

1. the FU Co-op Association of Marshall County with the Farmers Co-op Company of Amherst and Claremont (1991)
2. the Lake Preston Cooperative Association with the Oldham Farmers Elevator (1993)
3. the FU Cooperative Association of McCook County with the Salem-Rumpus Ridge Co-op and the Montrose Grain and Lumber Company (January 4, 1994)
4. the Howard FU Oil Company with the Howard FU Cooperative Association grain elevator (June 30, 1994).[8]

Until 1990, American cooperatives had been basically defensive in nature. They had been formed to keep food processors from becoming concentrated or to supply products farmers couldn't otherwise buy. By the mid-1990s, cooperatives had adopted an offensive posture by striving to add value to a producer's products. If farmers were to stay on the land as independent operators, they had to find alternative marketing for their products. The inspiration for this new approach came from Europe, where the majority of food production was handled by farmer-owned cooperatives and the success of "New Wave" co-ops in North Dakota and Minnesota. An office of co-op development to promote farmer owned co-ops had created more than twenty value-added cooperatives in North Dakota by the end of 1994. The new cooperative movement was a survival response to the trend towards industrial agriculture, increased international competition and less support from the federal farm program. At the national rural summit at Ames, Iowa, in late April 1995, a Wisconsin dairyman, Drew Johnson, symbolized the farmers' new attitude. Johnson said, "We've got to get away from just worrying about production and get into the marketing end of it. We've got to get that back." He went on to say, "Farmers have let it go, and we've got to get our markets back."[9] The success of the Dakota Growers Pasta Cooperative at Carrington, North Dakota, was indicative of what a value-added cooperative enterprise could do for farmers. The durum wheat farmers, who invested more than $40 million in the processing operation, received twenty-two cents for their durum on a box of pasta retailing for one dollar. This compared with the ten cents per box received by producers selling to an outside producer.[10]

The development of value-added facilities required investment by both the private and public sectors. Most important, the cooperation of the USDA was needed. The USDA spent most of its appropriation on research related to production agriculture, very little of the money went to the marketing of farm products. To nurture new markets, the USDA needed to provide low interest loans to help producers to take advantage of value-added opportunities.[11]

Members of the South Dakota Farmers Union recommended that an office of co-op development be created in the Governor's Office of Economic Development. J. D. Lynd, executive secretary of the South Dakota Association of Cooperatives, hoped for legislation that would provide for state financing of value-added cooperative ventures. A debate over farmer-owned co-ops occurred during the 1996 session of the South Dakota Legislature. Advocates of value-added agriculture stressed the need to process more farm products in the state. Several of the lawmakers established the Rural Coalition, seeking ways for the Legislature to help spur construction of more agricultural plants. Two champions of the value-added co-ops, Representative Robert Weber (Rep.) of Strandburg and Senator Bernie Hunhoff (Dem.) of Yankton, argued that the state could provide assistance through tax refunds and taking a proactive stance towards providing transportation and favorable utility rates.

Ground was broken for a $32.5 million soybean processing plant at Volga on August 21, 1995. The cooperative plant, capable of processing 16 million bushels of soybeans per year, was the forerunner of "New Wave" co-ops in South Dakota. The state's 11,000 soybean producers raised 65-75 million bushels of soybeans per year.[12] In 1995, soybeans were shipped out-of-state for processing with about 40 percent of the state's crop shipped back at the farmers' expense, mostly for animal feed. The proposed farmer-owned plant could process 50,000 bushels of soybeans a day. At the groundbreaking, Governor Janklow said, "This is the only way to go. Agriculture can't survive in South Dakota without projects like this." He went on to say that farmers must lead themselves to a brighter economy.[13]

The South Dakota Soybean Processors plant officially opened on September 8, 1996. The first new soybean plant built in twenty years had a storage capacity of 1.06 million bushels and produced 1,250 tons of soybean meal and 265 tons of soy oil daily.[14] The co-op bought the product from farmers at market prices. Sixty pounds of beans were routinely converted to forty-four pounds of meal, twelve pounds of soy oil and four pounds of husk.[15] The resulting soy meal and soy oil were sold directly to area farmers or to manufacturers as farmers became marketing agents for their own crops. Any profits from the processing and selling of the value-added products were distributed to the co-op's members and other investors. In addition to processing soybeans into food and animal feed, the farmer owned cooperatives had the opportunity to be on the leading edge in the production and marketing of fuels and lubricants made from soybeans. Such fuels and lubricants were biodegradable and made from a product which was renewed annually. The Volga plant became a showcase for the vertical integration of agriculture. It illustrated how farmers could control their own destiny.

Federal assistance for organizing new farmer-owned cooperatives was announced in March 1995. The USDA reported that a newly reorganized Rural Economic and Community Development (RECD) agency would assist in rebuilding rural communities. Former SDFU

President Dallas Tonsager was named Acting RECD Director in South Dakota. A co-op development specialist would be added to the RECD state office to work with new cooperatives.[16] The USDA's commitment to cooperatives dated back to the late '30 and early '40s. At that time, Emil Loriks was a co-op development specialist with the Farm Security Administration, the predecessor of the FmHA.

At the end of 1994, the USDA reported that South Dakota had 170 farmer-owned cooperatives. The combined business volume of all of these cooperatives was $1.8 billion. Because farmers often held membership in more than one cooperative, the total co-op membership in the state was estimated to be 140,095.[17]

SOUTH DAKOTA FARMERS UNION SERVICE ASSOCIATION

National Farmers Union Insurance encountered some financial difficulties during the farm crisis of the mid-1980s. Approximately fourteen years earlier, in 1969, the NFU had sold its three insurance companies — National Farmers Union Property and Casualty Company (founded in 1945), National Farmers Union Standard Company (founded in 1968) and National Farmers Union Life Insurance Company (founded in 1937) to the Baldwin-United Corporation. The ownership of the three companies shifted to Mortgage Guarantee Investment Corporation (MGIC) several years later, after Baldwin acquired MGIC. When Baldwin-United announced it was filing Chapter 11 reorganization under the Bankruptcy Act, in late 1983, the NFU immediately began efforts to reacquire ownership of National Farmers Union Insurance Companies.[18] The insurance companies were not a part of the Chapter 11 action.

A consortium, the Insurance Acquisition Corporation (IAC), located in Denver, was formed to purchase National Farmers Union Insurance Companies from MGIC. The consortium, composed of four Farmers Union State Insurance agencies, Farmers Union Central Exchange and Lamar Investors paid MGIC $15 million for substantially all of the assets of NFU's insurance companies.[19] The South Dakota Service Association committed $90,000 to the acquisition. CENEX committed $3 million to the buyout of Baldwin-United.[20]

The insurance business continued to be very difficult in the rural areas until 1987. Having experienced substantial losses, National Farmers Union Insurance Companies engaged in an ambitious restructuring program to capture a stronger market share in the countryside. Claudia Svarstad, senior vice president for marketing, an Aberdeen native and an Augustana College graduate, reported that the company had abandoned some unprofitable lines and had laid off personnel at the Denver office, many of the employees had had long tenure with the companies. On August 31, 1987, Stanley Moore, former president of the North Dakota Farmers Union, was named chief executive officer and chairman of the board of National Farmers Union Insurance companies.[21]

A change in leadership had also occurred in the SDFU Service Association. Clifford Ott, manager of the Service Association for twenty-one years, retired in June 1984. SDFU President Leland Swenson appointed Jacob J. "Jake" Krull as Ott's successor. Krull, a Watertown native, took over the management reins on July 1, 1984. Krull graduated from South Dakota State University with a degree in economics. After serving in the Army from 1961-1963, he had become a Farmers Union insurance agent.[22]

Five years later, on October 2, 1989, Krull was named president of the National Farmers Union Property and Casualty companies. He succeeded Stanley Moore, who continued to

Jake Krull, President of the NFU Property and Casualty Companies.

Wayne Bartscher, Manager of the SDFU Service Association.

serve as chairman of the board. Krull had an outstanding record in insurance sales and public service. In his twenty-one years as an insurance agent, he was honored twelve times with the companies' top national sales award. As a public servant, following his discharge from the Army, Krull had joined the National Guard. He was a Brigadier General and Assistant Adjutant General (Army) for South Dakota from 1986 until his move to Denver in 1989. Krull was a state senator from 1972-1982, serving as minority leader of the Senate from 1978-1982.[23] In his first years as president, Krull emphasized "ESP" — the need for all agents and employees to reduce expenses, improve service and exude professionalism.[24]

Wayne Bartscher of Rapid City was selected by SDFU President Dallas Tonsager to replace Krull. Bartscher, born and raised at Salem, graduated from South Dakota State University in 1977 with a degree in commercial economics. He immediately became an FU agent at Brookings, and the next year, he was appointed West River District Manager. He left that position to become manager of the Service Association on October 1, 1989. Bartscher was responsible for overseeing the work of thirty-one SDFU insurance agencies. In 1989, FU insurance generated an annual premium of $12-$13 million in South Dakota.[25]

In the early 1990s, the Farmers Union insured over 13,000 vehicles and 3,500 farm and city homes in South Dakota. Annual premiums totaled more than $15 million with $7 million of this amount in property and casualty insurance premiums.[26] In 1994, National Farmers Union Insurance Companies (NFUIC) had over 400 agents selling personal lines in twelve states and business accounts in twenty-three states. NFUIC sales totaled $105 million with an additional $26 million coming from crop insurance premiums.[27] Following the lean years prior to 1992, premium growth at both the state and national levels increased each year.[28] National Farmers Union Insurance celebrated its 50th anniversary on June 25, 1995, and looked toward the future with considerable confidence.

THE REGIONAL COOPERATIVES: CENEX

The regional cooperatives were not immune to the agrarian crisis of the 1980s. Both CENEX and Harvest States (the former GTA) were forced to retrench and restructure after 1985. Through merger, consolidation, divestiture, innovation and good management both cooperatives regained their economic stature and, once again, were thriving enterprises in the 1990s.

CENEX expanded its physical holdings before the farm crisis reached its pinnacle. In 1981, the cooperative purchased Solar Gas, a propane supplier with forty retail outlets. The acquisition increased CENEX's annual volume of propane to approximately 200 million gallons.[29] Late in the same year, the firm broadened its base in the Pacific Northwest through the purchase of Western Farmers Association, a regional cooperative with headquarters in Seattle that provided feed and seed to almost fifty retail units.[30]

Investment in petroleum production and distribution continued to be the primary goal of CENEX. In 1983, sixty miles of pipeline were purchased from the Bow Pipeline Company, facilitating the flow of oil from the Kevin producing area to the Laurel, Montana, refinery.[31] Fifty-nine producing wells, in the Powder River Basin, near Gillette, Wyoming, were purchased from Shell Oil in early 1986. The transaction involved nearly 900 barrels of daily crude oil production as well as between two and three million barrels of proven reserves. The new holding increased CENEX's oil and gas reserves to a company record of 14.3 million barrels.[32] As the rural economy gradually recovered, the regional cooperative's board of directors agreed to invest $30 million in the construction of a fuels' pipeline from Minot to Fargo, North Dakota, beginning in 1991. The 246 mile addition to the pipeline network improved the supply of diesel fuel and gasoline from the refinery at Laurel, Montana, to local coops and customers.[33]

Consolidation and cooperation among the major farm cooperatives was one result of the rural depression of the 1980s. On June 30, 1992, CENEX purchased Farmland Industries' (the former CCA) share of ownership in the National Cooperative Refinery Association (NCRA). The purchase increased CENEX's ownership in the 75,000 barrel per day McPherson, Kansas, refinery from 43.7 to 74.2 percent. It resulted in increased distribution efficiencies and balanced the supply and demand for both regional cooperatives. The purchase had been accompanied by a product exchange agreement. Farmland continued to draw from the NCRA refinery, and CENEX secured refined fuels from the Farmland refinery at Coffeyville, Kansas.[34]

The regional cooperative's venture into ethanol production lasted less than five years. In November 1987, CENEX purchased the Dawn Enterprises ethanol plant at Walhalla, North Dakota. It had an 11 million gallon capacity and the potential of using 4 million bushels of small grain and corn produced by farmers each year. Drought conditions, which substantially increased corn prices, and the bankruptcy suit of its former owners, caused the plant to be closed for a year in 1988-1989. CENEX returned the plant to operation after investing $1.3 million to upgrade the facility's systems.[35] On May 8, 1991, CENEX transferred the plant to Archer Daniels Midland (ADM), having decided to exit from the ethanol manufacturing business. Given the major capital expenditures that were ahead in its petroleum-related business, CENEX believed it appropriate to transfer the plant to ADM, which viewed ethanol production as one of its core enterprises. CENEX, nevertheless, remained committed to the marketing and distribution of oxygenated fuels.[36]

A second major commitment of CENEX was to continue to invest in and to promote agronomy supply systems. In fulfilling this goal, Darrell Moseson and the other administrative officers of CENEX, in 1987, entered into conversations with officials of Land O'Lakes, another regional cooperative, on how they could consolidate their operations, strengthen their respective firms, and increase their production, marketing and distribution efficiency.[37] Land O'Lakes, with its headquarters at Arden Hills, Minnesota, was a farm supply and food marketing cooperative, serving about 1,400 co-ops and 8,000 dairy members in 1988.

The joint venture between CENEX and Land O'Lakes eliminated duplication and inefficiencies, achieving $10 million in savings. Some of these savings resulted from closing facilities and dismissing employees. At the same time, each corporation maintained its identity and separate culture.[38] The success of the new relationship between the two cooperatives led to discussions of possible consolidation of the operations of CENEX, Land O'Lakes and Farmland Industries. CENEX made the right decision by not merging with Farmland Industries. When the merger was considered, Farmland would have been the dominant partner. Since the initial discussions, Farmland has suffered economic reversals, and today, CENEX is much stronger than Farmland.[39] Looking beyond their respective heritages and philosophical differences, the leaders of the three cooperatives concurred there must be constant readjustments in the cooperative system to ensure its survival. Some type of collaboration would enhance their ability to remain competitive in any economic environment.

The third major goal of CENEX was to return cash to its member-owner cooperatives. Throughout the first five years of the nineties, the cooperative annually set net income records. At the close of the fiscal year ending September 30, 1994, the firm announced a net income of $82.2 million on sales of $2.184 billion. Since 1931, CENEX, or the FUCE, had returned to patrons in the form of patronage, stock returns and other payments, a total of $404.9 million.[40] Early in 1995, the regional cooperative proudly announced a record $75.5 million in patronage distributions for the preceding fiscal year. The 116 co-ops doing business with CENEX in South Dakota earned preferred stock totaling $3.6 million, and they also received $1.5 million in cash.[41]

CENEX continued to realize the value of education. The co-op invested nearly $800,000 during 1994 in support of the cooperative education program. State Unions, however, were no longer allotted 5 percent of the net profit based upon annual patronage. The Unions, instead, were required to make requests for set aside monies.[42] The CENEX Foundation scholarship program, created in 1961, awarded seventy-eight college juniors and seniors cooperative studies scholarships in 1994. Since the program began, the foundation had awarded $1.5 million to more than 2,350 students pursuing agriculture and cooperative studies at schools in the CENEX trade area.[43]

CENEX had moved into a new headquarters building in January 1982. Located on a forty-seven acre site at Inver Grove Heights, Minnesota, the $16 million building had been in the planning stage for ten years. Darrell Moseson, who had presided over operations in the new structure since its occupancy, retired in 1987, after being with CENEX for thirty-six years. His successor was Noel K. Estenson, the senior vice president for Petroleum and Finance. South Dakota's representation on the board of directors changed during the '80s. Richard Traphagen of Columbia was elected to the thirteen member CENEX board for a three year term, beginning in 1983. He was the president of the board of the FU Co-op Association of Brown County and operated a 1,000 acre farm. The graduate of Northern State University soon became secretary-treasurer of the regional cooperative's board.[44] Roy Kopperud of

Lake Preston, who served five terms on the board, was replaced by Aaron Glanzer of Salem in 1988. Glanzer represented Nebraska and western Iowa as well as South Dakota.[45]

Some of the longtime members of the Union questioned the decision to change the corporate name of the cooperative from Farmers Union Central Exchange to CENEX Incorporated at the annual meeting on December 8-9, 1993. The greatest concern was expressed by delegates from North and South Dakota. They felt that the firm was attempting to disassociate itself from its historic ties with the Farmers Union. After a civil and rational discussion, the delegates, by a slim majority, voted that the cooperative's formal name should be the same as the name CENEX and the registered trademark formally adopted to identify the regional and member cooperative in 1972.[46] There was no sulking once the decision had been made.[47]

At the end of 1994, CENEX served more than 320,000 farmers and ranchers from the Prairie Plains to the Pacific Northwest. More than 1,600 locals in twenty-one states handled the regional cooperative's refined petroleum products, propane, lubricants, tires, batteries, and vehicle accessories. As a result of its joint venture with Land O'Lakes, it also supplied the local co-ops with plant food, crop protection products and a wide range of agronomic and information services.[48]

THE REGIONAL COOPERATIVES: HARVEST STATES COOPERATIVES (FUGTA)

The dramatic decline of the export grain trade in the early 1980s had a chilling effect on Harvest States Cooperatives, formerly called the Farmers Union Grain Terminal Association. In 1981, the regional co-op had handled a record 700 million bushels of grain. By 1985, the amount had declined to 380 million bushels. GTA experienced a net loss of $10.9 million for the fiscal year which ended May 31, 1982. The decline in grain exports, low grain prices and high interest rates caused a swing of $30.5 million from the prior year's net profit of $19.6 million. It was the first financial loss for GTA since 1968 and the second since the cooperative began operations in 1938.[49]

The regional cooperative had changed its name in 1983. Ninety-eight percent of GTA's members approved changes in the co-op's by-laws which 1) permitted acquisition of North Pacific Grain Growers of Portland, Oregon, 2) changed the name of the organization from the Farmers Union Grain Terminal Association to Harvest States Cooperatives, 3) increased the board of directors from thirteen to fifteen and 4) changed the procedure used to allocate delegates to annual meetings. The changes became official on June 1, 1983. North Pacific Grain Growers served forty-seven local co-ops with 293 substations in Washington, Idaho, Oregon and Utah. The Pacific Northwest regional owned terminal elevators at Spokane and Kennewick, Washington, and a 6.5 million bushel export facility on the Columbia River at Kalama, Washington.[50]

The Farmers Union Grain Terminal Association paid an enormous price for its efforts to expand its international export trade in grain from 1975 to 1981. Retrenchment became the order of the day, and by 1987, the co-op's work force had been reduced from 5,000 to 2,000 employees. Half of the Great Plains Supply Company lumberyards had been closed, and an 80 percent share of the Froedtert Malt Company had been sold to a French corporation.[51]

The Grain Terminal Association (Harvest States) began a retreat from providing education funds to state Unions in 1981. A controversial decision changed the bylaws of the cooperative, leaving the payment of education funds solely up to the discretion of GTA's board of directors.[52] As of 1995, money no longer comes to the SDFU from Harvest States for educational purposes.[53]

Harvest State's leadership was forced to reevaluate the role of grain in the cooperative's overall operations. With increasing international competition and rising world grain stocks, the grain industry faced a very challenging future. While the co-op continued an aggressive grain marketing effort, greater emphasis was placed on value-added diversification in creating new market opportunities for producer-owner grain and oilseed.[54] Because of the uncertainties created by a fast changing economic environment, the cooperative created several partnership relationships with its members, with other cooperatives and with non-cooperatives. Relationships were sought that would make the most of organizational strengths and that would contribute to long-term financial strength. These general guidelines were applied in Harvest States' capital expenditure program in the '90s.

In 1990, the durum mill of the Amber Milling Company, Huron, Ohio, a subsidiary of Harvest States, was expanded to meet the growing demand for quality pasta ingredients. Four years later, Amber Milling Company began construction of a flour mill at Mount Pocono in eastern Pennsylvania. The finished plant would produce bakery flour from hard, red, spring wheat and semolina and durum flour for the pasta industry. Amber Milling also began construction of another flour mill at Kenosha, Wisconsin.[55] With the purchase of two other flour mills, Harvest States expected to handle 50 million bushels of grain per year.[56] Holsum Foods, another Harvest States subsidiary, acquired Louis Albert and Sons Food Company, Omaha, Nebraska, and Private Brands, a leading Chicago service company in 1990.[57] Early in 1995, Harvest States joined with a California firm, Wilsey Foods, Incorporated, to acquire and operate the bulk mayonnaise, dressings, sauces, syrup and soup-base business of Portion Pac Incorporated (PPI), an affiliate of H. J. Heinz Company. Wilsey was owned by Mitsui and Company, a Japanese firm. Holsum Foods Division assumed direction of PPI. The acquisition of Portion Pac facilitated the export of value-added products to Pacific Rim countries.[58]

GTA Feeds, another division of Harvest States, purchased NORCO mills in 1992. The acquisition of this animal feed manufacturing firm in Iowa and Nebraska gave GTA Feeds six additional feed mills and entrance into the large cattle and hog production states of the western Corn Belt and Prairie Plains.[59]

A South Dakota native became president and chief executive officer of Harvest States Cooperatives on January 1, 1995. John D. Johnson of Baltic, South Dakota, was named president shortly after Allen D. Hanson announced his retirement. Johnson, a 1970 graduate of Black Hills State College, Spearfish, had joined GTA Feeds as a feed consultant in the late seventies. He was named central region sales and marketing manager in 1982, director of sales and marketing for GTA Feeds in 1986 and a vice president of Harvest States Cooperatives on January 1, 1990. In the latter capacity, he became general manager of Harvest States' GTA Feeds division. Immediately prior to becoming president. Johnson was vice president in charge of the new Farm Marketing and Supply area, created when the feed division and the company-owned local elevators were combined.[60]

Phil Testerman retired from the Harvest States board of directors at the end of 1993. First elected to the board in 1972, he had risen to be the first vice president of the group. His replacement, Merlin Van Walleghen of Letcher, represented the farmers of District Three, which included South Dakota, Nebraska and Kansas. Van Walleghen had been president of the South Dakota Association of Cooperatives for seven years, and for ten years, he had presided over the board of the Mitchell Farmers Cooperative Association.[61] In late 1997, he was 1st Vice Chairman of the Board of Directors. Tyrone Moos, Philip, was also a member of the board.

In November 1995, Eastern Farmers Co-op, with its headquarters in Garretson, joined Harvest States Cooperatives. Eastern enrolled approximately 800 members and had operations in nine southeastern South Dakota communities. As a division of Harvest States Cooperatives, Eastern Farmers Co-op made more efficient use of available capital and human resources. The merger provided a larger pool of transportation, access to Harvest States' export facilities on the Pacific Coast and cost efficiency in products.

Harvest States Cooperatives' total revenues were nearly $4 billion for the 1993-1994 fiscal year. It offered grain marketing and other services to some 500 member cooperatives, having approximately 125,000 producer-owners. Its own local agricultural service centers directly served an additional 25,000 farmers, primarily in Minnesota, Montana, North and South Dakota. The regional cooperative's producer-owners were located in thirteen states, from Wisconsin to the Pacific Northwest. Aside from its grain marketing and processing activities, Harvest States had three primary divisions in 1995: GTA Feeds, Holsum Foods and Honeymead, the latter division crushed soybeans and refined soybean oil.[62]

Beginning in October 1997, CENEX and Harvest States studied the advantages of an alliance or merger to strengthen their respective positions in the market place. Together, the two co-ops had combined annual sales of approximately $10 billion. CENEX had 1,400 local co-op members, 2,500 employees and sales of $2.7 billion. Harvest States had 600 local co-op members, 38,000 direct farmer members, 2,350 employees, and sales of $7.1 billion. It was the largest cooperative grain marketer in the United States.[63] On June 1, 1998, the two regional cooperatives that had emerged out of the Farmers Union became one. With a resounding vote by the delegates, CENEX and Harvest States Cooperatives formed CENEX-Harvest States Cooperatives.

EDUCATION: ADJUSTING TO ECONOMIC AND SOCIAL CHANGE

The SDFU education program did not escape the worst agricultural depression in fifty years. Funds were not available during the 1980s to conduct educational activities as vigorously as in preceding decades. Between 1982 and 1989, the South Dakota Farmers Union had six different Directors of Education. In an attempt to economize but, at the same time, maintain essential services, President Leland Swenson logically combined Education and Cooperative Services in the summer of 1985. It was not until August 1, 1989, with the appointment of Elaine Tobin, that the Union, once again, had a full-time Director of Education. Dorothea Geidel of Ethan and Verla Lindblad of Wolsey served as her assistants. In the spring of 1992, Linda Johnson of Lake Preston was appointed Assistant Director of Education.

DIRECTORS OF EDUCATION
1982-1995

Pamela Lubinus Kloos	1982-1984
Cheri Smith	1984-1985
Maynard Whitmyre	1985-1987 *

SDFU Junior Advisory Council, State Camp, 1981-1982.

Verla Lindblad	1987-1989 **
Dorothea Geidel	1989 ***
Elaine Tobin	1989-1995
Brian Jennings	1995-1997
Bonnie Geyer	1997-present

* Director of Education and Cooperative Services

** Part-time Director

*** Interim Director.

Elaine Tobin, a native of Huron, graduated from Northern State University in 1963, with majors in music and English education. From 1963 to 1974, she taught music and high school English in several South Dakota high schools. Prior to her appointment as Education Director, Tobin had been Family Services Coordinator for the University of South Dakota Head Start Program.[64] In naming Tobin as Education Director, SDFU President Dallas Tonsager said "Her appointment underscores our strong commitment to increasing understanding of farmer owned cooperatives and other key family farm and ranch issues."[65]

Following Tobin's departure in early 1995, SDFU President Dennis Wiese appointed Brian Jennings as the farm organization's new education director. Jennings graduated from South Dakota State University in May 1995, with a major in political science and a minor in agricultural business. He was a State Senate legislative intern during the 1994 session. Jennings

The Advisory Council takes a stroll following the announcement of new officers (late 1980s). L to R: Senior Advisory Council: Jamie Larson, Langford, Becky Putnam, Corsica; Junior Advisory Council: Jennifer Brower, Willow Lake, Kevin Stepka, Tyndall, Don Satiroff, Vermillion, Brian Kelly, Armour, Rebecca Petik, Keldron and Stephanie Symens, Amherst.

joined the state office staff on a full-time basis following his graduation from SDSU.[66] Jennings left the central office in June 1997 to join the staff of U.S. Senator Tim Johnson.

Bonnie Geyer, Iroquois, South Dakota, replaced Jennings in July 1997. She spent her childhood on a farm in western Kingsbury County. Geyer worked for the membership department of the SDFU from 1980-1988. Since 1988, she had been assistant to the president of the state Union. Her current responsibilities include coordinating and conducting farm and ranch youth organizations and adult and cooperative education programs.

Societal changes had a negative impact on the camping program. By the early 1990s, less than fifty young Union members enrolled annually for the state camp at Placerville. After reaching a low of twenty-six campers in 1991, the enrollment for the annual summer event increased to forty-seven in 1993. Several factors had radically reduced attendance, including 1) the decline in farm population during the eighties, 2) many high school youths being employed in off-the-farm summer jobs and 3) the distraction of other activities created via modern technology.

Although cooperation, economic and international, continued to be stressed in the camp curriculum, contemporary social issues became a part of the educational program. There were speeches and discussions that considered gender roles, date rape, the plight of the homeless, eating disorders, Mothers Against Drunk Drivers (MADD), organ donations and living wills. Role playing, a very effective teaching technique, was used in studies of the

Julie Schmeising, State Department of Health, making a presentation at a State Camp in the late 1980s.

State Legislature, the American political process, a mock United Nations conference and the plight of Third World refugees.

In 1969, a spokesman for the Farmers Union Central Exchange had said that "No other organization or group provides the kind of concentrated cooperative philosophy that we drill into thousands of farm youth during every camp and study session every year."[67] Activities and involvement proved more effective than "drill" or rote learning. In 1994, the SDFU sponsored a youth poster and essay contest to help celebrate the 150th anniversary of cooperation and the Rochdale Pioneers. The theme for the contest was "Celebrating 150 Years of Cooperation." First place winners in each category received $50 savings bonds.

The cooperative store at the various state camps remained the most effective way to teach campers the importance of cooperatives and how they functioned. The youngsters learned by doing; they became a part of the decision making process in their own store. Organizing the co-op store was one of the first activities for campers when they arrived at Placerville. Articles of incorporation and bylaws were formulated and a board of directors was elected. All transactions at the store were completed with credit union share drafts. At the end of the camp session, all of the recorded profits from the co-op were distributed as dividends to the campers or donated to the Farmers Union Education fund.

Two new camps were created in recent years. Maynard Whitmyre introduced a new intermediate camp for twelve to fourteen year olds in the summer of 1986. With emphasis on "Sharing Global Harvests," Union youths, grades six through eight, met on the campus of the Dakota Bible College at Huron.[68] Approximately fifty campers attended a Farmers Union

Camp Co-op Store.

Reunion Camp at Placerville in June 1993. Campers included current and former Farmers Union staff members, volunteer education leaders, former summer camp staff and former Junior and Senior Advisory Council members from the '60s through the '80s. In "think tank" sessions, they focused on the future of the Farmers Union, its youth program and rural South Dakota. When not considering the farm organization's future, they engaged in those activities associated with camps past and present — hiking, singing around a roaring campfire, dancing and participating in KP activities in the dining hall.[69] The Reunion Camp became an annual event, open to all individuals who attended the senior youth camp between 1960 and 1990.

At the 1983 state camp, the international project was global interdependence. Through reading, discussion and role playing, the campers learned how necessary cooperation was in the quest for world peace. Ten years later, two members of the SDFU, Tom Neuberger of Canistota and Edward Strand of Mitchell exemplified this Union ideal. They were participants in the "Farmer-to-Farmer" program organized by the National Farmers Union and funded by the U. S. Agency for International Development (AID).

With eight other volunteers, they spent three weeks in Hungary in early 1993, working one-on-one with family farmers in that former Communist-bloc country. The program was designed to provide Hungarian farmers and farm organizations with technical assistance as they moved from state-owned enterprises to private farms. Neuberger, president of the McCook County Farmers Union, and Strand, a member of the board of directors of the Mitchell Farmers Union Cooperative Association (CENEX), discovered that the Hungarian

Cartoon Theme — Farm programs and the Contract with America.

farmers did not need help with production agriculture. They needed assistance with value-added marketing and the development of true cooperatives and farm organizations. Strand was impressed with the Hungarian farmers' love of the land and their close knit family structure. It reminded him of community life in rural South Dakota in the 1950s.[70]

As Edward Strand observed, South Dakota was much different today than in 1955. American society, in general, was much different than it was forty years ago. Different times called for different approaches to solving problems. President Wiese, in his monthly column in the *Union Farmer*, the "President's File," wrote in the December 1994 issue that the SDFU should begin "Doing Things Different." Because of the many demands placed upon young farmers, he called for fewer meetings. He wanted meetings that truly emphasized the significant and substantive issues facing Union members. He called for meetings that not only centered on issues affecting producers today but meetings at which the young farmers could develop a vision of what they wanted rural South Dakota to look like ten years in the future.

The cooperatives were adjusting to the rapidly changing economic and social environment through consolidation, restructuring and a major emphasis upon marketing value-added products. While maintaining the traditions and values of the Farmers Union, the delivery of cooperative education, too, has to be adjusted to a society standing on the threshold of the 21st century. The pool of rural youth on which to draw for the future of the South Dakota Farmers Union is much smaller than it was in 1955. For those youngsters who still reside on the farm, there are many other activities competing for their time. Most critical, there currently is a lack of volunteers to operate a successful education program. Approximately 70 percent of farm women, the traditional source of volunteers, now hold off-farm jobs. The farmer, himself, may hold another job in the struggle for economic survival. Many adult volunteers leave the education program after their family members have graduated from high school. A major function of the Education Director as the century draws to a close is to secure volunteers.[71] If the Union is to play a significant role in the next century, the Farmers Union's program of education for farm youth has to be redesigned and reinvigorated at all levels. An awareness of what the Farmers Union and cooperatives have done for farmers in the past, what they are doing for farmers in the present, and what they can do for farmers in the future remains essential.

Chapter 21

THE STATE OF AGRICULTURE IN THE 90s

"Change is inevitable in a progressive country. Change is constant."
— Benjamin Disraeli

During the hyperactive 1960s, Bob Dylan, a Minnesota songwriter and musician, composed a song that spoke not only to that turbulent decade but also to the state of agriculture in the final decades of the 20th century. The American farmer realized more than ever that "The Times They Are A-Changin'" as the century drew to a close. Family farmers and their champion, the Farmers Union, were confronted with 1) Congressional legislation that ended six decades of crop subsidies and regulations, 2) vertical integration, 3) the consolidation of farmland in fewer hands, 4) the growing power of giant agribusiness firms and food processors, 5) the globalization of agricultural markets, 6) the privatization of technology transfer and 7) the generalization of competition — having to manage a more volatile, highly competitive global market without government price stabilization programs.

THE FREEDOM TO FARM ACT

Agricultural programs begun over sixty years earlier came to an end in the early spring of 1996. The Agricultural Adjustment Act of 1933 was enacted when farmers on the northern Prairie Plains were selling wheat for 28 cents a bushel and unable to pay taxes and interest. Surveys showed that half of farm mortgages were delinquent, half [of the] tractors were inoperable because farmers could not afford repairs, that agricultural lending had simply stopped, and many of the farmers needed blankets, clothes and shoes for their families.

The driving force behind the AAA had been Henry A. Wallace, a leading agricultural economist and journalist, a statistician and preeminent plant scientist. The former Iowan presided over the USDA for eight years of the New Deal. He was the Secretary of Agriculture during the Department's golden age. Farm programs and soil conservation were enacted, food reserves established, rural electrification became a reality, food stamps and school lunches were introduced, international food assistance was formalized, and the Department of Agriculture financed advanced work in research, plant genetics and agricultural economics. By 1980, the social aspects of the Wallace-generated farm programs overshadowed the benefits paid to producers. Food stamps, school lunch programs, food for the elderly, food aid for Indian Reservations and other food assistance programs cost $17 billion compared to $3.6 billion for price supports and subsidies.

A complete break with the longstanding safety net of federal crop subsidies and regulations occurred with the election of the 104th Congress in November 1994. Many of the candidates had promised to cut taxes and mandatory spending. The new lawmakers, the great majority representing urban and suburban constituents, were driven to reduce the deficit and to gain control over entitlements. Representative Dick Armey (Rep. Texas), the Majority Leader, was an outspoken critic of traditional farm programs. In his colleagues' eyes, subsidies had to go although farm programs accounted for only 20 percent of USDA's spending and 1 percent of overall federal spending. A five year renewal or revision of the federal government's farm policy was due in 1995. The new Congress had to determine the government's role in the economic well-being of America's 1.9 million farms.[1]

Throughout 1995, Congress debated the end of traditional subsidies and government controls. Attempting to move toward a more free-market based approach to agriculture, Congress devised the Freedom to Farm Act. At a conference celebrating the accomplishments of Henry A. Wallace, former Senator George McGovern said the advocates of the Freedom to Farm Act were men of Lilliputian minds in hailing the demise of Wallace's New Deal programs. McGovern said they were replacing the former Secretary's policies with a grubby little policy, largely the result of the always underestimated influence of greed.

In late 1995, President William J. Clinton vetoed the new farm bill because Republicans had attached it to a balanced budget bill. Three months later, on March 28, 1996, Congress approved a reworked Freedom to Farm Act. On April 4, 1996, Clinton quietly and reluctantly signed the bill, saying that it "fails to provide an adequate safety net for family farmers."

The new farm bill, the 39th major piece of agricultural legislation since 1933, included the following key provisions:

1. An end to government guaranteed prices for corn and other feed grains plus cotton, rice and wheat. A program that had been a staple of U. S. farm policy since the Great Depression.
2. Fixed but declining payments until 2002. A one-time sign-up for payments was scheduled. Farmers who did not sign-up would miss out on the seven year program. Payments would total $36 billion over the seven years, accounting for most of the spending of the $47 billion allowed under the new law.
3. The USDA could no longer order farmers to idle land or limit the kinds of crops grown to collect farm payments.
4. Growers of wheat, soybeans and corn remained eligible for government loans that enabled them to keep crops off the market until prices rose.
5. The Department of Agriculture was directed to monitor foreign compliance with the agricultural provisions of the General Agreement on Tariffs and Trades (GATT). The controversial Market Promotion Program, which favored major U. S. corporations was renamed the Market Access Program with participation limited to small businesses, farmer-owned cooperatives and agricultural groups.
6. The Food Stamp program was reauthorized for two years. Other food programs were reauthorized for seven years.
7. The creation of a $300 million Fund for Rural America over four years. One-third of the money would be spent on research, one-third on rural development and the remaining third would be expended at the discretion of the USDA.

8. The Conservation Reserve Program (CRP), which paid farmers to take highly erodible cropland out of production, was retained. The USDA, however, could not enroll more than 36.4 million acres, the current number under contract. Farmers could withdraw land if the acres were not classified environmentally sensitive. The Wetlands Reserve Program on farmland continued as well as the Swampbuster program that discouraged converting wetlands into cropland.

The administration of the CRP created controversy. Speaking for the South Dakota Farmers Union, President Dennis Wiese criticized bureaucrats who imposed harsh penalties on farmers if they did not fully comply with the program's regulations. He argued that the law's provisions should be based more on local conditions, involving farmers and area agricultural officers. Wiese said it was better to offer financial incentives for farmers to protect soil and water than to punish them for mistakes.

Environmentalists in the Eastern states argued that the Plains States were favored over the rest of the country. They complained that too much money was being spent on preventing wind erosion on the Plains and that inadequate attention was being given to retiring land along waterways to prevent pollution in the East. Of the approximate 36 million acres in the CRP, some 15 million acres were committed to the program in the Plains' States from Texas to North Dakota. The Secretary of Agriculture, Dan Glickman, on February 12, 1997, unveiled new conservation rules. He expanded the CRP's focus from taking erodible cropland out of production to including environmentally sensitive farmland that affected water quality and habitat. Under the new rules, two-thirds of the nation's 360 million cropland acres, including 8.4 million acres of wetlands that had been farmed, were potentially eligible. Glickman made Prairie Pothole Conservation a priority, potholes were a major breeding ground for waterfowl. Significantly, the Dakotas and Minnesota have been the heart of the pothole area.

Agribusiness corporations also attacked the renewed Conservation Reserve Program. They wanted a reduction of the overall acreage allowed to the program and reduced payments to the participants in the program. The more cropland available for production would be to their economic advantage. Agribusiness interests accused Glickman of supply management, the controlling of farm production, which could mean a decline in surplus grain and an increase in market prices.[2]

Proponents of the CRP argued that the program reduced the severity of dust storms and helped in providing game habitat. It also provided a stable source of income as participating farmers received guaranteed payments for ten years for taking the environmentally sensitive land out of production. CRP payments in the Plains' States often exceeded local cash rent rates, often making it more attractive for farmers to idle their land than grow crops on it. The new Freedom to Farm Act barred payments on newly enrolled acreage from being higher than local cropland rental rates. Land already in the program was exempt. In 1996, CRP rates in South Dakota averaged 56 percent higher than local cropland rents.

While supporters of the Freedom to Farm bill boasted that the legislation ended a costly, antiquated system under which the government dictated planting decisions through subsidies and acreage controls, the radical change in the farm program came under immediate fire from the Farmers Union. The new legislation appeared to be a bribe to wean farmers off farm programs. Dennis Wiese noted that the transition payments were actually guaranteed for two years, not seven. The payments were substantially less than the average annual deficiency payments received by farmers under the 1990 farm law. The payment schedule was based on 85 percent participation. Because participants did not have to plant crops to receive payments, participation would be close to 100 percent. Payments would be further reduced because they would be pro-rated based on available funds.[3] Nationwide, 1.7 million farmers signed contracts that would provide fixed but declining payments over the seven years. In 1996, the payments totalled $5.3 billion. South Dakota's farmers received $143 million of this amount.

Wiese pointed out that the payment plan tarnished the image of agriculture by providing hefty payments at a time when the market prices for corn, soybeans and wheat were high. He went on to say, "This is just blatant welfare, where you just throw money out there and the farmer doesn't even have to plant a crop .... It's a terrible image situation and it should be an even bigger concern to consumers and taxpayers." The payment program appeared totally incompatible with the fundamental interests of both the Democrats and Republicans to overcome the budget deficit. Their urban constituents would object to cutting other entitlement programs while giving money to farmers regardless of what the prices were. In addition, when crop prices were high, the largest handouts would be directed to the large, wealthy farmers. Almost 25 percent of the transition payments would go to large family operations at the expense of family farms.

The Freedom to Farm Act was closely connected to a proposed balanced budget amendment. Farmers who signed the "production flexibility contract" that purportedly guaranteed seven years of agricultural subsidies discovered that the contract contained an escape clause if a balanced budget amendment were ever approved. Attachment "Part 7B" under "modifications" to "Appendix to Form CCC-478" stated that "In the event a statute is enacted during the effective period of this contract under which (the USDA) would be required to materially change the terms of the contract, (the USDA) may require signatories to elect between acceptance of modifications consistent with provisions of such statute or termination of the contract." Thus, if the amendment became law, the seven year agricultural subsidies, the ten year CRP and disaster payments would be cut.

Although Farmers Union members liked the planting flexibility in the Freedom to Farm Act, they feared that all-out production would drive commodity prices to new lows. Overproduction threatened fragile land and, over time, would build surpluses threatening the survival of the traditional family farm. Of particular concern to the SDFU was the lack of a safety net. At a time of low payments and low prices, farmers would have to rely on private lenders. The reality would be that if farmers weren't making money, lenders would not loan them anything. In addition, lower commodity prices would lead to lower agricultural values.

Addressing the annual convention of the SDFU at Sioux Falls in October 1996, NFU President Lee Swenson stated that the Freedom to Farm bill did not offer adequate marketing options. It left farmers vulnerable to market-price changes. Swenson stated that "The impact that producers now see is $2.30 a bushel corn, when it was $5.00 a few months ago. Soybeans are down. Wheat is down. That's the impact of the so-called Freedom to Farm Act."[4] Wheat prices fell from $7.00 a bushel in the spring of 1996 to $3.60 per bushel in February 1997. The National Farmers Union wanted a bill that would have raised the loan rate farmers could receive from the government to the average cash price for corn in the past years. Higher loan rates would enable farmers to put some of their grain under loan, use their loan money to pay bills and hold the grain off of the market, hoping for a higher price.

Farmers could sell grain and pay off loans at a profit. The higher the loan rate, the more options farmers would have, tending to lift the market price.[5]

Farm legislation from 1933 to 1995 had stressed stability. Farm bills were designed to provide a stable, abundant supply of food for the public while conserving and maintaining the productivity of the land. The goal was to provide farmers a fair return for their production of key food commodities, preserving the family farm. The farm program had expanded to include water quality, wildlife and wetlands conservation. In the spring of 1996, Congress had confined much of the traditional program to the trash can. Small or average-sized farmholders faced an uncertain and possibly unstable future. There appeared to be no clear and permanent direction for agricultural policy. Abandoning the old program could lead to wild leaps and drops in crop supplies and prices. The new legislation provided for no adjustments if prices plunged or yields were low. Reliance on global markets, foreign demand for U. S. commodities as well as the protectionist policies of other countries, added to the uncertainty faced America's tillers of the soil.

Supporters of the Freedom to Farm Act argued continually that new and expanded foreign markets would keep commodity prices high. In the future, the health of American agriculture would be tied to supply and demand in the global market.

The reliance on foreign markets led to disaster in 1998. Commodity prices tumbled when the Pacific Rim countries of East Asia experienced a radical downturn in their economies. U.S. agricultural exports declined from about $4.4 billion to $3.7 billion. It was ironic that except for agriculture, the American economy was doing well, but if Asia's was not, agriculture suffered. The outlook for farm income became grim as prices declined drastically. By September 1998, based on a statewide average, the cash value of corn fell to $1.46, compared to $2.19 a year earlier and $3.69 the year before that. Other crop prices fell correspondingly following the passage of the Freedom to Farm law in 1996.

	1996	1997	1998
Wheat	$4.30	$3.40	$2.70
Sorghum	2.34	2.70	1.80
Barley	2.74	2.35	1.85
Soybeans	7.35	6.45	4.85

Sioux Falls Argus Leader, 15 July 1998.

The hog market fell to a monthly average of 19 cents a pound in December 1998, including a short dip to nine cents a pound.

Mike O'Connor, director of the federal Farm Service Agency in Huron, said "I know this, if prices continue at this level, it could make the 80s look like a walk in the park." The reduced farm income forced farmers to lower spending and increase borrowing. Loan renewals and extensions increased and loan repayments slowed. In assessing the situation, Dennis Wiese commented, "Debt loads are increasing. Independent farmers now face increasing competition from corporate farms and market control from a few dominant meat packers and grain companies. In some ways, you might be able to argue that it's worse than in 1985."

Government disaster assistance, surprising yields for corn and soybeans and payments for land put in conservation acres helped head off a complete disaster in agriculture in 1998. In the fall of 1998, Congress approved a $6 billion emergency aid plan for agriculture. In the early summer of 1999, President Clinton asked whether the Freedom to Farm Act "is fundamentally flawed" and hinted that it might be necessary to pass a new farm bill.

VERTICAL INTEGRATION AND AGRIBUSINESS

During the 90s, the food industry continued to move rapidly from open production and sales to vertical integration in which a company controlled the flow of commodities across various stages of output, processing and distribution. Changing food demands from consumers and technological advances in production techniques contributed to vertical integration. Consumers wanted lean and uniform beef, pork and poultry; they desired breakfast cereals with certain nutritional characteristics. "Market power," resulting from concentration in the food industry, raised an overall threat of the loss of free markets for farmers. Contract production and vertical integration implied that firms exert more control over the quality of output.

The United States had 2.1 million farms in 1992. Almost three-fourths of them were classified as small with annual gross sales of less than $50,000. In contrast, 558,000 "commercial" farms with sales of more than $50,000 accounted for 88 percent of total gross farm sales and 68 percent of the land in agricultural production. Fewer farms produced more food, and although farmers remained independent operators they formed contractual alliances with the big processing companies that made the economic decisions. With the increased industrialization of agriculture, farmers have been increasingly selling their products through contracts with processors and other buyers. Between 1970 and 1990, the share of hogs produced under contract rose from 2 to 21 percent. The processor or integrator typically owned the commodity being produced and supplied inputs. The farmers was paid for his services. In other cases, farmers had marketing contracts in which they owned the commodity being produced and made the production decisions but they sold to a buyer at a negotiated price. In either case, the real winners were the big companies. Middlemen received the economic benefits, not the farmer.

Economists attributed slumping cattle and hog prices in 1998 to oversupply, low consumption and weak exports. Producers said that packers manipulated prices. After all, 82 percent of all beef in the United States was processed by four companies. Similar circumstances existed in the pork industry. In 1998, in Sioux Falls, the cash market for typical slaughter hogs fell to $15 a hundredweight or $40 for a 250 pound hog, the lowest level in twenty-seven years. Packer manipulation was suspected when in three weeks time, the hog market went from 9 cents a pound to 30 cents per pound. Asking for more honesty from the meat packers, farmers and ranchers called for legislation to ensure full disclosure of prices the packers paid producers. Under the voluntary system currently used, about 60 percent of slaughter cattle purchases, for example, were reported.

During the 1930s, Henry A. Wallace, Secretary of Agriculture, asked "How are farmers to get equality of bargaining power without help from the government?" With major support from the South Dakota Farmers Union, the 1999 South Dakota Legislature passed Senate Bill 95. The bill required meatpackers, each day, to report all livestock purchases to federal and state agricultural officials. It also stipulated that a packer purchasing or soliciting livestock for slaughter in South Dakota could not discriminate in prices paid or offered to be paid to sellers of that livestock unless there were legitimate reasons for price differences. Those could include variation in quality and transportation costs. Senate Bill 95 was scheduled to go into effect on July 1, 1999.

Prior to the implementation of Senate Bill 95, the American Meat Institute (AMI) filed a legal action in federal court during the week of June 21, 1999, to block the law. The AMI claimed that the bill was unconstitutional because it regulated and discriminated against interstate commerce. AMI's attorneys contended that purchases of livestock that are slaughtered in South Dakota included those animals purchased in other states. The federal court was asked to find the law invalid and unenforceable. Prior to a court hearing, scheduled for July 20, John Morrell and Company and Iowa Beef Packers (IBP) stopped making cash livestock purchases. Instead, they purchased on a grade-and-yield basis, which meant the producer was paid after the animal was slaughtered. Dennis Wiese, speaking for the SDFU said "This isn't about producers. This isn't about legislation. This is about corporate greed."

On July 27, U.S. Circuit Court Judge Charles Kornmann struck down the section of the law that required meatpackers to pay the same price for equal-quality livestock. The judge left intact the portions that required packers to make public the prices they pay for all livestock.

One can argue with certitude that reoccurring economic setbacks and new technology contributed to the concentration of productive farm land into fewer and fewer hands. Several students of American agriculture believe that the farm programs begun in 1933 also worked to the benefit of large-scale producers. When Secretary Wallace advanced the AAA program in 1933, he indicated that it could contribute to the decline of the small family farm. The programs were designed to provide a safety net for farmers, lowering their risks and enabling them to adopt new equipment, technologies and facilities. Federal programs reduced reliance on diversified farming because of the insurance provided by price supports. The more that participants in the farm program specialized, the more their productivity increased and the more they expanded. Supporters of this thesis argue that farm credit programs encouraged the substitution of capital and management for labor.

Like John Batcheller in the 1920s, many family farmers remained suspicious of the Department of Agriculture. They remembered statements that emanated from the USDA during the 1970s such as Secretary Butz's "Adapt or die." Some young executives advised the Department to take no steps to help small farmers. Many farmers remained convinced that bureaucrats in the USDA were in league with corporate agriculture, that USDA financed research promoted agricultural industrialization. In the mid-90s, agricultural research did favor large-scale and corporate farms at the expense of family-size operations. Help was given to those corporate interests whose huge livestock facilities threatened water and soil resources. Non-corporate farmers pointed out that agribusiness interests could afford to do their own research and that they were the major cause of environmental problems. Disadvantaged farmers suggested that USDA research should be directed toward dispelling the myth that chemically based, land intensive agricultural systems were the key to feeding the world's growing population.

THE GLOBAL COMMODITY MARKET

The American farmer will play a crucial role in determining whether a stable global population of an estimated ten to twelve billion people fifty years in the future can be sustained. North America's importance as the world's breadbasket will increase in the 21st century. In determining how to feed the billions of new planetary residents, environmentalists and technologists vigorously disagree. Some planners, supported by the corporate sector, believe that increased technology would meet dietary demands. Environmentalists argue that the combination of a burgeoning population and an increased standard of living would cause a crash of the Earth's life support system in the coming century. The ecological-minded stress that America's advanced technology, emphasizing extensive agriculture, is responsible for high sediment loadings in streams, for pesticide and nitrate contamination of ground water and for the large social costs resulting from the displacement of independent farmers.

As a result of the Freedom to Farm legislation, independent farmers will have to rely to a much greater degree on the global commodity market. Secretary of Agriculture, Dan Glickman, asserted that long-term trade, not domestic commodity programs, would define agriculture's future. At the turn of the century, approximately 1 percent of the gross farm income will come from government programs. He believes that the outlook for the American farmers is good if the grain trade can be expanded to countries in Asia and Latin America. American farmers are twice as reliant on international markets than the U. S. economy as a whole. It is estimated that approximately 50 percent of South Dakota's production is exported. In fiscal 1996, government payments to farmers amounted to $750 million. At the same time, America's total farm exports for 1996 were estimated at $60 billion. By 2005, agricultural exports are projected to reach nearly $80 billion. In 1996, two out of every three bushels of wheat grown in the United States were sold overseas. China, by 2000, is expected to be a net importer of up to 8 million tons of U. S. grain. Farm exports must continue to grow. Widespread chaos could occur in the farm economy and related industries if this does not happen.

In relying on international markets, the American farmer must contend with several unknowns. If the past is a reliable indicator, a certain amount of the money due the producer will be drained off by agribusiness middlemen. Competition from other nations, individually or in a bloc (i.e. the European Union) may create a volatility in prices. Governments in the past have subsidized their producers to gain a competitive edge in the international market. The United States can counter foreign competition through the Export Enhancement Program (EEP). The Congressional budget permits $500 million to be used to subsidize American commodity exporters, allowing them to compete on the world market with those countries subsidizing their producers. The Department of Agriculture, however, has been reluctant to implement the EEP. There may also be a resurgence of economic nationalism in various parts of the globe causing the erection, once again, of tariff barriers. Finally, much of the population explosion is projected to take place in countries of the world that lack the financial resources to purchase foodstuffs from American producers. Aside from humanitarian agencies, who will reimburse the U. S. farmer for his production costs?

A CHANGED ECONOMIC AND SOCIAL ENVIRONMENT

The Freedom to Farm bill provided more freedom but also more risk. Market prices would favor the farmer as long as inventories were low. In the long run, good management would be the key to survival. More time and energy would have to be directed toward marketing, growing what would make the most money. South Dakota's farmers and ranchers in the 21st century will have to be talented business managers with an understanding of global food trends, computer science and cash flow.

Outmigration from rural towns became pronounced in the 1930s, accelerated in the 50s, and increased dramatically in the 1980s. In 1994, for the first time since 1850, the number of farms dwindled to fewer than two million; the total stood at 1,925,300. Farm numbers in the United States had peaked at 6.8 million in 1935. With the decrease in farm numbers, there has been an increase in farm size. Between 1987 and 1992, the average farm size increased from 462 to 491 acres.[6] Economic reverses, accelerating costs and new technology all contributed to the depopulation of the countryside. Technology also changed rural values. Improved transportation and communication after the 1930s contributed to farm families becoming more urbanized in their outlook. Family relations became more egalitarian because of more education for women and increased work opportunities. Modern appliances enabled an increased number of women to work off the farm. In 1994, 3.5 million people, over half of whom had off-farm jobs, lived on farms where 32 million once farmed. Small towns and villages that depended on farming have been forced to depend on non-farm activities if they are to survive. Rural communities are now interdependent. Taking a larger view, agriculture remained the nation's largest industry. In 1989, agriculture employed approximately 23 million persons but only 3.2 million of them were in farm production.

No one can deny the considerable changes that have taken place in American agriculture since World War II. Whether all of these changes have been for the good is debatable. Certainly the status of the small family farm has changed and is in grave danger. Hiram Drache, a proponent of industrialized farming, said that the small family farm is now a myth. He defined the late 20th century family farm as primarily an agricultural business in which the farmer has become a risk-taking manager, working the land with his family.[7] In sharp contrast, the South Dakota Farmers Union justifiably recognizes the continued existence of the small family farm. The Union continues to value these farms, to champion their virtues and works through its tripartite program of cooperation, education and legislation to ensure quality life in rural South Dakota.

Chapter 22

AT THE CROSSROADS — AN ASSESSMENT AND AN ESSAY

We must live with change to live at all.
— Thomas H. Steichen.

Study how a society uses its land, and you can come to pretty reliable conclusions as to what its future will be.
— E. F. Schumacher, *Small Is Beautiful.*

Family Farm — An agricultural production unit which can be efficiently operated by a full-time farmer and his family, that provides most of its labor, management and investment, and derives its major income from the farm.
— Leland Swenson, 1982.

Over eighty-five years have passed since J. K. Weinmaster stepped off the train at Mitchell and began organizing the Farmers Union in South Dakota. The recruiting ground was fertile as most South Dakotans resided on farms, and the problems that confronted them were similar to those earlier sodbusters who applied their plows to the Prairie Plains in the 1860s and 1870s. Because of twenty years of low farm prices, however, adverse climatic conditions and dissension within the Union's ranks, the South Dakota Farmers Union did not achieve sizable membership growth until after 1940. Numbers are only one way of evaluating an organization, its accomplishments are another.

Beginning in the 1930s, despite the daunting challenges enumerated above, the SDFU began to make significant contributions to improving the quality of life in South Dakota. Believing that social and economic good should be distributed generally throughout society, the Union, through its legislative program, led in the struggle for a national Farm Bankruptcy Act and the passage of an ore tax on the state level. When the Farmers Union realized that Congress would not legislate cost of production, it became a leading advocate of parity. It was the Farmers Union, through the Farmers' Holiday Movement, that created a greater awareness among Americans of the farmers' desperate economic plight in the early thirties.

In the post-World War II era, the South Dakota Farmers Union was the leading proponent for developing the Missouri River Basin, for promoting public power and rural electrification and for rural telephone cooperatives. The Union was one of the first organizations in South Dakota to recognize what the decline in railroad transportation would do to the rural areas and to advocate a South Dakota rail authority. It was the Farmers Union that led in the struggle to eliminate the detested personal property tax, or so-called "liar's tax." The SDFU secured passage of the Family Farm Act, and it continues to fend off attempts by corporate agriculture to establish a foothold in South Dakota.

Taxes remain the number one issue for the South Dakota Farmers Union. Believing that sales and real estate taxes are regressive and unfair to farmers, the Union, working with other interest groups, continues its longstanding advocacy of net personal and corporate income taxes. In promoting its legislative agenda, the SDFU has not taken an extreme ideological position. It accommodates a broad range of state and federal government actions on behalf of rural producers, and it has been flexible in the adoption of lobbying programs, adapting to the times.

From the beginning of the state's history, cooperative enterprises began to play an important role in the life and livelihood of the farmer. Co-ops helped shape the character, economic, social and political life of the state. Although the SDFU shied away from owning cooperatives, it became the strongest supporter of cooperatives in South Dakota, beginning in the 1930s. Members were encouraged to participate in those co-ops that bore the Union's mark of approval or its name. The Farmers Union became the foremost organization in the state working to perpetuate the tradition of cooperative buying and selling.

Co-ops provided the means by which farmers were able to work together for common objectives, to gain equity and justice for themselves and for others in the marketplace. The benefits of the cooperatives spread well beyond the marketplace. They contributed to a better lifestyle for rural Americans. President Lyndon B. Johnson, in the mid-1960s, provided an excellent summary of the value of co-ops when he said:

> The catalog of cooperative accomplishments is large. You brought credit where it seemed unobtainable. You brought light and power to rural Americans when it appeared out of reach. You brought modern telephone service to tens of thousands and lowered the price of gasoline and fertilizer to many thousands more. Your goal has been opportunity for all and not just success for some. You have achieved so much because you cared most about people.[1]

Cooperatives enhanced the spirit of community. A primary function of the Farmers Union was to develop the spirit of cooperation through education. In the late 1920s, the SDFU introduced an educational model that attracted the National Farmers Union. Through the Junior locals and an excellent camping program, which began in the mid-1930s, Farmers Union youth learned not only the values and benefits of cooperation but the importance of ethnic and racial understanding. In international relations, the Union has always emphasized peace and justice. The SDFU has been a leading proponent of the United Nations and a major participant in programs that brought farmers from other continents to South Dakota to enhance their cooperative and farming skills.

There have been several significant turning points during the SDFU's eighty-five years of existence. On the positive side there was the growing support of progressive members for the New Deal farm program. The role of NFU President James G. Patton in developing the

Farmers Union into a truly united national body and his relationship with governmental leaders in Washington, D.C., and urban liberals must also be underscored.

The SDFU, however, cannot rest upon its past accomplishments, significant as they have been. One cannot be unduly optimistic in the light of the societal changes that have been and are taking place. There is a lot less income security for today's farmers. The failure to adopt the Brannan Plan in the late 1940s and the rejection of the 1963 wheat referendum contributed to the destabilization of a supply management system in American agriculture. The farm crisis of the 1980s led many farmers to abandon the land.

There have been other changes, evolutionary changes, that the Farmers Union is confronting and will have to confront in the future. Since the 1920s, the farm bloc in Congress has been gradually reduced in strength. The traditional alliance with urban liberals is showing stress. Concern for balanced budgets has put certain programs (i.e. the food stamp program) used to garner support from the urban sector for farm programs in jeopardy. Many elements in the business community are not very enthusiastic about a producer-assured income. Farmers have encountered increasing bureaucratic red tape and more trips to the ASCS office in meeting government guidelines. The Congress authorized a new farm bill in 1996. In countering the negative aspects of this legislation, the SDFU, in consort with the Unions in other states, continues to work for the passage of farmer friendly programs.[2]

The most insidious enemy of the family farmer has been the American idolatry of giantism. It is one of the causes and certainly one of the effects of modern technology. Unreliable government farm policies, inconsistent and low farm prices with corporate giants determining market prices, and modern technology are primarily responsible for the migration of people from rural areas to urban centers. South Dakota, in July 1997, had 32,500 farms, a decline of 1,500 farms from July 1994. In 1931, the state had a high of 84,300 farms. Despite the Great Depression, South Dakota's farm population did not begin a radical decline until the 1960s. Federal farm policies and technology played a major role in the decline.

Many of the state's farm families barely survived in the 1980s, during the worst agricultural depression since the 1930s. Land values and prices collapsed in 1981. Drought conditions in the late 70s, low commodity prices and burdensome debt compelled many farmers to file for bankruptcy or to lose their property through mortgage foreclosures and farm auctions during the 80s. South Dakota had only 703 bankruptcies in 1981. At the height of the farm crisis, in 1987, 1,653 people filed for bankruptcy, an increase of over 130 percent in six years.[3] A rural wit from Nebraska expressed the financial plight of the farmer with the following question and answer. She asked, "What's the difference between a farmer and a crow?" Her ensuing reply was "A crow can still make a deposit on a John Deere."

In South Dakota, the average age of farmers and ranchers was 51.1 years in 1992, an all-time high. Fewer young people were willing or able to take a gamble on farm and ranch life. It was difficult to start farming because of the high cost of technology. A farmer could spend $200,000 for a new combine. Many young farmers were unable to economically weather the 80s. They moved out of the area or sought employment in nearby towns and cities.[4] Aging farmers did not want to expand because they did not want to incur debt. When it came time to quit, they sold to their neighbor or neighbors. In South Dakota, the family farm was not disappearing, it was becoming larger. The shift to larger sole proprietorships, partnerships and family corporations. however, meant a significant reduction in the number of rural consumers and the further decline of main street in the state's smaller communities.

In South Dakota, the land devoted to agriculture in 1997 totaled 44.2 million acres. The average size of a farm was 1,354 acres. There were 6,500 farms with sales from $1,000 to $9,999; 16,000 farms with sales ranging from $10,000 to $99,000; and 10,000 farms with sales of $100,000 or more. Nationwide, many of the people who moved from the land to urban communities took jobs in farm-related work. In North and South Dakota, however, farm production jobs continued to exceed jobs in the wholesale and retail sections of agriculture. In South Dakota, in 1992, 100,503 people worked directly in farm production, which was 26.7 percent of the state's work force.

Roland Leonhardt, a Farmers Union member from Kingsbury County, said:

> I can climb up on top of my silo — and it's high — and I can see 25 places that have been vacated since I started out there 25 years ago. Those 25 families would support four merchants in one of these small towns.[5]

He went on to list the businesses lost in Oldham, including two grocery stores, a theater, a drugstore and a barber shop.

Fewer and larger farms and fewer and larger processing industries has been an ominous trend over the past three decades. A report prepared by the Leopold Center for Sustainable Agriculture at Iowa State University stated, "We no longer have a competitive agriculture sector in the United States." At the end of 1993, four firms controlled 82 percent of beef slaughter, 45 percent of pork slaughter, 71 percent of flour milling and 76 percent of soybean processing. When four firms control 40 percent or more of the market, the market is no longer a competitive market. A few firms control the food system from seed to shelf.[6]

Modern technology has been a blessing and a curse for the family farmer in South Dakota. At his disposal, if he can afford them, are four wheel drive, air conditioned, diesel tractors, computer regulated and fully automated feed lots and the means to feed and milk dairy cows with automation. Farming is less laborious and more efficient. The American farmer outproduces every agriculturalist anywhere in the world. But there is a dark side to the use of this technology by the family farmer. In most instances, its acquisition means going into debt. In 1994, the income strictly from farm sources averaged $179,872. Farm operating expenses, however, averaged $154,499. These same farms averaged about $16,000 in nonfarm income. Many of the state's farm women entered the paid labor force in disproportionately high numbers during the economic upheaval of the farm crisis years of the 1980s. They had been helpers, mothers and wives who were pushed or jumped into off-farm work to help pay the bills. Rural women may spend the day at a job in town and then come home not only to family responsibilities but a mountain of bookkeeping and other paper work for the farm business. Without the nonfarm income, however, many farmers would have incurred a deficit. An analysis of the records of 200 South Dakota farms indicated that the total farm debt on December 31, 1994, averaged $194,269 per farm, despite better than average crop production.

Civilizations have risen on an agricultural base, and they have fallen when the agricultural base declined. The depopulation of rural America is a national tragedy. Many urban people do not realize how dependent they are on rural folk. The same is true of many politicians. Following current trends, they look only to the urban areas for their votes. The politicos in the nation's capital tend to ignore Secretary of Agriculture Glickman's statement that agriculture and defense relate directly to national security and that they must be supported. Agricultural subsidies in the mid-90s amounted to about $10 billion out of a $1.4 trillion budget. There were far, far more people on welfare than there were people who actually farmed. The Republican controlled Congress in 1994 planned to spend more on foreign aid than it planned to spend on subsidies for American farmers.

Recent research at Ohio State University confirmed that farmers continued to personify the honest, hardworking and moral ideals associated with a healthy national culture. The research indicated that farmers were more stable, more religious, happier and more satisfied with life. They were among the better adjusted elements of society. Farmers were more likely to have been raised with a mother and a father present, were less likely than any other group to believe that a person "must live for today and let tomorrow take care of itself." The research confirmed the 19th century view that a nation could flourish only if agriculture was prosperous and the family farm preserved.

The number of people who question whether bigger and biggest is best is increasing. Much of the emphasis on "biggest is best" has come from land grant institutions, lenders, certain farm publications and the USDA. Bigness creates impersonality, insensitivity and a lust to concentrate power. Joe Robbie, a former attorney for the South Dakota Farmers Union, had this in mind when he told a group of farmers shortly after World War II that "As a man's waistline expands, his mind contracts in about the same ratio."[7] Michael Duffy, an Economics Professor and Extension Economist at Iowa State University, after careful examination of available data, said it is a myth that big is efficient and small is inefficient. The cost of producing a bushel of corn in Iowa is less on farms of 100 to 1,000 acres than on those greater than 1,000 acres. The total cost of producing hogs is relatively the same for operators selling from 1,200 to 3,000 animals a year as those selling more than 3,000 hogs. Duffy concluded that farms are getting bigger not because it lowers their production costs to do so, but because they need more volume to generate income. E. F. Schumacher, in *Small Is Beautiful*, questioned the social structure of an agricultural system which obtained its justification from large-scale mechanization and the heavy use of chemicals. Such a system, he argued, lives on irreplaceable capital which it treats as income. The irreplaceable capital in this case is the topsoil, the topsoil which carries an immense variety of living beings, including man.[8]

A commentator on agriculture, Wendell Berry, wrote — "Land is not productive; people are productive." If the country's shrinking farm land was left to giant corporations, they would, as was wont with giant corporations, exploit, ruin the land and move on to other investments. Land has to be nurtured, otherwise it dies. To do the hard work of nurturing, people have to live on the land, to love it, to know it and have the skills needed to keep the land productive for generations.

Contemporary adherents of Schumacher's philosophy advocate sustainable agriculture. In attempting to address agriculture's environmental problems, the advocates of sustainable farms would reduce the use of costly fertilizers and chemicals through a return to crop rotations, and less specialization among farmers and livestock on virtually every farm. A six year study prepared by the Northwest Area Foundation of St. Paul, Minnesota, argued that sustainable agriculture could reverse the trend toward bigger and fewer farms and increase the overall rural population. The authors of the study acknowledge that sustainable agriculture may not prevail, but they are convinced that it offers a farming system that rewards land stewardship, provides economic opportunity for farmers and supports healthy rural communities.

The South Dakota Farmers Union is at the crossroads. It can submit to the forces that have created consolidation — the changes in technology which have reduced the ratio of labor to output — or adapt some of the same technology in the struggle to save family agriculture. One way in which farmers can take charge of their destiny is to unite in promoting their own "seed to shelf" endeavors. The base is present for doing this. There are fewer farms in South Dakota, but agriculture continues to be the bulwark of the economy. Agriculture in 1991 contributed $13.2 billion to the state's economy, more than three times the economic impact of any other single industry and larger than all other industries combined. Every resident of the state has a direct or indirect interest in the level of farm and ranch income. When commodity prices decline, the entire state suffers, especially communities that have a population under 20,000. SDFU President Dennis Wiese said that "When the farmer prospers, everybody prospers." This is very true in South Dakota.

Unity and commitment are keys to the survival of family agriculture. The state's farmers must unite and invest in value-added cooperatives. One of the Union's tasks should be to encourage new farmer-owned industries similar to those now in operation or proposed — such as ethanol production, soybean processing, a farmers' pork cooperative, a spring wheat processors co-op and a beef co-op. What is needed is a type of co-op fever as experienced in the state during the 1940s and 1950s. The new co-ops would be different from those of an earlier date. In the closed co-op, the farmer or rancher makes a substantial investment, is given the right to deliver the product, then collect the dividend. Discipline is required of the members because they must fulfill their obligation to deliver the product.

The South Dakota Farmers Union must continue to be a leader in encouraging the formation of cooperatives that will process and market the food grown by the state's farmers. The greater the farmers' involvement in the food chain the greater their chances for survival. South Dakota has only one processing cooperative, the soybean plant at Volga, whereas its neighbors, Minnesota and North Dakota, have several processing installations owned and operated by farmers. Former Governer George Sinner of North Dakota, speaking at the AGARAMA in Sioux Falls in February 1997, issued the following warning to his rural audience, "In my judgment, we are in danger of losing individual ownership of productive land in this country. Cooperatives may be one of the last ditch efforts to preserve the personal, private ownership of the land in America."[9] The traditional individualism of South Dakota's farmers must be set aside and a commitment made to create cooperatives for the processing of hogs, cattle and small grains.

The Rural Development Act of 1972 had been designed to help maintain farm numbers by providing economic opportunities for off-farm work. A former president of the SDFU, Dallas Tonsager, the current State Director of Rural Development for the USDA, reiterated the program's advantages when he spoke to assembled Union members at the 1996 annual convention. Tonsager said that low-interest loans and loan guarantees were available through his office for projects involving agricultural processing, small rural water systems, community buildings and home purchases. He emphasized that these loans could bring new life to struggling farms and rural communities. Essential to any program's success was strong leadership. Tonsager noted that a group in east-central South Dakota — representing Iroquois, Cavour, Yale, Tulare and Frankfort — had earned a federal designation as "enterprise communities," making them eligible for a $2.9 million federal grant. The grant was used to establish a revolving loan fund for development projects. In a year's time, the fund directors made almost $1 million in loans. The loans helped businesses to begin and expand, to construct housing developments and to conduct community cleanup campaigns.[10]

Unity is needed to make the farmers more visible in legislative endeavors. Farmers must set aside philosophical differences and unite in a common cause — the preservation of family agriculture and the small town. Politicians count numbers. The numbers increase with each ally the Union can secure in urban areas. The Farmers Union, by joining with city consumers, organized labor, educators and people with the same philosophical position (i.e. taxation), can revitalize its presence in the state and national legislative bodies. Political activism, a trademark of the SDFU since the 1920s, is direly needed to maintain the Union's values.

Another formidable challenge for the SDFU is to attract young people back to the farm and small town. The average age of the South Dakota farmer in 1992 was fifty-one years of age.[11] James Beddow, former president of Dakota Wesleyan University, said, "We have mined the young people out of our rural areas, just as we do gold and timber. We have to reinvest in those areas."[12] The return of the young to the rural areas will depend on what those areas have to offer in economic incentives and income security.

The South Dakota Farmers Union met the challenges of the past in defense of the family farm. Circumstances are very different today than they were forty to sixty years ago. Different circumstances call for different solutions. The basic parts of the Farmers Union triangle, however, still apply in achieving those solutions — legislation, cooperation and education. Built into these words are activism, unity and understanding and/or creativity.

There has to be a future for family agriculture. The family farmer recognizes the whole relationship between man and nature. The fundamental principle of agriculture is that it deals with life, with living substances. Ward Sinclair, a farmer and a journalist, recognized this relationship. Looking toward spring, Ward wrote:

> The farmer is much like the tree. As the days turn longer in the waning winter, the time of dormancy begins to ebb and the sap rises in anticipation of resurgence and renewal. No formality announces this beginning. It is a feeling in the bones, if you will, that culminates in a rush to gently drop the first seed into the soil.[13]

APPENDICES

Appendix A

NATIONAL FARMERS UNION
PRESIDENTS

Newton J. Gresham	Founder
Charles S. Barrett	1906-1928
C. E. Huff	1928-1930
John A. Simpson	1930-1934
E. H. Everson	1934-1937
John Vesecky	1937-1940
James G. Patton	1940-1966
Tony T. Dechant	1966-1980
George W. Stone	1980-1984
Cyril H. Carpenter	1984-1988
Leland H. Swenson	1988-

SOUTH DAKOTA FARMERS UNION
PRESIDENTS

S. W. Ricords	1915-1918
Clair D. Johnson	1918
John W. Batcheller	1918-1928
E. H. Everson	1928-1934
Emil Loriks	1934-1938
John Skage	1938-1944
Oscar Fosheim	1944-1947
Paul Opsahl	1947-1958
Harold Golseth	1958-1961
Ben Radcliffe	1961-1981
Leland H. Swenson	1981-1988
Dallas Tonsager	1988-1993
Dennis Wiese	1993-

SOUTH DAKOTA FARMERS UNION
VICE PRESIDENTS

Clair D. Johnson	1917
E. H. Everson	1918-1928
W. H. Preston	1929-1932
Emil Loriks	1933-1934
Oscar Fosheim	1934-1936
John Skage	1936-1938

Roy Brazell	1938-1943
David J. Tiede	1943-1944
Elmer Beck	1944-1947
Roy Brazell	1947-1949
Harold Golseth	1949-1958
Ben Radcliffe	1958-1961
Herman Pietz	1961-1969
LeRoy Wiitala	1969-1971
Adam Seidel	1971-1985
Michael O'Connor	1985-1987
Dallas Tonsager	1987-1988
Cecil Hall	1988-1995
Dennis Van Overshelde	1995-1997
Phil Cyre	1997-

Appendix B

THE BOARD OF DIRECTORS

1918

E. H. Everson
Frank Bradley
O. J. Murphy
J. Glenn Richards
Henry Turner

1919

E. E. Hunter, Letcher
Frank Bradley, Tripp
Erland Johnson, Plankinton
E. H. Everson, Bonesteel
George E. West, Presho

1920

E. E. Hunter, Letcher
Frank Bradley, Tripp
Frank Meiers, White Lake
William Sinkular, Dallas
George E. West, Presho (Replaced by Ira E. Frazier, Presho)

1921

Frank Meiers, White Lake
Ed Rhein, Ravinia
E. P. Lloyd, Rockham
William Sinkular, Dallas
Edward Brabander, Creighton

In 1921, the SDFU established district representation.
The listings that follow are by district.

DISTRICT I

1922-1925	P. G. Eckhoff, Platte
1926-1927	J. M. Nagle, Yankton

1928	W. B. Benson, Parker
1929	J. M. Nagle, Yankton
1930-1934	J. W. Batchellor, Mission Hill
1935-1942	Enoch Hofstad, Elk Point
1943-1946	D. J. Tiede, Parkston
1947-1951	Sam K. Ulrickson, Canton
1952-1958	Norman E. Olsen, Canton
1959-1961	Herman Pietz, Parkston
1962-1982	Lafe Lunder, Canton
1983-1998	Lavern Aisenbrey, Olivet

DISTRICT II

1922	J. E. House, Chamberlain
1923-1932	Joe Atkins, Chamberlain
1933	H.W. Phillips, Sioux Falls
1934-1939	Jesse Kistler, Mt. Vernon
1940-1941	Everett Heggen, Brandon
1942-1958	Henry Vinz, Spencer
1959-1974	Ewald Osterberg, Salem
1975-1981	Robert Fortner, Mt. Vernon
1982-1983	Darrell Roth, Spencer
1984-1996	Ervin Goldammer, Mitchell
1996-1998	Kenneth Artz, Stickney

DISTRICT III

1922	Albert Wharton, E. P. Lloyd, Rockham
1923-1930	E. P. Lloyd, Rockham
1931-1933	H. P. Richardson, Highmore
1934-1944	W. C. Hermann, Rockham
1945-1947	Roy S. Glover, Frederick
1948-1964	Loyd B. Larson, Brookings
1965-1971	David L. Hanson, Waubay
1972	LaVerne Jensen, Erwin
1973-1978	Ray Wiese, Gary
1979-1984	Paul Symens, Amherst
1985-1995	Robert Weber, Strandburg
1996-1998	Wayne Soren, Lake Preston

DISTRICT IV

1922-1923	William Sinkular, Dallas
1924	E. E. Owen, Colome
1925	William Sinkular, Dallas
1926	E. E. Owen, Colome

1927	Peter T. Nelson, Dallas
1928-1932	N. M. Peterson, Colome
1933	F. C. Wilson, Colome
1934-1939	William Sinkular, Dallas
1940	Karl Keck
1941	T. J. Jacobsen, Dallas
1942-1944	J. E. Broline, Winner
1945-1968	Harry E. Caslin, Kennebec
1969-1973	Dorothy Authier, Vivian
1974-1987	Larry Kehn, Batesland
1988-1998	Kenneth Miner, Gregory

DISTRICT V

1922-1923	Harry Caslin, Kennebec
1924	E. C. Griffith, Kadoka
1925	Harry Caslin, Kennebec
1926	E. C. Griffith, Kadoka
1927-1928	D. D. Collins, Belle Fourche
1929-1932	E. N. Hammerquist, Farmingdale
1933	W. F. McNenny, Sturgis
1934	E. N. Hammerquist, Farmingdale
1935-1937	Fred Root, Powell and Keystone
1938-1939	L. D. Lyman, Murdo
1940-1942	W. F. McNenny, Sturgis
1943-1947	Roland Veren, Sturgis
1948-1949	John Riedy, Thunder Hawk
1950-1953	Fred Ferwerda, Kadoka
1954	John Riedy, Thunder Hawk
1955-1966	LeRoy Hardy, Sturgis
1967-1984	George Levin, Hereford
1985-1988	John Youngberg, Newell
1989-1998	Richard Harwood, Union Center

In 1955, the South Dakota Farmers Union redistricted the state. Two new districts, VI and VII, were created. Districts I, II, and IV remained intact while districts III and V were divided along county lines to create the new districts.

DISTRICT VI

1955-1975	Frank Butler, Shade Hill
1976-1994	Carroll McKenna, Zeona
1995-1998	Jerry Petik, Meadow

DISTRICT VII

1955-1958	Ben Radcliffe, Hitchcock
1959-1970	Ray W. Nelson, Bowdle
1971-1984	Harvey Faulstich, Orient
1985-1994	Clifford Gascoigne, Yale
1995-1996	Mary Jo Nemec, Holabird
1996-1998	John Sumption, Frederick

Appendix C

SOUTH DAKOTA FARMERS UNION
SECRETARY-TREASURERS

R. B. Warne	1917
Albert Stevens	1918-1921
E. H. Sauber	1921-1937
Paul G. Erickson	1937-1941
Helen Shepherd	1942
Paul G. Erickson	1943-1944
Melvin Thomsen	1945
Ed Backlund	1945-1947
Roy S. Glover	1947-1950
Emil Loriks	1950-1952
Mrs. Gene (Anderson) Olson	1952-1959
Mrs. Joy L. (Johnson) McGregor	1959-1963
Mrs. Illene M. Harrison	1963-1967
Mrs. D. Maxine Peterson	1967-1970
Leland (Lee) Swenson	1970-1978
Norman Paul Tofflemire	1978-

SOUTH DAKOTA FARMERS UNION
DIRECTORS OF EDUCATION

Mrs. David (Minnie) Lovinger	1930-1932
Mrs. O. H. (Elsie) Olson	1932-1935
Paul G. Erickson	1936-1941
Mrs. Arnold (Marie) Styles	1941-1945
Mrs.Olga Radke, Acting Director,	1941-1942
Mrs. Lena Husman	1945-1947
Mrs. Hazel Brekke	1947-1950
Mrs. Henry (Leona) Ravenberg	1951-1957
Shirley Dickinson	1957-1958
Mrs. Berdyne Halse	1958-1959
Arlene Schley	1959-1967
Joan Olson	1967-1969
Barbara Heidenreich, Acting Director	1969-1970
Mrs. Wilma Putzier	1970-1978
Kenneth Birgen	1979-1982
Mrs. Pamela (Lubinus) Kloos	1982-1984
Cheri Smith	1984-1985
Maynard Whitmyre	1985-1987
Verla Lindblad (Part-time Director)	1987-1989

Dorothea Geidel (Interim Director)	1989
Elaine Tobin	1989-1995
Brian Jennings	1995-1997
Bonnie Geyer	1997-present

Appendix D

SDFU CONVENTION SITES
ANNUAL MEMBERSHIP TOTALS (INCOMPLETE)

YEAR	SITE	MEMBERSHIP	LOCALS
1915	Mitchell	—	—
1916	Mitchell	—	—
1917	Mitchell	—	—
1918	Mitchell	—	—
1919	Mitchell	—	—
1920	Mitchell	—	—
1921	Mitchell	Less than 4,000	—
1922	Mitchell		
1923	Chamberlain	—	—
1924	Chamberlain	—	—
1925	Chamberlain	—	—
1926	Chamberlain*	—	—
1917	Miller	7,092	
1928	Winner	7,017	333
1929	Rapid City	7,175	329
1930	Yankton	7,694	358
1931	Mitchell	7,073	
1932	Huron**	5,603	379
1933	Winner	4,660	356
1934	Rapid City	6,146	375
1935	Yankton	6,763	403
1936	Sioux Falls	5,611	392
1937	Huron	4,462	346
1938	Winner	4,430	329
1939	Rapid City	4,242	310
1940	Yankton	4,723	296
1941	Mitchell	4,738	265
1942	Aberdeen	5,297	284
1943	Huron	5,880	272
1944	Mitchell	11,647	374
1945	Yankton	11,410	390
1946	Rapid City	11,512	386
1947	Pierre	11,255	370
1948	Watertown	13,762	401
1949	Watertown	21,600	—
1950	Mitchell	18,920	—

YEAR	SITE	MEMBERSHIP	LOCALS
1951	Mitchell	17,147	—
1952	Mitchell	—	—
1953	Huron	17,724	—
1954	Huron	18,694	463
1955	Huron	18,900	444
1956	Huron	19,701	452
1957	Huron	19,002	460
1958	Huron	19,466	—
1959	Huron	19,011	452
1960	Huron	18,595	468
1961	Huron	17,455	451
1962	Huron	16,785	444
1963	Huron	16,627	432
1964	Huron	15,110	427
1965	Huron	14,555	401
1966	Huron	14,355	401
1967	Huron	14,303	362
1968	Huron	13,411	200
1969	Huron	12.731	—
1970	Huron	12,313	—
1971	Huron	11,959	190
1972	Huron	12,227	180
1973	Huron	12,319	174
1974	Huron	—	—
1975	Huron	13,751	160
1976	Huron	14,707	154
1977	Huron	15,058	146
1978	Huron	14,939	139
1979	Huron	14,602	136
1980	Huron	14,628	128
1981	Huron	15,081	125
1982	Huron	15,096	113
1983	Huron	14,906	—
1984	Huron	—	—
1985	Huron	—	—
1986	Aberdeen	10,913	—
1987	Aberdeen	10,745	—
1988	Mitchell	9,984	—
1989	Yankton	—	—
1990	Brookings	10,600	—
1991	Rapid City	—	—
1992	Huron	—	—
1993	Pierre	—	—
1994	Aberdeen	—	—
1995	Watertown	—	—

1996	Sioux Falls	—	—
1997	Watertown	—	—
1998	Rapid City	—	—
1999	Huron	—	—

*Drought — drastic membership decline in Districts 4&5

** Severe Drought

ENDNOTES

CHAPTER 1

1. Herbert T. Hoover, "Farmers Fight Back: A Survey of Rural Political Organizations, 1873 - 1983," *South Dakota History* 13 (Spring/Summer 1983): 126.
2. Charles and Joyce Conrad, *50 Years, The North Dakota Farmers Union* (n.p., 1976), 4.
3. Herbert S. Schell, *History of South Dakota* (Lincoln: University of Nebraska Press, 1961), 224.
4. Ibid.
5. Ibid.
6. Conrad, *50 Years*, 4.
7. Conrad, *50 Years*, 6.
8. Mildred K. Stoltz, *This Is Yours, The Montana Farmers Union and Its Cooperative Associates* (Minneapolis: The Lund Press, 1956), 3.
9. *South Dakota Union Farmer*, 19 May 1926.
10. Ibid.
11. Ibid., 29 September 1932.
12. Ibid.
13. Ibid.
14. William P. Tucker, "Populism Up-to-Date: The Story of the Farmers Union," *Agricultural History* XXI (October 1947): 199.
15. Ibid.
16. Ibid., 200.

CHAPTER 2

1. John L. Shover, *Cornbelt Rebellion, the Farmers' Holiday Association*, (Urbana: University of Illinois Press, 1965), 10.
2. Robert L. Morlan, *Political Prairie Fire, The Nonpartisan League, 1915 - 1922* (Minneapolis: Univesity of Minnesota Press, 1955), 16.
3. There is evidence that the first Farmers Union organizers from Nebraska arrived in southern Bon Homme County, South Dakota, in 1910. J. W. Batcheller, without providing documentation, mentioned a Roosevelt Local in the forenamed county. If the local existed, it did not last long. - *Union Farmer*, March 1990; Thompson, *History of the South Dakota Farmers Union*, p.15. Interview with J. W. Batcheller, 10 July 1948.
   In the "50 Years Ago" column of a January 1956 edition of the *Madison Daily Leader*, there appeared the following item: January 20, 1906: "...Fiske, state organizer of the state and national Farmers Union, has just organized a Union at Wentworth with the following officers: P. A. Johnson, President, C. N. Fader, Vice President; A. F. Dworak, Secretary and C. P. Johnson, Treasurer and Peter Warrens, local organizer.- *Union Farmer*, 20 February 1956.
4. *Union Farmer*, 28 December 1959.
5. Florence and Russell Tilberg, interview by author, Ethan, South Dakota, 26 July 1994.
6. *Union Farmer*, 25 January 1965.
7. Ibid., 12 April 1939.

8. Ibid., 28 December 1959.
9. Tucker, "Populism Up-to-Date: The Story of the Farmers Union," 200.
10. Florence and Russell Tilberg, 26 July 1994.
11. Stoltz, *This Is Yours*, 13.
12. Florence and Russell Tilberg, 26 July 1994.
13. *Union Farmer*, 28 December 1959.
14. *Union Farmer*, 26 April 1939.
15. Ibid., 12 April 1939.
16. *Union Farmer*, 14 September 1938.
17. Darrell Moseson, Interview by author, Sioux Falls, South Dakota, 17 July 1995.
18. S. Douglas Murphy, interview by author, Woonsocket, South Dakota, 2 August 1994; Harmony Hall was razed circa 1990 - 1991.
19. *Union Farmer*, July-August 1990.
20. *Union Farmer* , 13 August 1948.
21. Robert S. Thompson, *The History of The South Dakota Farmers Union, 1914-1952*, M. A. Thesis, The University of South Dakota, June 1953, 17.
22. *Union Farmer*, 9 July 1948.
23. Minutes of the Executive Board, South Dakota Farmers Union (SDFU), 9 July 1917; *Union Farmer*, 9 July 1948.
24. Minutes of the Executive Board, SDFU, 16 August 1917.
25. Ibid., 16 August 1917.
26. *Union Farmer*, 9 July 1948.
27. Minutes of the Executive Board, SDFU, 16-17 March 1917.
28. *Union Farmer*, 9 July 1948.
29. Ibid., 13 August 1948.
30. Minutes of the Executive Board, SDFU, 15 December 1919.
31. *Union Farmer*, 10 November 1952.
32. Thompson, *History of the South Dakota Farmers Union*, 18; Thompson's interview with J. W. Batcheller, 10 July 1948.
33. *Union Farmer*, 10 November 1952.

CHAPTER 3

1. *Union Farmer*, 16 June 1926.
2. Ibid., 6 April 1927.

   The SDFU adhered to the limitations placed on political partisanship. A Committee on Political Action asked delegates to the Union's 1924 annual convention at Chamberlain to vote on their choices for President, U. S. Senator and Congressman. The delegates said No. - *Union Farmer*, 5 November 1924.
4. Thompson, *History of the South Dakota Farmers Union*, 20.
5. Ibid., 85.
6. Ibid., Thompson's interview with J. W. Batcheller, 10 July 1948, 20-21.
7. Ibid., 23.
8. Ibid., pp. 22-23.
9. Theodore Saloutos and John D. Hicks, *Agricultural Discontent in the Middle West, 1900 - 1939* (Madison: University of Wisconsin Press, 1951), 258.

10. Robert L. Tontz, "Membership of General Farmers Organizations, United States, 1874 - 1860," *Agricultural History* 38 (July 1964):150.
11. Crampton, *The National Farmers Union, Ideology of a Pressure Group*. (Lincoln: University of Nebraska Press, 1965), 13.
12. Ibid.
13. *Union Farmer*, 21 December 1927.
14. Ibid., 20 October 1926; 18 July 1928.
15. Ibid., 16 June 1926.
16. Ibid., 26 October 1927.
17. Ibid., 24 October 1928.
18. Ibid., 23 November 1927.
19. Tontz, "Membership of General Farm Organizations," 156.
20. Ibid.
21. Conrad, *50 Years*, 15.
22. *Union Farmer*, 5 May 1926.
23. *Union Farmer*, 2 December 1925.
24. Conrad, *50 Years*, 14.
25. Conrad, *50 Years*, 13.
26. Ibid., 16.
27. Tucker, "Populism Up-to-Date: The Story of the Farmers Union," 204.
28. *Union Farmer*, 11 May 1927.
29. Ibid., 31 August 1927.
30. Ibid., 7 November 1928.
31. Ibid., 21 November 1928.
32. Ibid., 21 November 1928.
33. Ibid., 26 October 1927.
34. Minutes of the Executive Board, SDFU, 14 March 1928.
35. Shover, *Cornbelt Rebellion, The Farmers' Holiday Association*, 18.
36. In July 1928, the Farmers Union, Farm Bureau and the Grange agreed to launch the Farmers National Grain Corporation, incorporated that fall as the first national marketing cooperative under the Federal Farm Board. There was dissent in the Union along geographical lines regarding the Farm Board. Members liked or disliked the board based on the commodities produced in their area.-Tucker, "Populism Up-to-Date: The Story of the Farmers Union," 204.
37. *Union Farmer*, 27 April 1927.
38. Ibid., 1 June 1927.
39. Ibid., 24 October 1928.
40. Ibid., 29 February 1928.
41. Ibid., 24 October 1928.
42. Thompson, *History of the South Dakota Farmers Union*, 26 -27.
43. Henry W. Landswehr, Winfred, South Dakota, to Loriks, 30 September 1937, Box 4, File 26, General Correspondence L, 1937 - 1938. Loriks Papers, CWS.
44. *Union Farmer*, 24 October 1928. C. S. Barrett retired at the NFU convention at Denver in November 1928. His successor was Charles Huff of Salina Kansas. Joe Atkins of Wessington Springs, South Dakota, became vice president. Ibid., 4 June 1930.
45. S. Douglas Murphy, Interview by author, Woonsocket, South Dakota, 2 August 1994.

46. Thompson, *History of the South Dakota Farmers Union*, 26.
47. Official Minutes, FECUA, South Dakota Division, 1941, 32.
48. Official Minutes of FECUA, South Dakota Division, 1941, 32.
49. Until 1915, there had been little protection for the farmer in the market place. Within four years, six Farmers Union Livestock Commission houses were established in the Midwest and the Prairie Plains. Shortly after their establishment, Congress favored the "Packers Consent Decree," which broke up the big packers' interest in owning stockyards, the "Co-op Bill of Rights" and the Capper-Volstead Act. - *Union Farmer*, 25 October 1965.
50. Thompson, *History of the South Dakota Farmer Union*, 98.
51. Minutes of the Executive Board, SDFU, 31 November 1918.
52. *Union Farmer*, 30 December 1925.
53. Minutes of the Executive Board, SDFU, 11 November 1924.
54. *Union Farmer*, 21 October 1925.
55. Ibid., 27 October 1925.
56. Minutes of the Executive Board, SDFU, 17 February 1925.
57. Ibid., 1 February 1929.
58. Ibid., 13 February 1929.
59. *Union Farmer*, 27 February 1929.
60. Official Minutes, FECUA, South Dakota Division, 1927, 15.
61. Minutes of the Executive Board, SDFU, 6-7 January 1926.
62. Official Minutes of FECUA, South Dakota Division, 1927, 15.
63. Ibid.
64. *Union Farmer*, 3 August 1927.
65. Ibid., 17 August 1927.
66. *Union Farmer*, 10 October 1928.

CHAPTER 4

1. Shover, *Cornbelt Rebellion*, 22.
2. Tucker, "Populism Up-to-Date: The Story of the Farmers Union,"207.
3. Shover, *Cornbelt Rebellion*, 24-25.
4. Gilbert C. Fite, "John A. Simpson: The Southwest's Militant Farm Leader," *The Mississippi Valley Historical Review* 35 (March 1949): 573.
5. Crampton, *The National Farmers Union*, 123.
6. Fite, "John A. Simpson," 568.
7. Ibid., 575.
8. Ibid., 582.
9. Shover, *Cornbelt Rebellion*, 121.
10. *Union Farmer*, 12 February 1930.
11. Ibid., 16 July 1930.
12. Ibid., 21 October - 4 November 1931.
13. R. D. Lusk, "Life and Death of 470 Acres," Saturday Evening Post, 13 August 1938.
14. Studs Terkel, *Hard Times, An Oral History of the Great Depression* (New York: Pantheon Books, 1970), 258.

15. At the SDFU's annual convention in 1925, a motion was made that after January 1926, all farm produce would be withheld from the market until wheat sold for $1.75 a bushel and corn for $1.00 a bushel.- *Union Farmer*, 4 November 1925.
16. Shover, *Cornbelt Rebellion*, 35.
17. Homer Ayres, interview by Earl Hausle, Sturgis, South Dakota, 28 June 1971, Tape # 187. South Dakota Oral History Center (SDOHC), University of South Dakota, Vermillion, South Dakota.
18. William C. Pratt, "Rethinking The Farm Revolt of the 1930s," *Great Plains Quarterly*, 8 (Summer 1988), 132.
19. *Union Farmer*, 10 August 1932.
20. Emil Loriks, "The Great Depression - The Early Thirties & the Farm Holiday Movement," *Selected Papers of the Dakota History Conference, 1969 - 1977*. Series # 8. (Madison: The Karl E. Mundt Historical and Educational Foundation, 1981), 498.
21. Emil Loriks, interview by Paul O'Rourke, June 1970, MS Tape #006, Reel 2 of 3. SDOHC.
22. Ibid.
23. Loriks to E. N. Hammerquist, Farmingdale, South Dakota, 23 August 1932, Box 2, File 13, Loriks Correspondence, 1927 - 1975. Loriks Papers, CWS.
24. Ibid., Minutes of a Holiday meeting, Cataract Hotel, Sioux Falls, South Dakota, 25 September 1932.
25. Emil Loriks, Interview by Paul O"Rourke, June 1970, Arlington, South Dakota, MS #0007. SDOHC.
26. Shover, *Cornbelt Rebellion*, 52.
27. Emil Loriks, 20 September 1932, Box 2, File 47, Farmers Holiday. Loriks Papers, CWS.
28. Ibid.
29. Trudeau was on a committee to organize the Modern Seventy-Sixers. They wanted to revitalize the free enterprise system by breaking, by force if necessary, the stranglehold of corporate capitalism. Joseph A. Trudeau to Loriks, 27 September 1932, Box 2, File 13, Loriks Correspondence, 1927 - 1975. Loriks Papers, CWS.
30. *Union Farmer*, 19 October - 2 November 1932; E. H. Everson to Loriks, 3 October 1932, Box 2, File 13, Loriks Correspondence, 1927 - 1975. Loriks Papers, CWS.
31. Ibid., Loriks to John Bosch, Atwater, Minnesota, 9 December 1932.
32. Emil Loriks, interview by Paul O"Rourke, June 1970, Tape MS #0007. SDOHC.
33. An action that protected the farm itself was called a "penny auction." A "Sears-Roebuck sale" protected tangible property other than land, buildings and other things annexed to the land.
34. Shover, *Cornbelt Rebellion*, 72.
35. Pratt, "Rethinking the Farm Revolt of the 1930s," 134.
36. Thompson, *History of the South Dakota Farmers Union*, 61.
37. Emil Loriks, "The Great Depression - The Early Thirties & The Farm Holiday Movement," 500.
38. Oscar Brekke, Albee, South Dakota to Loriks, 5 September 1933, Box 6, File 13, National Farm Holiday, 1932 - 1935. Loriks Papers, CWS.
39. Ibid., Mose L. Lindau, Milbank, South Dakota to Milo Reno, 9 September 1933.
40. Emil Loriks, interview by Paul O'Rourke, June 1970, Arlington, South Dakota, Tape MS#0006. SDOHC.

41. Oscar Brekke to Loriks, 5 September 1933, Box 6, File 13, National Farm Holiday, 1932 - 1935. Loriks Papers, CWS.
42. Ibid.
43. Emil Loriks, interview by Paul O'Rourke, June 1970, Tape MS #006. SDOHC.
44. Shover, *Cornbelt Rebellion*, 77.
45. Form letter from Loriks to all Holiday Association members, n.d., Box 6, File 13, National Farm Holiday, 1932 - 1935. Loriks Papers, CWS.
46. *Union Farmer*, 19 April 1933.
47. Form letter from Loriks to South Dakota National Farmers Holiday Association membership, 12 May 1933, Box 6, File 13, National Farm Holiday, 1932 - 1935. Loriks Papers, CWS.
48. Ibid., Milo Reno to Loriks.
49. *Union Farmer*, 1 November 1933.
50. Newsletter from Loriks to Holiday members, 3 October 1933, Box 2, File 13, Loriks Correspondence, 1927 - 1975. Loriks Papers, CWS.
51. Telegram from Governor Berry to Loriks, 27 October 1933, Box 6, File 13, National Farm Holiday. Loriks Papers, CWS.
52. John E. Miller, "Restrained, Respectable Radicals: The South Dakota Farm Holiday," *Agricultural History*, 59 (July 1985): 442.
53. *Union Farmer*, 18 October - 1 November 1933.
54. Shover, *Cornbelt Rebellion*, 157.
55. Emil Loriks, interview by Paul O"Rourke, June 1970, Tape MS #0007. SDOHC.
56. *Sioux Falls Argus Leader*, 5 November 1933.
57. *Union Farmer*, 18 October - 1 November 1933.
58. Loriks to Irvin Heggestad, Carter, South Dakota, 27 November 1933, Box 6, File 13, National Farm Holiday, 1932 - 1935. Loriks Papers, CWS.
59. Ibid.
60. *Union Farmer*, 13 December 1933.
61. Milo Reno to Loriks, 6 March 1934, Box 6, File 13, National Farm Holiday, 1932 - 1935. Loriks Papers, CWS.
62. Shover, *Cornbelt Rebellion*, 93.
63. *Union Farmer*, 18 October - 1 November 1933.
64. Shover, *Cornbelt Rebellion*, 169.
65. Loriks to L. M. Peet, *Farm Holiday News*, Des Moines, Iowa, 17 April 1934, Box 6, File 13, National Farm Holiday, 1932 - 1935. Loriks Papers, CWS.
66. Emil Loriks, Interview by Paul O'Rourke, June 1970, Tape MS #006. SDOHC.
67. Loriks to Irvin Heggestad, 27 November 1933, Box 6, File 13, National Farm Holiday, 1932 - 1935. Loriks Papers, CWS.
68. On January 14, 1938, the merger of the Farm Holiday into the Farmers Union was considered. Its members would be accepted into the Union up to September 1, 1938, upon the payment of regular annual dues for the current year. No initiation fee would be required.- *Union Farmer*, 2 February 1938. Those Holiday members who joined the Union helped make it a more progressive organization.
69. Shover, *Cornbelt Rebellion*, 216.
70. Pratt, "Rethinking the Farm Revolt of the 1930s," 139.

CHAPTER 5

1. Crampton, *The National Farmers Union*, 140; NFU Convention Minutes, 1933.
2. Crampton, *The National Farmers Union*, 141; NFU Convention Minutes, 1934.
3. *Union Farmer*, 17 October 1934.
4. Ibid., 18 October-1 November 1933.
5. Ibid., 10 January 1934.
6. Ibid., 15-29 November 1933.
7. Ibid., 17 October 1934.
8. Resolutions of the National Farmers Holiday Association Convention, 3 May 1934, Box 6, File 13, National Farm Holiday, 1932 -1935. Loriks Papers, CWS.
9. *Union Farmer*, 18 October - 1 November 1933.
10. Loriks to John Bosch, Atwater, Minnesota, 29 April 1934, Box 6, File 13, National Farm Holiday, 1932 - 1935. Loriks Papers, CWS.
11. Edward C. Blackorby, *Prairie Rebel: The Public Life of William Lemke* (Lincoln: University of Nebraska Press, 1963), 186-187.
12. Ibid., 198-199.
13. *Union Farmer*, 17 October 1934. Lemke was introduced by Gutzom Borglum at the 1934 convention.
14. Frazier was elected the first Nonpartisan League governor of North Dakota in 1916.
15. *Union Farmer*, 27 June 1934.
16. Ibid., 17 October 1934.
17. Ibid., 27 June 1934. The Farm Bankruptcy Act was declared unconstitutional by the Supreme Court in 1935. It was repassed by Congress in August 1935 and sustained by the Supreme Court in 1937. The new act provided for a three-year moratorium against seizure for farmers who secured court permission, thus enabling debt-burdened farmers to keep possession of their properties by paying a fair and reasonable rental determined by the court.
18. Ibid., 5 April 1933.
19. Emil Loriks, interview by Paul O'Rourke, Arlington, South Dakota, 9 June 1971, Tape # MS224, SDOHC, USD.
20. *Union Farmer*, 18 September 1935. Through correspondence and an interview with Elizabeth Williams, Professor of Journalism, South Dakota State University, during the early 1980s, Loriks said: "Huey was a spellbinder on the platform and had more `guts' than any public official of his time to deal with the problems of his state and its people in the most tragic economic era of our history." - Elizabeth Evenson Williams, "Emil Loriks and the South Dakota Farmers Union View Huey Long," Papers of the 16th Dakota History Conference, April 12 - 14, 1984 (Madison: The Karl E. Mundt Historical and Educational Foundation, 1984), 546. Loriks further stated that "Huey Long and the Farmers Union were mutually supportive of each other, and that Long was worthy of admiration because of his support from diverse sources." - Elizabeth Evenson Williams, "A South Dakota Agrarian Looks at Huey Long," Papers of the 17th Dakota History Conference, April 11 - 13, 1985 (Madison: The Karl E. Mundt Historical and Educational Foundation, 1985), 289.
21. Crampton, *The National Farmers Union*, 141. NFU Convention Minutes, 1936.
22. Crampton, *The National Farmers Union*, 143.
23. Ibid., 143.

24. H. G. Keeney to Loriks, 11 June 1937, Box 9, File 9, *Union Farmer*, 1937. Loriks Papers, CWS.
25. Blackorby, *Prairie Rebel: The Public Life of William Lemke*, 232.
26. Crampton, *The National Farmers Union*, 228.
27. Ibid., 228.
28. Shover, *Cornbelt Rebellion*, 2.
29. Emil Loriks, interview by Paul O'Rourke, 1 June 1970, Tape # MS224, SDOHC, USD.
30. Emil Loriks, interview by Paul O'Rourke, June 1970, Tape # MS0007, Reel 3 of 3, SDOHC, USD.
31. Schell, *History of South Dakota*, 283.
32. Paul O'Rourke, "South Dakota Politics During the New Deal Years,"*South Dakota History* 1 (Summer 1971): 238.
33. The farm progressives were able to write into the bill a slightly graduated rate for higher incomes.
34. Thompson, *History of the South Dakota Farmers Union*, 85.
35. Schell, *History of South Dakota*, 286.
36. O'Rourke, "South Dakota Politics During the New Deal Years," 237.
37. Oscar Fosheim, interview by Paul O'Rourke, Howard, South Dakota, June 1970, Tape # MS004, Reel 3 of 3, SDOHC, USD.
38. O'Rourke, "South Dakota Politics During the New Deal Years,"238.
39. Ibid., 242.
40. *Union Farmer*, 17 October 1934. Emil Loriks, interview by Paul O'Rourke, 9 June 1971, Tape # MS224, SDOHC, USD.
41. Oscar Fosheim, interview by Paul O'Rourke, June 1970, Tape # MS002, Reel 1 of 3, SDOHC, USD.
42. O,Rourke, "South Dakota Politics During the New Deal Years," 242.
43. Loriks to *Sioux Falls Argus Leader*, Box 7, File 4, Loriks, Personal, 1933 - 1934. Loriks Papers, CWS.
44. Ibid.
45. Emil Loriks, interview by Paul O'Rourke, 9 June 1971, Tape # MS224, SDOHC, USD.
46. Schell, *History of South Dakota*, 286.
47. Emil Loriks, interview by Paul O'Rourke, June 1970, Tape # MS006, Reel 2 of 3, SDOHC, USD.
48. Oscar Fosheim, interview by Paul O'Rourke, June 1970, Tape # MS004, Reel 3 of 3, SDOHC, SDU. A senator's salary during the 1930s was $300 for two years. They voted themselves a $150 expense allowance. - Emil Loriks, Interview by Paul O'Rourke, 9 June 1971, Tape # MS224, SDOHC, USD.
49. Ibid., June 1970, Tape # MS006, Reel 1 of 3, SDOHC, USD.
50. Loriks to H. A. Cowden, Union Oil Company, Kansas City, Missouri, 20 June 1934, Box 7, File 55, SD Co-op Assn., 1934 - 1937. Loriks Papers, CWS.
51. Saloutos and Hicks, *Agricultural Discontent in the Middle West, 1900 - 1939*, 494.
52. *Sioux Falls Argus Leader*, 25 October 1934.
53. Oscar Fosheim, interview by Paul O'Rourke, June 1970, Tape # MS002, Reel 1 of 3, SDOHC, USD.
54. Emil Loriks, Interview by Paul O'Rourke, June 1970, Tape # MS006, Reel 2 of 3, SDOHC, USD.

55. Oscar Fosheim, interview by Paul O'Rourke, June 1970, Tape # MS002, Reel 1 of 3, SDOHC, USD.
56. Oscar Fosheim, June 1970, Tape # MS004, Reel 3 of 3, SDOHC, USD.
57. O'Rourke, "South Dakota Politics During the New Deal Years," 252.
58. *Union Farmer*, 16 October 1935.
59. Emil Loriks, interview by Paul O'Rourke, June 1970, Tape # MS006, Reel 1 of 3, SDOHC, USD.
60. Studs Terkel, *Hard Times, An Oral History of the Great Depression* (New York: Pantheon Books, 1970), 257.
61. Oscar Fosheim, interview by Paul O'Rourke, June 1970, Tape # MS003, Reel 2 of 3, SDOHC, USD.
62. O'Rourke, "South Dakota Politics During the New Deal Years," 260.
63. D. B. Gurney to Loriks, 19 March 1934, Box 7, File 4, Personal, 1933 - 1934. Loriks Papers, CWS.
64. Oscar Fosheim, interview by Paul O'Rourke, June 1970, Tape # MS003, Reel 2 of 3, SDOHC, USD. *Union Farmer*, 6 March 1935.
65. Ibid., 6 March 1935.
66. The ore tax became an ongoing issue. In the 1937 legislative session, the tax was increased to 6 percent, but agitation was already underway to repeal or reduce the tax.
67. O'Rourke, "South Dakota Politics During the New Deal Years," 253.
68. Emil Loriks, interview by Paul O'Rourke, June 1970, Tape # MS0007, Reel 3 of 3, SDOHC, USD.

CHAPTER 6

1. Emil Loriks to John C. Erp, Canby, Minnesota, 20 October 1937, Box 4, File 19, General Correspondence E, 1937 - 1938. Loriks Papers, CWS.
2. Loriks' report to the convention — *Union Farmer*, 13 October and 27 October 1937.
3. *Union Farmer*, 1 January 1930; 27 January 1932.
4. John A. Crampton, *The National Farmers Union, Ideology of a Pressure Group* (Lincoln: University of Nebraska Press, 1965) 140.
5. *Union Farmer*, 23 October 1929.
6. *Ibid*, 2 January 1929.
7. *Ibid*. 19 June 1929.
8. *Ibid*., 21 October and 4 November 1931.
9. One South Dakota Union member suggested "that a picture of convenient size of our deceased national president, John A. Simpson, should be in every American farm home." James W. Vopat, White River, South Dakota, to Loriks, 26 March 1934, Box 7, File 4, Personal, 1933 - 1934. Loriks Papers, CWS.
10. *Union Farmer*, 12 February 1930.
11. In a letter to C. N. Rogers of Des Moines, Iowa, on 23 May 1938, Loriks agreed with Rogers that the Farm Bureau "which you refer to as a company union, has been very much a government agency." - Loriks to C. N. Rogers, Des Moines, Iowa, 23 May 1938, Box 4, File 31, General Correspondence R, 1937-1946. Loriks Papers, CWS.
12. Emil Loriks, interview by Paul O'Rourke, June 1970, Tape #006, Reel 1, SDOHC, USD.
13. Session Laws of South Dakota, 1935, Ch. 15, Section 3, page 38. (S.B. 120).
14. *Union Farmer*, 21 October and 4 November 1931.

15. *Ibid.*, 17 July 1929.
16. Loriks to Leslie Grace, Secretary, Sanborn County Farmers Union, Woonsocket, South Dakota, 2 October 1937, Box 4, File 22, General Correspondence H, 1926 - 1938. Loriks Papers, CWS.
17. B.A. Van Hatten, Conde, South Dakota, to Loriks, 30 January 1937, Box 5, File 43, Miscellaneous, 1936 - 1947. Loriks Papers, CWS.
18. Ibid.
19. S. H. Lassen, Brookings, South Dakota, to Loriks, 16 January 1937, Box 5, File 43, Miscellaneous, 1936 - 1947. Loriks Papers, CWS.
20. Loriks to Lawrence Staley, Secretary, Hyde County Farmers Union, 29 September 1937, Box 9, File 9, *Union Farmer*, 1937. Loriks Papers, CWS.
21. Homer Ayres to Oscar Fosheim and Loriks, 8 September 1937, Box 4, File 22, General Correspondence H. Loriks Papers, CWS.
22. Emil Loriks, interview by Paul O'Rourke, Tape No. 224, June 1970, SDOHC, USD.
23. Ibid.
24. Paul A. O'Rourke, "South Dakota Politics During the New Deal Years," *South Dakota History*, 1 (Summer 1971): 257.
25. *Sioux Falls Daily Argus-Leader*, 24 July 1935.
26. *Ibid*, 29 January 1937.
27. Box 9, File 9, *Union Farmer*, 1937. Loriks Papers, CWS.
28. Loriks to Morris Erickson, Secretary of the North Dakota Farmers Union, Jamestown, North Dakota, 27 October 1937, Box 4, File 19, General Correspondence E, 1937 - 1938. Loriks Papers, CWS.
29. Homer Ayres to Fosheim and Loriks, 8 September 1937, Box 4, File 22, General Correspondence H. CWS.
30. Homer Ayres, Zeona, South Dakota, to Loriks, 26 November 1937, Box 4, File 22, General Correspondence H. Loriks Papers, CWS.
31. *Union Farmer*, 21 December 1926.
32. Robert S. Thompson, "The History of the South Dakota Farmers Union, 1914-1952," 34-35.
33. *Union Farmer*, 13 and 27 October 1937.
34. *Ibid.*, 21 July 1937.
35. *Ibid.*, 26 June 1935.
36. Box 9, File 9, *Union Farmer*, 1936-1937. Loriks Papers, CWS.
37. *Union Farmer*, 23 October 1929.
38. Loriks to Oscar Brekke, Albee, South Dakota, 17 July 1937; Ibid., Loriks to Miss Magdalene Everson, 30 July 1937, Box 9, File 8, *Union Farmer*, 1936 - 1937. Loriks Paper, CWS.
39. *Union Farmer*, 17 October 1934.
40. Homer Ayres to Loriks, 30 March 1937, Box 5, File 15, General Correspondence A. Loriks Papers, CWS.
41. E. H. Bremer to Loriks, 23 March 1937, Box 7, File 19, Politics and Personal, 1936-1937. Loriks Papers, CWS.
42. M. W. Thatcher to Loriks, 19 March 1937, Box 7, File 19, Politics and Personal, 1936-1937. Loriks Papers, CWS.

43. E. H. Bremer to Loriks, 23 March 1937, Box 7, File 19, Politics and Personal, 1936-1937. Loriks Papers, CWS.
44. Loriks to Homer Ayres, Box 1, File 21, Ayres, 1937-1947. Loriks Papers, CWS.
45. Loriks to Morris Erickson, Jamestown, North Dakota, 15 September 1937, Box 9, File 8, *Union Farmer*, 1936 - 1937. Loriks Papers, CWS.
46. Loriks to Leslie Grace, Secretary, Sanborn County Farmers Union, Woonsocket, South Dakota, 2 October 1937, Box 4, File 22, General Correspondence H, 1926 - 1938. Loriks Papers, CWS.
47. H. G. Keeney to Loriks, 4 September 1937, Box 4, File 25, General Correspondence K, 1937-1938. Loriks Papers, CWS.
48. Morris Erickson, Secretary of the North Dakota Farmers Union, to Loriks, 28 September 1937, Box 4, File 19, General Correspondence E, 1937 - 1938. Loriks Papers, CWS.
49. Loriks to Henry A. Wallace, Secretary of Agriculture, 3 October 1937, Box 9, File 9, *Union Farmer*, 1937. Loriks Papers, CWS.
50. Loriks to Andrew Cunningham, 7 October 1937, Box 4, File 17, General Correspondence C, 1932-1938. Loriks Papers, CWS.
51. Loriks to Leslie Grace, Secretary, Sanborn County Farmers Union, Woonsocket, South Dakota, 2 October 1937, Box 4, File 22, General Correpondence H, 1926 - 1938. Loriks Papers, CWS.
52. Loriks to E. H. Sauber, 7 October 1937, Box 4, File 32, General Correspondence S, 1937-1938. Loriks Papers, CWS.
53. Loriks to Oscar Fosheim, 6 October 1937, Box 4, File 20, General Correspondence F, 1937-1948. Loriks Papers, CWS.
54. Loriks to Morris Erickson, Secretary of the North Dakota Farmers Union, Jamestown, North Dakota, 4 October 1937, Box 9, File 8, *Union Farmer*, 1936 - 1937. Loriks Papers, CWS.
55. Oscar Fosheim to Loriks, 23 September 1937, Box 4, File 20, General Correspondence F, 1937-1948. Loriks Papers, CWS.
56. Loriks to Paul Opsahl, Brentford, South Dakota, 27 October 1937, Box 4, File 29, General Correspondence O, 1937 - 1954. Loriks Papers, CWS.
57. E. H. Everson to Loriks, 2 November 1937, Box 4, File 19, General Correspondence E, 1937-1938. Loriks Papers, CWS.
58. Enoch Hofstad to John Vesecky, Salina, Kansas, 5 November 1937, Box 4, File 22, General Correspondence H, 1926 - 1936. Loriks Papers, CWS.
59. C. H. Hyde to Loriks, 28 October 1937, Box 4, File 22, General Correspondence H, 1926-1936. Loriks Papers, CWS.
60. *Union Farmer*, 10 November 1937.
61. Irvine Levine to Loriks, 26 November 1937, Box 4, File 26, General Correspondence L, 1937-1938. Loriks Papers, CWS.

CHAPTER 7

1. In 1925, a local youth program was established in Haakon County. Three years later, there were thirty active Junior locals. - *Union Farmer*, December 1990.
2. *Union Farmer*, 21 October 1925.
3. Minutes of FECUA, South Dakota Division, October 9-11, 1928.

4. Ibid., 31 July 1929.
5. History of the Farmers Union Junior Work, Papers of Florence Tilberg, Ethan, South Dakota.
6. Ibid.
7. Ibid.; *Union Farmer*, 23 October 1929; Minutes of FECUA, South Dakota Division, October 8-10, 1929.
8. History of the Farmers Union Junior Work, Papers of Florence Tilberg, Ethan, South Dakota.
9. *Union Farmer*, 23 October 1929.
10. History of the Farmers Union Junior Work, Papers of Florence Tilberg, Ethan, South Dakota.
11. *Union Farmer*, 12 February 1930.
12. Emil Loriks to Mrs. Elsie Olson, 24 May 1938, Box 4, File 29, General Correspondence O, 1937 - 1954. Loriks Papers, CWS.
13. *Union Farmer*, August 1982.
14. History of the Farmers Union Junior Work, Papers of Florence Tilberg, Ethan, South Dakota.
15. Ibid.
16. *Union Farmer*, 18 October - 1 November 1933.
17. Florence and Russell Tilberg, interview by author, Ethan, South Dakota, 26 July 1994.
18. Ibid.
19. *Union Farmer*, 18 November 1931.
20. Ibid., 17 October 1934.
21. Ibid., 12-26 October 1938.
22. Gladys Talbott Edwards, *The Farmers Union Triangle* (Jamestown, North Dakota: Farmers Union Educational Service, 1941), 74.
23. *Union Farmer*, 12 - 26 October 1938.
24. Ibid., November-December 1983.
25. Ibid., July-August 1984.
26. Ibid., 13 May 1936.
27. Ibid., 18 October-1 November 1933.
28. Ibid., 13 December 1933.
29. Ibid., 14 October-28 October 1936.
30. Ibid., July-August 1984.
31. Ibid., July-August 1984.
32. Ibid., 13 October-27 October 1937.
33. Loriks, Box 4, File 13, General Correspondence R, 1937 - 1948. Loriks Papers, CWS.
34. The presence of the president of the SDFU on the final evening of the state camp became a tradition.
35. Florence and Russell Tilberg, interview by author, Ethan, South Dakota, 26 July 1994.
36. Ibid.
37. *Union Farmer*, 25 September 1925.
38. Ibid., 14 October-28 October 1936.
39. Ibid., 21 July 1937.
40. Gladys Talbott Edwards, *The Farmers Union Triangle*, 86 - 87.

CHAPTER 8

1. John A. Crampton, *The National Farmers Union, Ideology of a Pressure Group*, 38.
2. Gladys Talbott Edwards, *The Farmers Union Triangle* (Jamestown, North Dakota: Farmers Union Educational Service, 1941), 110-111.
3. Larry Remele, "The North Dakota Farmers Union and the Nonpartisan League: Breakdown of a Coalition," *North Dakota Quarterly* 46 (Autumn 1978): 45.
4. Theodore Saloutos and John D. Hicks, *Agricultural Discontent in the Middle West, 1900-1939*, 58-59.
5. Ibid., 290.
6. Ibid., 291.
7. *Union Farmer*, July 1975.
8. Ibid., June 1930.
9. Ibid., October 1967.
10. IBid., 20 May 1931.
11. Robert S. Thompson, *The History of the South Dakota Farmers Union, 1914 - 1952*, 104.
12. Official Minutes, Farmers Educational and Cooperative Union of America (FECUA), South Dakota Division, 1933, 25-26.
13. Official Minutes, FECUA, South Dakota Division, 1937, 31; Emil Loriks to W. C. Hermann, Rockham, South Dakota, 30 September 1937, Box 9, File 8, *Union Farmer*, 1936-1937. Loriks Papers, CWS.
14. *Union Farmer*, 14 October and 28 October 1936.
15. Ibid., 4 March 1936.
16. Ibid., 14 October and 28 October 1936.
17. Ibid., 20 January 1937.
18. Emil Loriks to O. S. Eastvold, Canton, South Dakota, 12 April 1938, Box 4, File 19, General Correspondence E, 1937 - 1938, Loriks Papers, CWS.
19. Loriks to Oscar Fosheim, 2 November 1937, Box 4, File 20, General Correspondence F, 1937 - 1948, Loriks Papers, CWS.
20. Darrell Moseson, interview by the author, Sioux Falls, South Dakota, 17 July 1995.
21. *Union Farmer*, 12 October and 26 October 1938.
22. Ibid, 16 August 1939.
23. Ibid., 21 October and 4 November 1931.
24. Ibid., 7 September 1932.
25. Ibid., 19 October and 2 November 1932.
26. 50th Anniversary, A History of the Sanborn County Farmers Union Oil Company, Woonsocket, South Dakota, 1928-1978.
27. S. Douglas Murphy, interview by author, Woonsocket, South Dakota, 2 August 1994.
28. 50th Anniversary, A History of the Sanborn County Farmers Union Oil Company, Woonsocket, South Dakota, 1928 - 1978.
29. Ibid.
30. S. Douglas Murphy, interview by author, 2 August 1994; 50th Anniversary, A History of the Sanborn County Cooperative Oil Company.
31. *Union Farmer*, July-August 1989.
32. *Union Farmer*, 21 October and 4 November 1931.
33. Ibid., 10 June 1936.

34. *Union Farmer*, 14 October and 28 October 1936. Loriks pointed out that the Scandinavians (Denmark, Norway and Sweden), through the cooperative principle, had created a new social and economic order within the old capitalist structure. He said that the SDFU had "accomplished much of what we are trying to do — [we have] secured `the Greatest Good for the Greatest Number.'"—Minutes of FECUA, South Dakota Division, 13-15 October 1936.
35. Ibid., 15 April 1936.
36. 25th Anniversary History of the Farmers Union Co-Op Association of McCook County, Salem and Spencer, 1937 - 1962.
37. *Union Farmer*, May-June 1986.
38. Ibid., 15 April 1936.
39. Official Minutes, FECUA, South Dakota Division, 1934, 18.
40. Official Minutes, FECUA, South Dakota Division, 1935, 25; Robert S. Thompson, The History of the South Dakota Farmers Union, 1914-1952, 107-108.
41. Official Minutes, FECUA, South Dakota Division, 1935, 25.
42. Ibid.
43. *Union Farmer*, 5 February 1936.
44. Henry Bell, interview by author, Flandreau, South Dakota, 22 August 1994.
45. *Union Farmer*, 15 April 1936.
46. Loriks to H. G. Keeney, 18 February 1938, Box 4, File 25, General Correspondence K, 1937 - 1938, Loriks Papers, CWS.
47. Ibid., H. G. Keeney to James G. Patton, Denver, Colorado, 2 March 1938.
48. M. R. Miller to Loriks, 22 February 1938, Box 4, File 27, General Correspondence M, 1937 - 1938, Loriks Papers, CWS.
49. Loriks to Emil Syftestad, 26 February 1938, Box 4, File 32, General Correspondence S, 1937 - 1938, Loriks Papers, CWS.
50. Official Minutes, FECUA, South Dakota Division, 1930, 22; *Union Farmer*, 26 February 1930.
51. *Union Farmer*, 7 September 1932.
52. Ibid., 21 October and 4 November 1931.
53. Enoch Hofstad, Elk Point, to Loriks, 4 January 1938, Box 4, File 22, General Correspondence H, 1926 - 1938, Loriks Papers, CWS.
54. *Union Farmer*, 12 February 1930.
55. Official Minutes, FECUA, South Dakota Division, 1930, 22; 1933, 12 and 36. Robert S. Thompson, The History of the South Dakota Farmers Union, 1914 - 1952, 100-101.
56. Minutes of the South Dakota Farmers Union Board of Directors, 8-9 January 1934; *Union Farmer*, 4 April 1934.
57. *Union Farmer*, 9 September 1931.
58. Ibid., 7 September 1932.
59. Ibid., 25 September 1929; 4 June 1930.
60. Ibid., 7 September 1932.
61. In his report to the 28th annual convention in 1943, President John Skage recommended that every cooperative in the state remit 5 percent of their net earnings to the state office to finance educational work.— Official Minutes, FECUA, South Dakota Division, 12-14 October 1943.

62. Loriks to Homer Ayres, 14 April 1937, Box 1, File 21, Ayres, 1937 - 1947, Loriks Papers, CWS.
63. Classified Directory of Cooperative Associations in South Dakota, June 1937, Box 5, File 22, Lundy, Moseson, Mahoney, Ferwerda, Ayres, Stern. Loriks Papers, CWS.
64. Paul O'Rourke's Interview with Emil Loriks, Tape #224, SDOHC, USD.
65. *Union Farmer*, January-February 1986.
66. Theodore Saloutos and John D. Hicks, *Agricultural Discontent In The Middle West, 1900-1939*, 248.
67. *Union Farmer*, 15 January 1930.
68. Ibid., 21 October and 4 November 1931.
69. Conrad, *50 Years*, 67.
70. Ibid., 67.
71. Ibid., 71.
72. Theodore Saloutos and John D. Hicks, *Agricultural Discontent In The Middle West, 1900 - 1939*. 252.
73. Leo N. Rickertsen, *To Gather Together, CENEX, The First Fifty Years* (Minneapolis: Cooperative Printing Association, 1980), 36.
74. Syftestad's managerial style was very different from that of M. W. Thatcher. Thatcher was a politician who used the limelight to create change. Syftestad was a mathematician who worked quietly to secure changes of his own. Ultimately, the two men would clash over who would manage the Farmers Union Central Exchange.— Rickertsen, *To Gather Together*, 75.
75. Theodore Saloutos and John D. Hicks, *Agricultural Discontent In The Middle West, 1900-1939*, 250.
76. E. A. Syftestad, General Manager, Central Exchange to Loriks, 1 December 1937, Box 4, File 32, General Correspondence S, 1937-1938. Loriks Papers, CWS.
77. Henry Bell, interview by author, Flandreau, South Dakota, 22 August 1994; Norman E. Olsen, interview by author, Canton, South Dakota, 21 July 1994.
78. L. S. Herron to Loriks, 21 October 1937, Box 9, File 8, *Union Farmer*, 1936 - 1937. Loriks Papers, CWS.

CHAPTER 9

1. Schell, *History of South Dakota*, 290.
2. Ibid., 291.
3. *Union Farmer*, 19 November 1924.
4. Ibid., 16 October 1935.
5. Emil Loriks to Bessie Rice, Winfred, South Dakota, 7 October 1937, Box 4, File 13, General Correspondence R, 1937 - 1948. Loriks Papers, CWS.
6. Ibid.
7. *Union Farmer*, 10 November 1937.
8. Ibid., 8 November 1939.
9. Ibid., 8 November 1939.
10. Ibid., 31 July 1940.
11. Ibid., 25 September 1940.
12. Ibid., 16 July 1941.
13. Crampton, *National Farmers Union*, 27.

14. Ibid., 28.
15. *Union Farmer*, 5 November 1941.
16. Crampton, *National Farmers Union*, 28.
17. Ibid., 32.
18. Ibid., 47.
19. *Union Farmer*, 3 February 1943.
20. Florence Tilberg, interview by author, Ethan, South Dakota, 26 July 1994.
21. Ibid., 20 January 1943.
22. Ibid., 4 November 1942.
23. Ibid., 8 September 1944.
24. Ibid., 23 January 1945.
25. Schell, *History of South Dakota*, 302-303.
26. *Union Farmer*, 23 April 1946.
27. Crampton, National Farmers Union, 25.
28. *Union Farmer*, January 1948.
29. Chester A. Graham to Emil Loriks, 15 December 1947, Box 9, File 11, Universal Military Training, 1947 - 1948. Loriks Papers, CWS.
30. Loriks to Roy Brazell, 2 April 1938, Box 4, File 16, General Correspondence B, 1937 - 1948. Loriks Papers, CWS.
31. *Union Farmer*, 30 March 1938.
32. Loriks to H. D. Rolph, President, Montana Farmers Union, Joplin, Montana, 24 May 1938, Box 4, File 31, General Correspondence R, 1937 - 1948. Loriks Papers, CWS.
33. *Union Farmer*, 25 May 1938.
34. Morris Erickson to Loriks, 14 May 1938, Box 4, File 28, General Correspondence N, 1937 - 1955; James Patton, Executive Secretary, NFU, to Loriks, 1 June 1938, Box 4, File 30, General Correspondence P, 1937 - 1954. Loriks Papers, CWS.
35. A. W. Ricker to Loriks, 16 April 1938, Box 4, File 31, General Correspondence R, 1937 - 1948. Loriks Papers, CWS.
36. *Union Farmer*, 20 July 1938.
37. Paul O'Rourke's interview with Emil Loriks, June 1970, Tape # 0006, Reel 1 of 3, SDOHC, USD. Ervine Levine, State Secretary of the Workers Alliance of South Dakota, had written to Loriks on 24 January 1938, stating that a treaty of mutual assistance had been signed between Labor's Nonpartisan League and the Farmers Union. The agreement called for joint-legislative action and cooperation in the 1938 election. — Ervine Levine to Emil Loriks, 24 January 1938, Box 4, File 26, General Correspondence L, 1937-1938. Loriks Papers, CWS.
38. Paul O'Rourke's interview with Emil Loriks, June 1970, Tape # 0007. SDOHC, USD.
39. Gerald Lange, "Americanism Over Radicalism: Mundt's First Election," Selected Papers of the First Nine Dakota History Conferences, 1969 - 1977, No. 8, H. W. Blakely, Editor. (Madison: Karl E. Mundt Historical and Educational Foundation, 1981).
40. Gerald Lange, ""Americanism Over Radicalism."
41. Albert J. Maag, Huron, to Loriks, 30 December 1938, Box 3, File 28, Fosheim, Luck, Maag, Mundt, Croll, 1934 - 1957. Loriks Papers, CWS.
42. *Union Farmer*, 25 November 1942; Crampton, *National Farmers Union*, 223.
43. Emil Loriks, Box 5, File 44, Miscellaneous, 1938 - 1948. Loriks Papers, CWS.

44. Paul O'Rourke's interview with Oscar Fosheim, June 1970, Tape # 0003, Reel 2 of 3, SDOHC, USD.
45. *Union Farmer*, 20 July 1938.
46. Ibid., January-February 1989.
47. Ibid., 15 February 1939; 6 November 1940.
48. Ibid., 12 February 1941.
49. Ibid., 26 February 1941; 12 March 1941.
50. Ibid., 5 November 1941.
51. Ibid., 15 February 1939.
52. Ibid., 12 February 1941.
53. Ibid., 23 April 1941.
54. Official Minutes of FECUA, South Dakota Division, 1941, 5-6.
55. Robert S. Thompson, *A History of the South Dakota Farmers Union, 1914-1952*, 81.
56. *Union Farmer*, 5 November 1941; Offical Minutes of FECUA, South Dakota Division, 1941, 6.
57. *Union Farmer*, 18 January 1939.
58. Ibid., 26 March 1941.
59. Ibid., 17 February 1943.
60. Ibid., 20 January 1943.
61. Official Minutes of FECUA, South Dakota Division, 1942, 27.
62. Robert S. Thompson, *A History of the South Dakota Farmers Union, 1914-1952*, 82; Official Minutes of FECUA, South Dakota Division, 1943, 28.
63. *Union Farmer*, 13 March 1945.
64. Robert S. Thompson, A History of the South Dakota Farmers Union, 1914-1952, 82.
65. Schell, A *History of South Dakota*, 312.
66. Studs Terkel, *Hard Times, An Oral History of the Great Depression* (New York: Pantheon Books, 1970), 56.
67. *Union Farmer*, 16 October 1935.
68. Ibid., 7 July 43.
69. James G. Patton, Statement before the Missouri River Basin Commission, Denver, Colorado, 26 June 1952, Box 7, File 37, River Development, 1952. Loriks Papers, CWS.
70. Ibid.
71. *Union Farmer*, 7 July 1943.
72. James G. Patton, Statement Before the Missouri Basin Commission, 26 June 1952.
73. Loriks to Robert Westergaard, Viborg, SD, 20 June 1952, Box 5, File 23, MVA, 1952. Loriks Papers, CWS.
74. Ibid.
75. *Union Farmer*, April-May 1945.
76. Ibid., May 1947.
77. Ibid., 10 June 1947.

CHAPTER 10

1. Crampton, The National Farmers Union, 191.
2. Ibid., p. 195.
3. *Union Farmer*, 1 January 1941.
4. Ibid., 27 October 1943.

5. Ibid., 23 October 1947.
6. Ibid., 27 October 1943.
7. Ibid., 5 November 1941.
8. Ibid., 27 October 1943.
9. Ibid., October 1945.
10. Crampton, *The National Farmers Union*, 175,
11. *Union Farmer*, 12 December 1947.
12. Ibid., 13 February 1945.
13. Ibid., 25 November 1943.
14. Ibid., 25 September 1945.
15. Ibid., 24 December 1946; 26 December 1947.
16. Ibid., 24 September 1946.
17. The net profit for the Brokerage increased dramatically between 1934 and 1935. The volume of business for the two years follows: 1934: Total Sales $236,809.95 Cost of Sales 208,793.18 Net Gain $28,016.77 1935: Total Sales $423,867.19 Cost of Sales 381,419.35 Net Gain $42,447.84. — Cooperative League of U.S.A. Yearbook, 1936, 244.
18. Official Minutes, FECUA, South Dakota Division, 1941, 28, 29.
19. Emil Loriks to Gust Moseson, 15 October 1940, Box 7, File 6, Personal, 1940 - 1970. Loriks Papers, CWS.
20. *Union Farmer*, 4 December 1940.
21. Brokerage Board Minutes, 14 August 1939, Box 4, File 49, Hofstad, 1938 - 1940. Loriks Papers, CWS.
22. *Union Farmer*, 27 September 1939.
23. Hofstad to Loriks, 9 June 1939, Box 4, File 49, Hofstad, 1938 - 1940. Loriks Papers, CWS.
24. W. C. Hermann, Rockham, South Dakota to Loriks, 28 April 1941, Box 7, File 7, Personal, 1940. Loriks Papers, CWS.
25. Darrell Moseson, interview by author, Sioux Falls, South Dakota, 17 July 1995. Oscar Johnson's son, Victor Johnson, had been assistant manager of the Brokerage when Moseson was the manager.
26. *Union Farmer*, 18 June 1941; 30 July 1941.
27. Ibid., 12 August 1941.
28. Ibid., 8 October 1941.
29. Ibid., 5 November 1941.
30. Ibid., 7 May 1941.
31. Ibid., 5 November 1941.
32. Official Minutes, FECUA, South Dakota Division, 1941, 49.
33. *Union Farmer*, 3 December 1941.
34. Ibid., 31 December 1941.
35. Ibid.,28 January 1942.
36. Ibid., 25 February 1942.
37. Ibid., 6 May 1942.
38. Ibid., 22 April 1942.
39. Ibid., 17 March 1943; Official Minutes, FECUA, South Dakota Division, 1943, 5, 9.
40. *Union Farmer*, 25 February 1942.
41. Ibid., 25 September 1945.

42. Ibid., 2 February 1944.
43. Ibid., 9 October 1944; 23 October 1944.
44. Ibid., 23 October 1944.
45. Ibid., 22 November 1944.
46. Ibid., 23 October 1944.
47. Ibid., 5 November 1941.
48. Ibid., 4 November 1942.
49. Official Minutes, FECUA, South Dakota Division, 1942, 8-9.
50. Ibid., 1944, 39.
51. *Union Farmer*, 22 November 1944.
52. Ibid., 8 November 1944.
53. Ibid., 22 September 1944.
54. Ibid., 22 September 1944; 9 October 1944.
55. Ibid., 22 November 1944.
56. Official Minutes, FECUA, South Dakota Division, 1944, 44.
57. Ibid.
58. *Union Farmer*, 23 October 1944.
59. Ibid., August 1981.
60. Oscar Fosheim, interview by Paul O'Rourke, June 1970, Tape MS 002, Reel 1 of 3. SDOHC, USD.
61. *Union Farmer*, 8 November 1944.
62. Oscar Fosheim, Interview by Paul O'Rourke, June 1970, Tape MS 002, Reel 1 of 3. SDOHC, USD.
63. *Union Farmer*, 23 October 1945.
64. Ibid., 23 October 1945.
65. Ibid., 22 October 1946.
66. Ibid., 22 October 1946.
67. Ibid., 23 October 1947.
68. Ibid., 23 October 1945.
69. Robert S. Thompson, *A History of the South Dakota Farmers Union, 1914-1952*, 86.
70. *Union Farmer*, 23 October 1945.
71. Ibid., 23 October 1947.
72. Ibid., 4 November 1942.
73. Ibid., 28 November 1947.
74. Ibid., 14 November 1947.
75. Ibid., 29 January 1941.
76. Ibid., 4 November 1942.
77. Ibid., 27 October 1943.
78 .Ibid., 8 July 1944.
79. Ibid., 10 July 1945.
80. 75 Years of Building for Today and Tomorrow, A Junior Youth Book, Youth and Young Adults Activities,edited by Linda Siderius. (National Farmers Union, n.d.), 21.
81. Minutes of the Board of Directors, SDFU, 11 October 1947.
82. Farmers Union, 22 January 1946.
83. Ibid., 23 October 1947.
84. Ibid., 10 May 1939.

85. Ibid., 12 May 1947.
86. Ibid., July - August 1984.
87. Ibid., 10 July 1945.
88. Florence and Russell Tilberg, interview by author, Ethan, South Dakota, 26 July 1994.
89. Skage believed that cooperation was the mainstay of the FU program. Economic strength for the farmer would increase his legislative strength and, in turn, make the United States more prosperous.— Official Minutes, FECUA, South Dakota Division, 11-13 October 1938.
90. Official Minutes, FECUA, South Dakota Division, 8-10 October, 1946.
91. Darrell Moseson, interview by author, Sioux Falls, South Dakota, 17 July 1995.
92. G. L. Moseson to Loriks, 18 April 1939, Box 5, File 22, Lundy-Moseson-Mahoney-Ferwerda-Ayres-Stern. Loriks Papers, CWS.
93. *Union Farmer*, 10 November 1943.
94. Ibid., 9 October 1944.
95. Ibid., 9 April 1946.
96. Ibid., 23 December 1942.
97. Ibid., 5 November 1941.
98. Official Minutes, FECUA, South Dakota Division, 1946, 10.
99. Minutes of the Board of Directors, SDFU, 12 October 1945.
100. *Union Farmer*, 28 January 1942, 6 May 1942, 9 September 1942, 7 October 1942, 4 November 1942.
101. Robert S. Thompson, *History of the South Dakota Farmers Union, 1914-1952*, 96.
102. Ibid., 96.
103. Crampton, The National Farmers Union, 150.
104. Ibid.
105. *Union Farmer*, 9 July 1948.
106. Robert S. Thompson, *History of the South Dakota Farmers Union, 1914-1952*, 96.
107. *Union Farmer*, 11 September 1961.
108. Ibid., July-August 1984.
109. Charles E. Nieman to Loriks, 21 December 1946, Box 1, File 53, Co-ops, 1936 - 1952; Richard C. Joyce to Loriks, 26 December 1946, Box 1, File 53, Co-ops, 1936 - 1952. Loriks Papers, CWS. South Dakota.
110. *Union Farmer*, 18 December 1940.
111. Ibid.
112. Ibid., 28 April 1943.
113. Conrad, *50 Years*, 97.
114., 112.
115. *Union Farmer*, 12 February 1946.
116. Tucker, "Populism Up-to-Date: The Story of the Farmers Union," 206.
117. *Union Farmer*, 4 August 1943.
118. Conrad, *50 Years*, 93.
119. *Union Farmer*, 4 August 1943.
120. Tucker, "Populism Up-to-Date: The Story of the Farmers Union," 206.
121. Ibid. 206.
122. *Grain Quarterly, National Federation of Grain Cooperatives* 5 (Fall 1947): 72.

CHAPTER 11

1. Schell, *History of South Dakota*, p. 302.
2. Ibid., 357.
3. Ibid., 359.
4. *Union Farmer*, 17 September 1956.
5. Ibid., 16 October 1961.
6. Ibid., 28 May 1956.
7. Ibid., 11 June 1948.
8. Radio Address of Fulton Lewis, Jr., 27 May 1952, over the Mutual Broadcasting System, Box 5, File 52, Miscellaneous, 1946 - 1954. Loriks Papers, CWS.
9. *Union Farmer*, 28 October 1957.
10. Loriks to August Dahme, 7 June 1949, Box 2, File 17, Dahme, 1949 -1956. Loriks Papers, CWS.
11. *Union Farmer*, 7 November 1955.
12. Homer Ayres to Loriks, 9 November 1955, Box 5, File 22, Lundy-Moseson-Ferwerda-Ayres-Stern. Loriks Papers, CWS.
13. Archie T. Erickson to Loriks, 22 July 1947, Box 6, File 39, Opsahl - Haugland, et.al. Loriks Papers, CWS.
14. *Union Farmers*, 12 November 1948.
15. Ibid., 23 October 1947.
16. Ibid., 24 November 1952.
17. Ibid., 26 October 1953.
18. Press Release, 10 August 1956, Box 4, File 47, History, 1955 - 1962. Loriks Papers, CWS,
19. *Union Farmer*, 11 August 1958.
20. Ibid., July - August, 1986; Minutes of the Board of Directors, South Dakota Farmers Union, 21 September 1947.
21. Ibid., 13 April 1959.
22. Loriks to August Dahme, 21 April 1950, Box 2, File 17, Dahme, 1949 - 1956; Loriks to August Dahme, 7 September 1949, Box 2, File 17, Dahme 1949 -1956. Loriks Papers, CWS.
23. Loriks to August Dahme, 21 April 1950, Box 2, File 17, Dahme, 1949 - 1956. Loriks Papers, CWS.
24. *Union Farmer*, 29 December 1958.
25. Ibid., 14 September 1959.
26. Ibid., 1 November 1954.
27. Ibid., 23 September 1949; 21 October 1949.
28. Ibid., 4 October 1959.
29. Ibid., 4 October 1954.
30. Ibid., 1 November 1954.
31. Ibid., 17 October 1955.
32. Ibid., 27 October 1943.
33. Minutes of the Board of Directors, SDFU, 2 June 1948.
34. Ibid., 21 September 1948.
35. *Union Farmer*, 12 November 1948.
36. Minutes of the Board of Directors, SDFU, 1 September 1949.
37. *Union Farmer*, 21 October 1949.

38. Minutes of the Board of Directors, SDFU, 11 July 1950.
39. *Union Farmer*, 8 December 1950.
40. Minutes of the Board of Directors, SDFU, 2 April 1951.
41. *Union Farmer*, 11 May 1951.
42. Loriks to Employment Security Department, Aberdeen, 21 February 1952, Box 5, File 17, Loriks, 1949 - 1957. Loriks Papers, CWS.
43. *Union Farmer*, 14 April 1952.
44. Ibid., 1 November 1954.
45. Minutes of the Board of Directors, SDFU, 24 June 1955; *Union Farmer*, 4 July 1955.
46. *Union Farmer*, 7 November 1955.
47. Ibid., 13 October 1958.
48. Ibid., 14 September 1953.
49. Schell, *History of South Dakota*, 305.
50. *Union Farmer*, 11 May 1953.
51. Ibid.
52. Ibid., 9 November 1953.
53. Ibid., 27 August 1962.
54. The Public Works Administration sponsored legislation to clear the way for public power projects. The legislation authorized municipal operation of power enterprises and permitted the formation of non-profit power corporations for rural areas.— *Sioux Falls Argus Leader*, 1 March 1935.
55. Ron Davis, "The Lights Come On in South Dakota," *East River Guardian*, 24 (May 1985), 15.
56. *East River Guardian*, May 1985, 21.
57. Ron Davis, "The Lights Come On in South Dakota," 14.
58. Patrick Dahl, "REA and the People — Powerful Partners," *East River Guardian*, 24 (May 1985), 6.
59. Ron Davis, "The Lights Come On in South Dakota," 16.
60. Ibid.
61. Ron Davis, "The Lights Come On in South Dakota," 16.
62. *Union Farmer*, 12 November 1948.
63. Ibid., 25 March 1949.
64. Ibid., 21 October 1949.
65. Ibid., 14 April 1950.
66. Ibid., 13 October 1958.
67. Ibid., 14 April 1950.
68. Joel Glover, "Roy Glover, Defender of the Family Farm,"22. A paper prepared for the Political Science Department, the University of South Dakota, 30 April 1988. South Dakota Farmers Union Archives, Huron, South Dakota.
69. *Union Farmer*, 30 January 1963.
70. Joel Glover, "Roy Glover, Defender of the Family Farm," 17.
71. *Union Farmer*, 28 January 1949.
72. Ibid., 25 March 1949.
73. Ibid., 28 July 1950, 26 October 1950, 9 March 1951.
74. Loriks, 28 February 1957, Box 3, File 28, Fosheim - Lusk - Maag - Croll, 1934-1957. Loriks Papers, CWS.

75. *Union Farmer*, 25 March 1957; 14 October 1957.
76. Ibid., 23 March 1959.
77. Ibid., 14 October 1957; 23 March 1959.
78. Ibid., 26 October 1953.
79. Ibid., 13 October 1958; Joel Glover, "Roy Glover, Defender of the Family Farm," 22.
80. *Union Farmer*, 13 January 1958.
81. Schell, *History of South Dakota*, 359.

CHAPTER 12

1. National Farmers Union Newsletter, Vol. 1, No. 11, 17 September 1954.
2. Ibid.
3. *Union Farmer*, 8 December 1952.
4. Ibid., 8 June 1953.
5. Ibid., 22 June 1953.
6. Ibid., 24 August 1953.
7. Ibid., 23 July 1953; 27 July 1953; 10 August 1953.
8. Press release from the National Farmers Union, Dallas, Texas, 10 March 1952, Box 5, File 17, Loriks, 1949-1957. Loriks Papers, CWS.
9. *Union Farmer*, 7 June 1954.
10. Ibid., 18 October 1954.
11. Ibid., 26 October 1953.
12. Ibid., 17 January 1955.
13. Ibid., 4 July 1955.
14. Ibid.
15. Ibid., 18 July 1955.
16. Ibid., 17 September 1956.
17. Ibid. 29 April 1957.
18. Ibid., 18 July 1955.
19. Ibid., 30 April 1956.
20. Ibid., 15 October 1956.
21. Ibid., 28 May 1956.
22. Ibid., 28 October 1954.
23. Ibid.; Opsahl apology, 11 October 1957, Box 6, File 39, Opsahl - Haugland, et. al., 1936-1958. Loriks Papers, CWS.
24. Loriks to Art Stern, Emery, South Dakota, 15 October 1957, Box 5, File 22, Lundy - Moseson, et. al. Loriks Papers, CWS.
25. Loriks to Paul Opsahl, 14 October 1957, Box 6, File 39, Opsahl -Haugland - Emerson - Skage - Eggs - Livestock, 1936-1958. Loriks Papers, CWS.
26. *Union Farmer*, 11 August 1958.
27. Ibid., 13 October 1958.
28. The G.O.P. Psalm, author unknown, Box 5, File 19, Loriks, 1946 -1966. Loriks Papers, CWS.
29. *Union Farmer*, 26 September 1960.
30. Ibid., 26 December 1960.
31. Crampton, *The National Farmers Union*, 171.

32. Ibid., 173.
33. Minutes, Board of Directors, SDFU, 20 March 1950.
34. *Union Farmer*, 26 October 1950.
35. Ibid., 22 October 1951.
36. Henry Bell, interview by author, Flandreau, South Dakota, 22 August 1994.
37. Norman Olsen, interview by author, Canton, South Dakota, 21 July 1994.
38. *Union Farmer*, 11 January 1954.
39. Ibid., 7 March 1955.
40. Loriks to Chan Gurney, 27 February 1948, Box 5, File 48, Miscellaneous Correspondence, 1948-1958; Homer Backlund, Mitchell, South Dakota, to Francis Case, 22 June 1951, Box 5, File 17, Loriks, 1949-1957. Loriks Papers, CWS.
41. *Union Farmer*, 6 December 1954.
42. Crampton, *The National Farmers Union*, 33.
43. Ibid.
44. *Union Farmer*, 13 October 1958.
45. Crampton, *The National Farmers Union*, 30.
46. *Union Farmer*, 28 July 1950.
47. Ibid., 10 March 1952.
48. Ibid., 10 March 1952; 26 October 1953.
49. Ibid., 12 April 1954.
50. Ibid., 10 May 1954.
51. Ibid., 16 May 1960.
52. Crampton, *The National Farmers Union*, 24.
53. Art Stern, Emery, South Dakota to Loriks, 20 November 1957, Box 5, File 22, Lundy - Moseson - Mahoney - Ferwerda - Ayres - Stern. Loriks Papers, CWS.
54. Loriks to John Skage, 30 July 1958, Box 6, File 39, Opsahl, Haugland, et. al., 1936 - 1958; Skage to Loriks, 28 August 1958, Box 6, File 39, Opsahl, Haugland, et. al., 1936 - 1958. Loriks Papers, CWS.
55. *Union Farmer*, 28 July 1958.
56. August Dahme to Loriks, 7 August 1958, Box 2, File 18, Dahme, 1940-1966. Loriks Papers, CWS.
57. Political Broadside, addressed to the delegates of the South Dakota Farmers Union Convention, prepared by August J. Hauck, Henry Ravenberg and F. L. Ericsson, Box 6, File 39, Opsahl, Haugland, et. al., 1936-1958. Loriks Papers, CWS. Emil Loriks said that M. W. Thatcher had a "terrifically high regard for John Skage. It was Skage who brought GTA and Central Exchange into South Dakota." - Loriks to Dahme, 19 September 1958, Box 2, File 18, Dahme, 1940-1966. Loriks Papers, CWS.
58. Campaign literature about Norman Olsen from Julian Holter, Merton Oliver and Henry Breugem, 1958, Box 6, File 39, Opsahl, Haugland, et. al, 1936-1958. Loriks Papers, CWS; *Union Farmers*, 15 September 1958.
59. Political Broadside, addressed to the delegates of the South Dakota Farmers Union Convention, Box 6, File 39, Opsahl, Haugland, et. al., 1936-1958. Loriks Papers, CWS.
60. Campaign literature about Norman Olsen, Ibid.
61. Arthur Sterns, Emery, South Dakota, to Loriks, 19 September 1958, Box 5, File 22, Lundy - Moseson - Mahoney - Ferwerda - Ayres - Stern. Loriks Papers, CWS.

62. John Skage to Loriks, 30 September 1958, Box 6, File 39, Opsahl, Haugland, et. al., 1936-1958. Loriks Papers, CWS.
63. Ibid.
64. *Union Farmer*, 11 August 1958; 17 November 1958.
65. Loriks to John Skage, 10 October 1958, Box 6, File 39, Opsahl, Haugland, et. al., 1936-1938. Loriks Papers, CWS.

CHAPTER 13

1. *Union Farmer*, 22 March 1954.
2. Ibid., 25 March 1957.
3. Ibid., 13 October 1958.
4. Ibid., 31 July 1961.
5. Ibid., 11 June 1948.
6. Ibid., 24 August 1953; 11 August 1958.
7. Ibid., 11 January 1960.
8. Ibid., 11 September 1961.
9. Ibid., 27 October 1952.
10. Ibid., 17 October 1955.
11. Ibid., 14 October 1957.
12. Ibid., 15 December 1958; 11 December 1961.
13. Ibid., 11 December 1961.
14. Ibid., 26 June 1959.
15. Ibid., 11 September 1961.
16. Ibid., 22 October 1951.
17. Ibid., 30 December 1957.
18. Ibid., 15 October 1956.
19. Ibid., 17 October 1955.
20. Ibid., 13 October 1958.
21. Ibid., 15 October 1956.
22. Ibid., 15 September 1958; 30 January 1961.
23. Ibid., 30 January 1961.
24. Ibid., June 1959.
25. Ibid., 29 December 1958.
26. Ibid., 29 June 1959; 13 July 1959.
27. Ibid., 30 November 1959.
28. Ibid., 26 September 1960.
29. Ibid., 23 March 1959.
30. Ibid., 31 October 1960.
31. Ibid., 13 June 1960.
32. Ibid., July-August 1990.
33. Ibid., 28 March 1960.
34. Ibid., 27 February 1948.
35. Ibid., 23 December 1949.
36. Ibid., 12 January 1951; 8 December 1950.
37. Ibid., 29 October 1956.

38. Ibid., 17 October 1955.
39. Ibid., 29 October 1956.
40. Ibid.
41. Ibid., 26 November 1951.
42. Ibid., July-August 1984.
43. Ibid., 26 January 1953.
44. Ibid., 10 December 1951.
45. Ibid., 25 August 1958.
46. Ibid., 15 September 1958.
47. Ibid., 28 September 1959.
48. Ibid., 28 November 1958.
49. Ibid., 9 January 1959.
50. Ibid., 9 November 1953.
51. Ibid., 15 November 1954.
52. Ibid., 11 August 1958.
53. Ibid., 11 February 1957.
54. The South Dakota Farmers Union entertained twenty farm organization leaders from India at Huron on November 12-13, 1958. The foreign guests were studying American agriculture and farm organizations. — *Union Farmer*, 17 November 1958.
55. Ibid., 28 November 1958.
56. Ibid., 26 September 1960; 30 October 1961.
57. Ibid., 30 September 1957.
58. Ibid., 15 December 1958; 11 January 1960.
59. *One World, All States Annual - 1948* (Denver: National Farmers Union, 1948). — Courtesy of Marian Kleinsasser Towne, Indianapolis, Indiana, a participant in the 1948 All States Camp.
60. Ibid., July-August 1984.
61. Ibid., 30 November 1959.
62. Ibid.
63. Ibid., 17 October 1960.
64. Ibid., 16 October 1961.
65. Ibid., 15 August 1960.

CHAPTER 14

1. Crampton, *The National Farmers Union*, 35.
2. *Union Farmer*, 16 May 1960.
3. Ibid., 14 December 1959.
4. Ibid., 28 December 1959.
5. Ibid., 14 December 1959.
6. *Union Farmer*, July-August 1994.
7. Ibid., 29 September 1958.
8. Ibid., 15 August 1955.
9. Walter L. Lassen, Executive Secretary, SD Association of Cooperatives, to Loriks, 4 November 1948, Box 2, File 1, Coop Association, 1948-1969. Loriks Papers, CWS.
10. *Union Farmer*, 18 October 1954.
11. Ibid., 14 January 1957.

12. Ibid., 8 December 1950; 25 May 1951.
13. Ibid., 8 September 1952.
14. Newsclipping, 4 March 1960, Box 2, File 18, Dahme, 1940-1966. Loriks Papers, CWS.
15. Gilbert C. Fite, "The Transformation of South Dakota Agriculture: The Effects of Mechanization, 1939-1964," *South Dakota History* 19 (Fall 1989): 291.
16. Ibid., 24 May 1954.
17. Ibid., 9 March 1959; 12 October 1959.
18. Karl Limvere, *Economic Democracy for the Northern Plains: Cooperatives and North Dakota* (Jamestown: North Dakota Farmers Union, 1980), 44.
19. Ibid., 9 March 1959.
20. Bulletin of the National Tax Equality Association, August 1960, Box 5, File 45, Miscellaneous, 1960-1963. Loriks Papers, CWS.
21. *Minneapolis Star* editorial, 5 August 1958, Box 5, File 45, Miscellaneous, 1960-1963. Loriks Papers, CWS.
22. Ken Holum, Executive Secretary of the South Dakota Association of Cooperatives, 24 February 1958, Box 2, File 1, Cooperative Association, 1948-1969. Loriks Papers, CWS.
23. John Sieh to Loriks, Box 6, File 39, Opsahl - Hougland, et. al. Loriks Papers, CWS.
24. Ibid.
25. Loriks to Howard A. Cowden, Kansas City, Missouri, 1 February 1948, Box 7, File 55, SD Co-op Assn., 1934-1957. Loriks Papers, CWS.
26. Loriks to Paul Opsahl, 2 February 1948, Box 6, File 39, Opsahl -Hougland, et.al. Loriks Papers, CWS.
27. Skage to Loriks, 9 February 1948, Box 6, File 39, Opsahl-Skage, et.al; Loriks to John Riedy, Thunderhawk, South Dakota, 9 February 1948, Box 7, File 55, SD Co-op Assn., 1934-1947; Loriks to Opsahl, 12 February 1948, Box 6, File 39, Opsahl - Hougland, et. al. Loriks Papers, CWS.
28. Skage to Loriks, 27 January 1956, Box 6, File 39, Opsahl - Hougland, et.al., 1936-1958. Loriks Papers, CWS.
29. Loriks to Walter Lassen, Secretary, SD Assoc. of Cooperatives, Huron, 18 February 1948; Loriks to Archie T. Erickson, 20 February 1948, Box 6, File 39, Opshal-Hougland, et.al. Loriks Papers, CWS.
30. Loriks to Archie T. Erickson, 20 February 1948, Ibid.
31. Loriks to Skage, 1948, Box 6, File 39, Opsahl-Skage, et.al. Loriks Papers, CWS.
32. Loriks to Thatcher and Syftestad, 1 October 1952, Box 5, File 56, Moseson, 1952-1957. Loriks Papers, CWS.
33. M. W. Thatcher to Loriks, 29 March 1957, Box 2, File 1, Coop. Assn., 1948-1969. Loriks Papers, CWS.
34. Conrad, *50 Years*, 190.
35. Loriks to August Dahme, 9 October 1958, Box 2, File 18, Dahme, 1940-1966. Loriks Papers, CWS.
36. *Union Farmer*, 11 May 1959.
37. Ibid., 31 August 1959; 15 February 1960.
38. Ibid., 14 December 1959.
39. Ibid., 13 November 1961.
40. Ibid., 25 August 1950; Bulletin of the National Tax Equality Association, August 1960, Box 5, File 45, Miscellaneous, 1960-1963. Loriks Papers, CWS.

41. Conrad, *50 Years*, 161.
42. *Union Farmer*, 29 December 1958.
43. Ibid., 5 December 1955.
44. Ibid., 29 September 1958.
45. Ibid., 15 August 1960.
46. Ibid., 28 December 1959; Bulletin of the National Tax Equality Association, August 1960, Box 5, File 45, Miscellaneous, 1960-1963. Loriks Papers, CWS.
47. Emil Loriks, Interview by Paul O'Rourke, June 9, 1971, Tape MS #224. SDOHC, USD.
48. *Union Farmer*, 15 December 1958.
49. Crampton, *The National Farmers Union*, 123.
50. Crampton, *The National Farmers Union*, 172.
51. *Union Farmer*, 15 August 1960.
52. Ibid., 16 October 1961.
53. Elizabeth E. Williams, *Emil Loriks, Building A New Economic Order* (Sioux Falls: The Center for Western Studies, 1987), 115.
54. *Union Farmer*, 12 October 1959.
55. Ibid., 15 December 1958.
56. President L. M. Stavig, Augustana College, to Loriks, 11 November 1952, Box 5, File 40, Miscellaneous, 1952. Loriks Papers, CWS.
57. Rickertsen, *To Gather Together, CENEX, The First Fifty Years*, 164-166.
58. Ibid., 171.
59. *Union Farmer*, 10 November 1952.
60. Ibid., 15 August 1960.
61. Ibid., 14 November 1960.
62. Ibid., 28 December 1959.
63. Ibid., 12 December 1960.
64. Ibid., 29 February 1960.
65. Ibid., 9 March 1959.
66. Ibid., 27 October 1958.
67. Conrad, *50 Years*, 181.
68. *Union Farmer*, 10 November 1952; 13 April 1953.
69. Ibid., 29 July 1957.
70. Ibid., 28 May 1948.
71. Ibid., 15 August 1960.
72. Two months before the November election, Alfred L. (Al) Johnson of Huron, a special representative for the FUGTA, was selected to head a seven state "Farmers for Kennedy-Johnson Committee." Johnson, a longtime figure in Farmers Union and federal farm programs, accepted the position after resigning as assistant to M. W. Thatcher. Before joining the GTA in 1953, he was chairman of the South Dakota Production and Marketing Administration.— *Union Farmer*, 12 September 1960.

CHAPTER 15

1. *Union Farmer*, 25 September 1960.
2. Ibid., 29 September 1962.
3. Ibid., 22 October 1962.
4. Ibid., 14 August 1961.

5. Ibid., 28 August 1961.
6. Ibid., 19 September 1962.
7. Conrad, *50 Years*, 208.
8. *Union Farmer*, 27 March 1963.
9. Ibid., 13 February 1963.
10. Ibid., 13 March 1963.
11. Ben Radcliffe, interview by author, Huron, South Dakota, 2 August 1994.
12. *Union Farmer*, 27 May 1963.
13. Ibid., 30 October 1963.
14. Ibid., 11 May 1964, 13 July 1964.
15. Ibid., 12 October 1964.
16. Ben Radcliffe, Interview by author, Huron, South Dakota, 2 August 1994.
17. *Union Farmer*, 13 March 1961.
18. Ibid., November 1981.
19. Ibid., 16 October 1961, April 1967. The improved agricultural economy of the 1970s slowed the decline of the number of farms in South Dakota. The South Dakota Agricultural Statistics Service reported the following data: Year No. of Farms Average Acres 1959 55,727 805 1969 45,726 997 1978 38,741 1,147.
20. Ibid., April 1969.
21. Ibid., 16 October 1961.
22. Ibid., November 1969.
23. Ibid., August 1971.
24. Ibid., September 1971.
25. Ibid., February 1971.
26. Ibid., April 1972, March 1972.
27. Ibid., October 1973.
28. Ibid., September 1972.
29. Ibid., June 1973.
30. Ibid., August 1974.
31. Ibid., April 1974.
32. Ibid., February 1974.
33. Ibid., September 1972.
34. Ibid., January 1974.
35. Ibid., April 1974.
36. Ibid., July 1973.
37. Ibid., September 1974.
38. Ibid., March 1975.
39. Ibid.
40. *Union Farmer*, July 1976; November 1976.
41. Ibid., February 1973.
42. Ibid., September 1978.
43. Ibid., September 1978.
44. Ibid., November 1979.
45. Ibid., July 1980.
46. Ibid., May 1981.
47. *Union Farmer*, February 1965.

48. Ben Radcliffe, interview by author, Huron, South Dakota, 2 August 1944.
49. Ibid.
50. *Union Farmer*, January 1979.
51. Ben Radcliffe, interview by author, Huron, South Dakota, 2 August 1994.
52. *Union Farmer*, February 1963.
53. Lynwood E. Oyos, ed. *Over A Century of Leadership, South Dakota Territorial and State Governors* (Sioux Falls, Center for Western Studies, 1987), 178-179.
54. *Union Farmer*, 13 March 1963.
55. Oyos, *Over a Century of Leadership*, 184.
56. Ibid., 193; *Union Farmer*, February 1969.
57. *Union Farmer*, April 1969.
58. Ibid., February 1970; Oyos, *Over a Century of Leadership*, 194.
59. *Union Farmer*, 10 February 1964.
60. Ibid., 22 November 1965.
61. Ibid., April 1971.
62. Ibid., November 1968, February 1969.
63. Ibid., August 1969.
64. Ibid., 11 December 1969; Loriks to Lafe Lunder, 29 December 1969, Box 2, File 53, Farmers Union, 1969. Loriks Papers, CWS.
65. *Union Farmer*, February 1970.
66. Ibid., July 1970.
67. Ibid., December 1970.
68. Ibid., December 1970.
69. Ibid., November 1971.
70. Ibid., January 1972.
71. *Union Farmer*, March 1973.
72. Ibid., February 1972.
73. Oyos, *Over a Century of Leadership*, 200.
74. The SDFU did not agree with Kneip on one issue. The Board of Directors opposed the Governor's plan for an appointed Public Utilities Commission (Senate Bill 270). Kneip proposed the creation of a Department of Energy, moving from an elected to an appointed Public Utilities Commission Board. The Union believed that the state's citizens needed to vote for officials who regulated gas and electric utilities, telephone rates, railroads and trucking within South Dakota. - Minutes of the Board of Directors, SDFU, 1-2 February 1978.
75. *Union Farmer*, February 1976.
76. Ibid., February 1977.
77. Oyos, Over a Century of Leadership, 200; *Union Farmer*, February 1978.
78. Ibid., November 1980.
79. Ibid., October 1967.
80. Ibid., October 1968.
81. Ibid., February 1970.
82. Ibid., April 1971.
83. Ibid., March 1974.
84. Ibid., March 1979.
85. Ibid., December 1974.

86. Ibid., February 1972.
87. Ibid., October and November 1973.
88. Ibid., June 1973.
89. Ibid., November 1973.
90. George Mann to Emil Loriks, 31 January 1975, Box 5, File 29, McGovern - Radcliffe, 1956 - 1976. Loriks Papers, CWS.
91. *Union Farmer*, June 1978.
92. Oyos, *Over a Century of Leadership*, 212.
93. *Union Farmer*, November 1979.
94. Ibid., December 1979.
95. Ibid., December 1979.
96. Ibid., April 1980.
97. Ibid., February 1980.
98. Oyos, Over a Century of Leadership, 213.
99. The National Farmers Organization had been established in 1955. Its members wanted to bring back farm holidays to pressure state legislators for favorable farm legislation. In 1958, the NFO tried collective bargaining by negotiating contracts with processors to raise and stabilize prices. Efforts to withhold hogs, cattle, corn and soy beans from the market in 1962, 1964 and 1967 failed when non-members refused to cooperate with NFO members. A revived agricultural economy during the 1970s reduced the activities of the NFO.
100. Ben Radcliffe, interview by author, Huron, South Dakota, 2 August 1994.
101. *Union Farmer*, August 1981.

CHAPTER 16

1. *Union Farmer*, 22 October 1962.
2. Ibid.
3. Ibid., 26 April 1963.
4. Ibid., May 1968.
5. Russell Tilberg, interview by author, Ethan, South Dakota, 26 July 1994.
6. *Union Farmer*, 28 June 1965; 10 June 1966; June 1969; July-August 1984.
7. Ibid., 15 August 1966; August 1969.
8. Ibid., May 1971; June 1977; November 1978; November 1980.
9. Ibid., October 1968.
10. Ibid., September 1969.
11. Ibid., October 1972.
12. Ibid., October 1968.
13. Ibid., May 1969; September 1980.
14. Ibid., November 1981.
15. Ibid., June 1969.
16. Ibid., July 1969.
17. Ibid., June 1973.
18. Ibid., December 1968.
19. Ibid., January 1972.
20. Ibid., January 1969.
21. Ibid., March 1974.

22. Ibid., May 1974.
23. Ibid., September 1974.
24. Ibid., May 1975.
25. Rickertsen, *To Gather Together*, 207.
26. *Union Farmer*, June 1976.
27. Ibid., April 1978.
28. Ibid., June 1977.
29. Ibid., November 1978.
30. Ibid., December 1971.
31. Ibid., November 1972.
32. Ibid., January 1977.
33. Ibid., May 1980.
34. Ibid., July 1980.
35. Ibid., 15 June 1964.
36. Ibid., 11 July 1966.
37. Ibid., May 1967.
38. Ibid., July 1969.
39. Ibid., August 1969, October 1970.
40. Ibid., October 1970.
41. Ibid., March 1979.
42. Ibid., December 1975.
43. Ibid., March 1972.
44. South Dakota Farmers Union Youth Activity Handbook (Huron: Department of Education, South Dakota Farmers Union, 1963),
45. *Union Farmer*, March 1971.
46. Ibid., August 1976.
47. Ibid., June 1981.
48. Ibid., July 1977.
49. Ibid., July 1975.
50. Ibid., December 1966.
51. Ibid., May 1968.
52. Ibid., January 1973.
53. Ibid., June 1974.
54. Ibid., September 1975.
55. Charles Groth, interview by author, Huron, South Dakota, 20 July 1995.
56. *Union Farmer*, May 1974.

CHAPTER 17

1. Patton suffered a stroke on December 10, 1984, causing paralysis of his left side. He died of another stroke on February 17, 1985.
2. Loriks, Box 4, File 47, History, 1955-1962. Loriks Papers, CWS.
3. Lafe Lunder to Loriks, 4 October 1969, Box 2, File 52, Farmers Union, 1969. Loriks Papers, CWS; Ibid., Lafe Lunder to August Dahme, October 1969.
4. *Union Farmer*, October 1969.
5. Lafe Lunder to Loriks, 16 December 1969, Box 2, File 53, Farmers Union, 1969. Loriks Papers, CWS.

6. Ben Radcliffe, Interview by author, Huron, South Dakota, 2 August 1994.
7. *Union Farmer*, November 1971.
8. Ben Radcliffe, Interview by author, Huron, South Dakota, 2 August 1994.
9. *Union Farmer*, November 1978.
10. Ibid., May 1971, May 1979.
11. Ibid., February 1975.
12. Ibid., April 1975, June 1975.
13. Ibid., November 1971.
14. Ibid., September 1971.
15. Ibid., July 1979.
16. Ibid., January-February 1991.
17. Ibid., 17 May 1965.
18. Ibid., December 1979.
19. Ibid.
20. Oyos, *Over A Century of Leadership*, 170.
21. *Union Farmer*, May 1971.
22. Oyos, *Over A Century of Leadership*, 202.
23. Ben Radcliffe, Interview by author, Huron, South Dakota, 2 August 1994.
24. *Union Farmer*, November 1981.
25. Ibid., October 1981.
26. Ibid., October 1981.
27. Ibid., 13 November 1963.
28. "Congress Has Settled Co-op Taxation," Box 2, File 3, Co-ops and the Farmers Union, 1953-1965. Loriks Paper, CWS.
29. Loriks to Gustav L. Moseson, 10 March 1965, Box 8, File 1, SD Coop Assn., 1959-1966. Loriks Papers, CWS.
30. *Union Farmer*, 17 May 1965.
31. Ibid., 12 April 1965.
32. Ibid., 13 December 1965.
33. Ibid., 13 December 1965.
34. Ibid., 10 January 1966.
35. Ibid., June 1967.
36. Ibid., December 1972.
37. Ibid., 28 July 1965.
38. Ibid., 27 May 1963.
39. Ibid., 22 February 1966
40. Ibid., July-August 1984.
41. Ibid., 12 November 1962.
42. Ibid., 22 November 1965.
43. Ibid., March 1976.
44. Ibid., July 1979.
45. Ibid., May-June 1983.
46. Rickertsen, *To Gather Together*, 208.
47. Ibid., 236-237.
48. *Union Farmer*, March 1978.
49. Ibid., July 1969, November 1970.

50. Ibid., November 1967.
51. Ibid., March 1974.
52. Ibid., July 1981.
53. Ibid., 8 March 1973.
54. Ibid., April 1972, July 1972.
55. Ibid., April 1974.
56. Ibid., April 1974.
57. Ibid., August 1972.
58. Darrell Moseson, interview by author, Sioux Falls, South Dakota, 17 July 1995. Gustav L. Moseson died in 1973 at age seventy-nine.
59. *Union Farmer*, January 1981.
60. Ibid., 30 August 1965.
61. Ibid., 30 August 1965.
62. Ibid., July 1977.
63. Ibid., August 1975.
64. Ibid., November 1975.
65. Ibid., October 1980.
66. Report of President B. J. Malusky, Grain Terminal Association Annual Report, 1978, 21.
67. Grain Terminal Association Annual Report, 1977, p. 19.
68. Grain Terminal Association Annual Report, 1978, 37.
69. *Union Farmer*, June 1970.
70. Loriks to Paul Opsahl, 31 December 1967, Box 5, File 46, Loriks, Miscellaneous, 1967-1968. Loriks Papers, CWS.
71. Norman Olsen, Interview by author, Canton, South Dakota, 21 July 1994.
72. Grain Terminal Association Annual Report, 1978, 22.

CHAPTER 18

1. *Union Farmer*, November 1981.
2. Ibid., December 1981.
3. Ibid., October 1982.
4. Ibid., April 1983.
5. Ibid., January - February 1985.
6. Charles Groth, interview by author, Huron, South Dakota, 20 July 1995. It may have been the only time in the state's history when a row of port-a-potties was set up on the Capitol grounds to accomodate a mass gathering.
7. Ibid., March-April 1985.
8. Ibid., January-February 1986.
9. Ibid., July-August 1986, September-October 1986.
10. Ibid., May-June 1986.
11. Ibid., November-December 1986.
12. Land prices in South Dakota peaked at $349 an acre in 1982.
13. Ibid., January-February 1987.
14. Ibid., October 1986.
15. Ibid., January-February 1987.
16. SDSU Data Center, South Dakota Agricultural Statistics Service.
17. *Sioux Falls Argus Leader*, 30 September 1986.

18. Ibid., May-June 1987.
19. Ibid., November-December, 1987.
20. Ibid., December 1992.
21. Ibid., November-December 1986.
22. Ibid., March-April 1988.
23. Ibid., March-April 1988.
24. Ibid., March-April 1988.
25. Ibid., May 1990.
26. Ibid., September-October 1989.
27. Ibid., May-June 1994. The major American exporters of grain made their greatest profit through handling the grain rather than through selling it to foreign consumers. — Charles Groth, interview by author, Huron, South Dakota, 20 July 1995.
28. Ibid., December 1992.
29. Ibid., March 1993.
30. *Union Farmer*, March 1993. Information was supplied by Lavern Schmidt of Ramona and Douglas Murphy of Woonsocket.
31. Ibid., April 1993.
32. Ibid., March 1994.
33. Ibid., March 1994.
34. Ibid., March 1994.
35. Ibid., September-October 1994.

CHAPTER 19

1. Ibid. January 1982.
2. Ibid., December 1994.
3. Ibid., March-April 1985.
4. Charles Groth, interview by author, Huron, South Dakota, 20 July 1995.
5. *Union Farmer*, July-August 1987.
6. Ibid., September-October 1987.
7. Ibid., January-February 1988.
8. Ibid., September-October 1988.
9. Ibid., March 1992.
10. Ibid., January-February 1992.
11. Ibid., November 1994.
12. The task force had proposed the Family Farm Act Corporate Hog Amendment. Patterned after legislation in Kansas, the amendment would have allowed counties to decided whether or not to have corporate hog facilities.
13. *Union Farmer*, March 1995.
14. The Dakota Pork plant in Huron closed in 1997.
15. *Sioux Falls Argus Leader*, 23 October 1996.
16. Ibid., 16 August 1996.
17. Charlie Johnson, "Readers' Forum," *Sioux Falls Argus Leader*, 15 November 1996.
18. Editorial, *Sioux Falls Argus Leader*, 25 August 1996.
19. Charlie Johnson, "Readers' Forum," *Sioux Falls Argus Leader*, 15 November 1996.
20. *Sioux Falls Argus Leader*, 15 August 1996.
21. Ibid., September 1996.

22. Ibid., 23 October 1996.
23. Ibid., November-December 1985.
24. Ibid., March-April 1985.
25. Ibid., September-October 1987.
26. Ibid., March-April 1989.
27. Ibid., March-April 1989.
28. Ibid., March 1990.
29. Ibid., May-June 1992.
30. Ibid., January-February 1994.
31. Ibid., March 1994.
32. IBid., March 1995.
33. Ibid., January-February 1995.
34. Ibid., 5 October 1995.
35. Ibid., 5 October 1995.
36. Ibid., 25 July 1996.
37. Ibid., 15 August 1996.
38. Ibid., January-February 1989.
39. Ibid., January-February 1989.
40. It was difficult to use REDI funds to establish cooperatives because the law required that for a certain amount of REDI money a certain number of jobs had to be produced.— Charles Groth, interview by author, Huron, South Dakota, 20 July 1995.
41. *Union Farmer*, March 1994.
42. Charles Groth, interview by author, Huron, South Dakota, 20 July 1995.
43. *Union Farmer*, December 1992.
44. Ibid., January-February 1989.
45. Ibid., April 1991.
46. Ibid., March 1992.
47. *Sioux Falls Argus Leader*, 3 July 1985.
48. *Sioux Falls Argus Leader*, 5 September 1996.
49. Ibid., March 1990.
50. Ibid., March 1991. AGP decided not to build in South Dakota. In 1995, however, the South Dakota Soybean Processors planned to build a $30 million plant near Volga. Paul Casper, a Farmers Union member from Lake Preston, is president of the soybean co-op.
51. Ibid., November 1982.
52. Ibid., December 1992.
53. Ibid., November-December 1987.
54. Ibid., March-April 1986.
55. Ibid., March 1992.
56. Ibid., June 1987.
57. Ibid., March-April 1989.
58. Ibid., November-December 1983.
59. Ibid., May, 1982.
60. George Anthan, Catholic Church Opposes Agriculture "Industrialization," *Sioux Falls Argus Leader*, July 1997.
61. Ibid., May-June 1983.

62. Ibid., September-October 1989.
63. Ibid., January-February 1990.
64. Ibid., November 1994.
65. Ibid., April 1982.
66. Ibid., March 1990.
67. Ibid., November-December 1985.
68. Ibid., January-February 1984.
69. Ibid., March 1992.
70. Ibid., January-February 1995; July-August 1995.

CHAPTER 20

1. *Union Farmer*, May-June 1994.
2. Ibid., June 1982.
3. *Union Farmer*, September-October 1985.
4. Ibid., December 1991.
5. Ibid., May-June 1993.
6. Ibid., May-June 1993.
7. Ibid., December 1991.
8. Questions were raised about the propriety of merging co-op service stations with grain elevators because of the differences in management. The management may not be good at both endeavors. - Darrell Moseson, interview by author, Sioux Falls, South Dakota, 17 July 1995.
9. *Sioux Falls Argus Leader*, 30 April 1995.
10. *Union Farmer*, March 1995; January-February 1995.
11. *Sioux Falls Argus Leader*, 30 April 1995.
12. Ibid., January-February 1994.
13. *Sioux Falls Argus Leader*, 22 August 1995.
14. Ibid.
15. Ibid., 11 November 1996.
16. Ibid., March 1995.
17. Ibid., January-February 1995.
18. Ibid., November-December 1983.
19. Ibid., July-August 1984.
20. Paul E. Huff and Raymond E. Novak, A Rich Heritage of Community, *A History of The National Farmers Union Insurance Companies* (Denver: National Farmers Union Property and Casualty Companies, 1995), 54.
21. *Union Farmer*, November-December 1987.
22. Ibid., July-August, 1984; September-October 1989.
23. Ibid., September-October 1989.
24. Huff and Novak, *A Rich Heritage of Community*, 60.
25. *Union Farmer*, December 1989.
26. Ibid., December 1991.
27. Ibid., March 1995.
28. Ibid., November 1994.
29. Ibid., November 1981.
30. Ibid., January 1982.

31. Ibid., September-October 1983.
32. Ibid., March-April 1986.
33. Ibid., September-October 1990.
34. Ibid., May-June 1992.
35. Ibid., January-February 1989.
36. Ibid., April 1991.
37. Ibid., March-April 1987.
38. Darrell Moseson, interview by author, Sioux Falls, South Dakota, 17 July 1995.
39. Charles Groth, interview by author, Huron, South Dakota, 20 July 1995.
40. *Union Farmer*, April 1994.
41. Ibid., March 1995.
42. Charles Groth, interview by author, Huron, South Dakota, 20 July 1995.
43. *Union Farmer*, November 1994.
44. Ibid., March 1983; January-February 1992.
45. Ibid., March 1991.
46. Ibid., November 1983.
47. Darrell Moseson, interview by author, Sioux Falls, South Dakota, 17 July 1995.
48. *Union Farmer*, December 1994.
49. Ibid., December 1982.
50. Ibid., May-June, 1983; July-August 1983.
51. Ibid., November-December 1987.
52. Ibid., October 1981.
53. Charles Groth, interview by author, Huron, South Dakota, 20 July 1995.
54. *Union Farmer*, December 1990.
55. Ibid., January-February 1994.
56. Ibid., July-August 1995.
57. Ibid., December 1990.
58. Ibid., March 1995.
59. Ibid., April 1992.
60. Ibid., January-February 1990; November 1994.
61. Ibid., December 1993.
62. Ibid., November 1994.
63. *Sioux Falls Argus Leader*, 8 October 1997.
64. Ibid., July-August 1989; September-October 1994.
65. Ibid., July-August 1989.
66. Ibid., March 1995.
67. Ibid., July 1969.
68. IBid., May-June 1986.
69. Ibid., July-August 1993.
70. Ibid., March 1993; April 1993.
71. Charles Groth, interview by author, Huron, South Dakota, 20 July 1995.

CHAPTER 21

1. *Sioux Falls Argus Leader*, 13 November 1994. The farm program cost $13.4 billion in 1993.
2. *Sioux Falls Argus Leader*, 13 February 1997.

3. Ibid., 14 March 1996.
4. Ibid., 23 October 1996.
5. Ibid., 23 October 1996.
6. Ibid., 10 November 1994.
7. Hiram M. Drache, *Legacy of the Land: Agriculture's Story to the Present.* (Danville, Illinois: Interstate Publishers, Inc., 1996).

CHAPTER 22

1. Rickertsen, To Gather Together, 216.
2. South Dakota received $431 million in farm subsidies in 1993. Actual dollars received according to programs follow: Feed Grain Deficiency Program $164,555,946 (Corn, oats and sorghum) Wheat Deficiency Program 83,266,999 Conservation Reserve Program 71,466,791 Disaster Aid 93,294,319 National Wool Act Program 8,514,507 Other Programs 10,235,795. — *Sioux Falls Argus Leader*, 11 December 1994.
3. *Sioux Falls Argus Leader*, 29 July 1996.
4. *Sioux Falls Argus Leader*, 2 November 1995.
5. Minneapolis Tribune, April 1968.
6. George Anthan, "Sociologist Perceived as Troublemaker or Visionary," *Sioux Falls Argus Leader*, March 1994.
7. Jack Bailey, "The Soap Box," Mitchell Daily Republic, circa. 21 July 1947.
8. E. F. Schumacher, Small Is Beautiful, Economics As If People Mattered (New York: Harper and Row, 1973), 19, 95, 107.
9. *Sioux Falls Argus Leader*, 14 February 1997.
10. *Sioux Falls Argus Leader*, 22 October 1996.
11. *Sioux Falls Argus Leader*, 23 October 1994.
12. *Sioux Falls Argus Leader*, 6 November 1993.
13. *Sioux Falls Argus Leader*, 19 March 1995.

ACKNOWLEDGMENTS

I began research for a comprehesive history of the South Dakota Farmers Union (SDFU) in 1993, and during the past six years many people have contributed to its preparation. I am particularly grateful to Charles (Chuck) Groth, Communications Director for the South Dakota Farmers Union. He provided access to books, documents and photographs in the archives of the Union's headquarters building in Huron. He gave me the names of several longtime members of the SDFU that would be available for interviews, and he recounted his experiences with the Union over the past twenty-five years. It was Chuck Groth who sought and found the funds necessary for the book's publication. A debt of gratitude is also owed to President Dennis Wiese and the Board of Directors of the SDFU for their approval and support of the project.

In addition to the documents in the Union's archives. the papers of Emil Loriks in the Center for Western Studies, Sioux Falls, the papers of Mrs. Russell Tilberg, Ethan, relating the history of the SDFU's Junior education program, Robert S. Thompson's unpublished masters thesis, The History of the South Dakota Farmers Union, 1914-1952, and Marian Kleinsasser Towne's information and photographs describing Union camp life were of inestimable value. A special thank you is in order for personnel of East River Electric Power Cooperative, Madison, for providing information on the impact of electric power on rural South Dakota.

For interviews and informative correspondence, I am indebted to Henry Bell, Charles Groth, Darrell Moseson, S. Douglas Murphy, Norman B. Olsen, Ben Radcliffe and Florence and Russell Tilberg. Unfortunately, four of the foregoing people have died since being interviewed. I am equally beholden to the staff of the South Dakota Oral History Center, the University of South Dakota. The Oral History Center contains interviews with SDFU leaders from the 1930s to 1970. The transcripts of all of the conversations and interviews proved invaluable to me.

Many thanks to the staff of the H. M. Briggs Library, South Dakota State University, Brookings, for giving me access to minutes of SDFU meetings and to back issues of the South Dakota *Union Farmer*. The organization's newspaper was a prime source of information. I am grateful to the staff of Mikkelsen Library, Augustana College, Sioux Falls, for permitting me to use books and journal articles that I found most useful in developing the necessary background information on farm organizations, particularly *The National Farmers Union.*

I owe a particular debt of gratitude to Dr. Harry Thompson, Director of Research Collections and Publications, the Center for Western Studies and to Dr. Arthur Huseboe, Executive Director of the Center for Western Studies and project director for this endeavor for their support and advice in readying the manuscript for publication.

Finally, I would say thank you with all my heart, once again, to my wife and fellow researcher, Bedia, for her constant vigilance, encouragement, and her patience in reading and rereading the manuscript.

BIBLIOGRAPHY

Sources

ARCHIVAL COLLECTIONS

Emil Loriks Papers. The Center for Western Studies, Augustana College, Sioux Falls, South Dakota

Minutes of the Executive Board, South Dakota Farmers Union, 1917-1929, 1944-1959, 1968-1989. South Dakota Farmers Union, Huron, South Dakota.

Official Minutes, Farmers Educational and Cooperative Union of America, South Dakota Division, 1926-1988. South Dakota Farmers Union, Huron, South Dakota.

Minutes of the Farmers Educational and Cooperative Union of America, National Conventions, 1933, 1934, 1936. South Dakota Farmers Union, Huron, South Dakota.

Glover, Joel. "Roy Glover, Defender of the Family Farm." A paper prepared for the Political Science Department, the University of South Dakota, 30 April 1988. South Dakota Farmers Union, Huron, South Dakota.

Thompson, Robert S., *The History of the South Dakota Farmers Union, 1914-1952*. Unpublished M.A. Thesis, University of South Dakota, June 1953. South Dakota Farmers Union, Huron, South Dakota.

History of the Farmers Union Junior Work. Papers of Mrs. Russell (Florence) Tilberg, Ethan, South Dakota.

UNPUBLISHED INTERVIEWS

Ayres, Homer. Interview by Earl Hausle, 28 June 1971. Interview # 187, transcript. South Dakota Oral History Center, University of South Dakota, Vermillion, South Dakota.

Fosheim, Oscar. Interview by Paul O'Rourke, June 1970. Interviews MS002, MS003, MS004, transcripts. South Dakota Oral History Center, University of South Dakota, Vermillion, South Dakota.

Loriks, Emil. Interview by Paul O'Rourke, June 1970. Interviews MS 006, MS007, transcripts. South Dakota Oral History Center, University of South Dakota, Vermillion, South Dakota.

Loriks, Emil. Interview by Paul O'Rourke, 9 June 1971. Interview MS224, transcript. South Dakota Oral History Center, University of South Dakota, Vermillion, South Dakota.

Unpublished Interviews by Author

Bell, Henry, active member of the Farmers Union for over fifty years, a cooperative builder. Interview by author, 22 August 1944, Flandreau, South Dakota. Tape Recording.

Groth, Charles, Communications Director of the South Dakota Farmers Union. Interview by author, 20 July 1995, Huron, South Dakota. Tape recording.

Moseson, Darrell, former president of CENEX, Inc. Interview by author, 17 July 1995, Sioux Falls, South Dakota. Transcript.

Murphy, S. Douglas, active member of the Farmers Union for over fifty years, a cooperative builder in Sanborn County. Interview by author, 2 August 1994, Woonsocket, South Dakota. Tape recording.

Olsen, Norman B., former vice president of the Farmers Union Grain Terminal Association. Interview by author, 21 July 1994, Canton, South Dakota. Transcript.

Radcliffe, Ben, President of the South Dakota Farmers Union, 1961-1981. Interview by author, 2 August 1994, Huron, South Dakota. Tape recording.

Tilberg, Florence and Russell, active members of the Farmers Union for over fifty years, major role players in the education program. Interview by author, 26 July 1994, Ethan, South Dakota. Tape recording and transcript.

PUBLISHED WORKS

Anthan, George. "Sociologist Perceived as Troublemaker or Visionary" *Sioux Falls Argus Leader*, March 1994.

Bailey, Jack. "The Soap Box" *Mitchell Daily Republic*, circa. 21 July 1947.

Blackorby, Edward C. *Prairie Rebel: The Public Life of William Lemke*. Lincoln: University of Nebraska Press, 1963.

Conrad, Charles and Joyce. *50 Years, The North Dakota Farmers Union*. n.p. 1976.

Crampton, John A. *The National Farmers Union, Ideology of a Pressure Group*. Lincoln: University of Nebraska Press, 1965.

Cooperative League of U.S.A. Yearbook, 1936.

Davis, Ron. "The Lights Come On In South Dakota" East River Guardian 24 (May 1985).

Dahl, Patrick. "REA and the People — Powerful Partners" *East River Guardian* 24 (May 1985).

Drache, Hiram M. *Legacy of the Land: Agriculture's Story to the Present*. Danville, Ilinois: Interstate Publishers, Inc., 1996.

Edwards, Gladys Talbott. *The Farmers Union Triangle*. Jamestown, North Dakota: Farmers Union Educational Service, 1941.

Fite, Gilbert C. "John A. Simpson: The Southwest's Militant Farm Leader" *Mississippi Valley Historical Review* 35 (March 1949).

________. "The Transformation of South Dakota Agriculture: The Effects of Mechanization, 1939-1964" South Dakota History 19 (Fall 1989).

Grain Quarterly, National Federation of Grain Cooperatives 5 (Fall 1947).

Grain Terminal Association Annual Report, 1977-1978.

History of the Farmers Union Co-op Association of McCook County, Salem and Spencer, 1917-1962.

History of the Sanborn County Farmers Union Oil Company, Woonsocket, South Dakota, 1928-1978.

Hoover, Herbert T. "Farmers Fight Back: A Survey of Rural Political Organizations, 1873-1983" South Dakota History 13 (Spring- Summer 1983).

Huff, Paul E. and Novak, Raymond E. *A Rich Heritage of Community, A History of The National Farmers Union Insurance Companies.* Denver: National Farmers Union Property and Casualty Companies, 1955.

Lange, Gerald. "Americanism over Radicalism: Mundt's First Election." Selected Papers of the First Nine Dakota History Conferences, 1969-1977, Series # 8. Madison: The Karl E. Mundt Historical and Educational Foundation, 1981).

Limvere, Karl. *Economic Democracy for the Northern Plains: Cooperatives and North Dakota.* (Jamestown: North Dakota Farmers Union, 1980).

Loriks, Emil. "The Great Depression - The Earl Thirties and the Farm Holiday Movement." Selected Papers of the Dakota History Conference, 1969-1977. Series # 8. (Madison: The Karl E. Mundt History and Educational Foundation, 1981).

Lusk, R.D. "Life and Death of 470 Acres" Saturday Evening Post, 13 August 1938.

Miller, John E. "Restrained, Respectable Radicals: The South Dakota Farm Holiday" *Agricultural History* 59 (July 1985).

Minneapolis Star, 5 August 1958.

Minneapolis Tribune, April 1968.

Morlan, Robert L. *Political Prairie Fire, The Nonpartisan League, 1915-1922.* Minneapolis: University of Minnesota Press, 1955.

National Farmers Union Newsletter 1, No.11, 17 September 1954.

One World, All States Annual - 1948. (Denver: National Farmers Union, 1948). Courtesy of Marian Kleinsasser Towne, Indianapolis, Indiana.

O'Rourke, Paul. "South Dakota Politics During the New Deal Years" *South Dakota History* 1 (Summer 1971).

Oyos, Lynwood E., Ed. *Over A Century of Leadership, South Dakota Territorial and State Governors.* Sioux Falls: Center for Western Studies, 1987.

Pratt, William C. "Glenn J. Talbott, the Farmers Union, and American Liberalism after World War II" *North Dakota History* 55 (Winter 1988).

________. "Rethinking the Farm Revolt of the 1930s" *Great Plains Quarterly* 8 (Summer 1988).

Remele, Larry. "The North Dakota Farmers Union and the Nonpartisan League: Breakdown of a Coalition" *North Dakota Quarterly* 46 (Autumn 1978).

Report of President B. J. Malusky, *Grain Terminal Association Annual Report*, 1978.

Rickertsen, Leo N. *To Gather Together, CENEX, The First Fifty Years.* Minneapolis: Cooperative Printing Association, 1980.

Saloutos, Theodore and Hicks, John D. *Agricultural Discontent in the Middle West, 1900-1939.* Madison: University of Wisconsin, 1951.

Schell, Herbert S. *History of South Dakota.* Lincoln: University of Nebraska Press, 1961.

Schumacher, E. F. *Small Is Beautiful, Economics As If People Mattered.* New York: Harper and Row, 1973.

Shover, John L. *Cornbelt Rebellion, The Farmers' Holiday Association.* Urbana: University of Illinois Press, 1965.

Siderius, Linda. *75 Years of Building For Today and Tomorrow, A Junior Youth Book, Youth and Young Adult Activities.* Denver: National Farmers Union, n.d.

Sioux Falls Argus Leader, 1933-1935, 1985, 1993-1997.

South Dakota Farmers Union Youth Activity Handbook. Huron: Department of Education, South Dakota Farmers Union, 1963.

South Dakota Union Farmer, 1918 - 1998.

Stoltz, Mildred K. *This Is Yours, The Montana Farmers Union and Its Cooperative Associates*. Minneapolis:The Lund Press, 1956.

Terkel, Studs. *Hard Times, An Oral History of the Great Depression*. New York: Pantheon Books, 1970.

Tontz, Robert L. "Membership of General Farm Organizations, United States, 1874-1960" *Agricultural History* 38 (July 1964).

Tucker, William P. "Populism Up-to-Date: The Story of the Farmers Union" *Agricultural History* XXI (October 1947).

Williams, Elizabeth Evenson. "A South Dakota Agrarian Looks at Huey Long." Papers of the 17th Dakota History Conference, April 11- 13, 1985. Madison: The Karl E. Mundt Historical and Educational Foundation, 1985.

________. "Emil Loriks and the South Dakota Farmers Union View Huey Long." Papers of the 16th Dakota History Conference, April 12- 14, 1984. Madison: The Karl E. Mundt Historical and Educational Foundation, 1984.

________. *Emil Loriks, Building A New Economic Order*. Sioux Falls: The Center for Western Studies, 1987.

INDEX

T

U